Chemical Power Sources

Chemical Power Sources

V. S. Bagotzky and **A. M. Skundin**

*The Institute of Electrochemistry of the
USSR Academy of Sciences*

Translated from the Russian by
O. Glebov and **V. Kisin**

1980

Academic Press

A Subsidiary of Harcourt Brace Jovanovich, Publishers

London · New York · Toronto · Sydney · San Francisco

ACADEMIC PRESS INC. (LONDON) LTD
24/28 Oval Road,
London NW1

United States Edition published by
ACADEMIC PRESS INC.
111 Fifth Avenue
New York, New York 10003

British Library Cataloguing in Publication Data
Bagotzky, V. S.
Chemical power sources.
1. Electrical batteries   2. Electrolytic cells
I. Title   II. Skundin, A. M.   III. Glebov, Oleg   IV. Kisin, Vitaliï
621.35   TK2901   80-40832

ISBN 0-12-072650-5

Filmset in 'Monophoto' Times New Roman by
Eta Services (Typesetters) Ltd., Beccles, Suffolk
Printed in Great Britain by
St Edmundsbury Press
Bury St Edmunds, Suffolk

Preface

Electrochemical power sources are devices for direct conversion of chemical energy into electricity. They find wide applications in various fields, for example, as power units for electronic devices, in vehicles of all types, spacecraft, portable appliances, etc. In the future their role will increase with the progress of the electric car, load-levelling and medical devices. Though the history of the cell began almost two centuries ago, continuous effort for further improvements is still under way. Fundamental changes in design and manufacturing processes are being introduced, new types of batteries developed and new electrochemical systems and reactions used in cells and batteries.

In the literature dealing with electrochemical cells, no single monograph covers the field of interest as a whole. The manufacture of cells and batteries involves electrochemical technology, as well as chemical engineering, machinery design and related skills. The scientific principles involved in the development, manufacture and use of batteries form an interdisciplinary field covering electrochemistry, physical chemistry, electrical engineering, the theory of thermal processes, and so on. Most of the existing books on batteries deal only with selected topics.

The aim of this book is to give a brief, but comprehensive and up-to-date treatment of most of the problems of cell research, of the efforts to improve the existing cell types and to develop new types, and of manufacturing, testing and application of electrochemical power sources.

The main function of the book is therefore to explain the most important features concerning each type of cell. This book is not a handbook of types and sizes of cells and batteries supplied by different manufacturers. Tables of cell parameters are given only to illustrate the performance of the corresponding cells and batteries. Some general ideas concerning battery use and maintenance are also discussed, though this does not make the book a guide in these fields.

v

A great variety of terms and definitions is found in the relevant literature. The same terms often have different meanings in different languages. We have tried throughout the book to use a system of terms which we consider consistent. When a term appears for the first time, its synonyms are given in parentheses following it.

The book is divided into two parts. The first is devoted to general features and fundamental principles of electrochemical processes and devices. In the second part, various systems of chemical power sources are considered, including their design, mechanism of reactions, performance, and specific features of their applicatons. Systems which are widely manufactured and employed are discussed, as well as systems under development. Problems of future development are analysed for each battery system. It must be mentioned that the relative volumes devoted to specific cell types in this book do not always reflect the importance of these types or the ratio of the respective research efforts.

The book is intended to be of use both to scientist and engineer engaged in cell development, and to all those who use batteries as regular practice.

Contents

Part One. GENERAL ASPECTS

Chapter One. WORKING PRINCIPLES OF CHEMICAL POWER SOURCES

Chapter Two. CELL TYPES

Chapter Three. PERFORMANCE

Chapter Four. ELECTROCHEMICAL ASPECTS OF CELL OPERATION

Symbols

a	constant in Tafel equation	$\mathscr{E}^{(T)}$	electromotive force, e.m.f. (thermodynamic value)
a_j	thermodynamic activity of jth species	$\mathscr{E}^\circ$	standard e.m.f.
$a_\pm$	mean ionic activity	$\bar{\mathscr{E}}$	thermoneutral voltage
b	width; constant in Tafel equation	E	strength of electric field
		F	Faraday number
C	Ah capacity	G	free energy
C°	rated Ah capacity	G°	standard free energy
C_{res}	residual Ah capacity	g	concentration, per cent
c_j	concentration of jth species	g_Q	specific consumption of reagent per unit charge
c_s	concentration near the surface	$g_Q^{(T)}$	theoretical specific consumption per unit charge
c_b	bulk concentration	g_W	specific consumption per unit energy
D	diffusion coefficient		
D_{eff}	effective diffusion coefficient	$g_W^{(T)}$	theoretical specific consumption per unit energy
d	thickness		
E	electrode potential	H	enthalpy
E°	standard electrode potential	h	height; efficiency of porous electrode
E_r	electrode potential referred to r.h.e. in the same solution		
$E_{(+)}$	electrode potential of positive electrode	I	electric current
		I_c	charge current
$E_{(-)}$	electrode potential of negative electrode	I_d	discharge current
		I_{adm}	maximum admissible discharge current
$E^{(i)}$	electrode potential at current density i	I_{sh}	short-circuit current
$\mathscr{E}$	open circuit voltage, o.c.v. (thermodynamic value)	I°	exchange current
		i	current density (apparent)

i_σ	true current density (per unit of true surface)	Q_c°	charge capacity
i°	exchange current density	$\bar{q}$	heat effect of reaction
i_l	limiting diffusion current density	$\bar{q}_{entr}$	unavailable heat
		R	gas constant
i_0	current density at the front face of a porous electrode	R_{ohm}	internal cell resistance (ohmic)
J_j	flux of jth species	R_{eff}	effective internal cell resistance
J_{diff}	diffusion flux	R_c	apparent internal cell resistance
J_{migr}	migration flux		
J_{conv}	convection flux	R_{pol}	polarization resistance
j	normalized current	R_e	electrolyte resistance
j_d	normalized discharge current	R_{pore}	electrolyte resistance in pores
j_c	normalized charge current	R_{ex}	resistance of the external circuit
k_r	reaction rate constant		
L	interelectrode distance	r	(pore) radius
L_{ohm}	characteristic ohmic length	r_{crit}	critical pore radius
L_{diff}	characteristic diffusion length	r_{max}	most probable pore radius
		r_s	resistance per unit electrode area
l	length, distance		
M	mass	r_{eff}	normalized effective internal cell resistance
M_r	mass of reactants		
M_e	mass of electrolyte	S	entropy; surface area
M_s	mass of structural materials	S_{rem}	area of heat removal surface
N	natural number	s_m	surface area per unit mass
n	number of electrons in single step	s_v	surface area per unit volume
P	power (electric)	T	temperature (absolute)
P_{max}	maximum power	T_{max}	maximum temperature
P_{adm}	maximum admissible discharge power	T_{min}	minimum temperature
		T_s	steady-state temperature
P_{pre}	maximum prescibed discharge power	T_{amb}	ambient temperature
		t	transference number
P_{therm}	power of heat generation	U	voltage (on-load)
p	pressure	U_d	discharge voltage
p_m	specific power per unit mass	U_c	charge voltage
p_v	specific power per unit volume	U_0	discharge voltage extrapolated to zero current
Q_d	electric charge for cell discharge	U_{init}	initial voltage
		U_{fin}	cut-off (final) voltage
Q_c	electric charge for cell charging	U_{max}	maximum voltage
		U_{min}	minimum voltage

$\bar{U}$	mean voltage	η_a	activation polarization
U_{crit}	critical voltage	$\eta^{(T)}$	thermodynamic efficiency
u_j	mobility of jth ion	η_U	voltage efficiency
V	bulk flow rate	$\eta^{(f)}$	net efficiency
v_j	rate of migration of jth ion	θ	normalized discharge time; wetting angle
W	electrical energy (Wh capacity); Warburg constant	θ_d	depth of discharge
		θ_c	charge state
W_{max}	maximum energy	θ_p	porosity
W_{pre}	prescribed electrical energy	κ	specific conductivity
w_m	specific energy per unit mass	κ_{eff}	effective (specific) conductivity
w_v	specific energy per unit volume	λ	utilization coefficient of reactants
x	co-ordinate; length of diffusion zone	μ_Q	Ah efficiency (of capacity or current utilization)
Z	impedance	μ_W	Wh efficiency (of energy utilization)
z_j	number of elementary charges of jth ion	μ	molar concentration
		ν	stoichiometric coefficient
α	heat transfer coefficient	π	perimeter
β	pore tortuosity coefficient	ρ	resistivity (specific)
γ	mass of fuel cell generator per unit power; surface tension	ρ_{eff}	effective resistivity
		Σ	true surface area
δ	particle size; thickness of diffusion layer	σ_v	true surface area per unit volume
δ_I	primary particle size	σ_m	true surface area per unit mass
δ_{II}	secondary particle size		
δ_{max}	most probable particle size	σ_s	surface roughness
$\bar{\delta}$	mean particle size	τ	time (current)
ε	coefficient of resistance enhancement (conductivity attenuation)	τ°	time of total discharge
		τ_{pre}	prescribed operational time
		τ_d	transient time of diffusion
ε_D	coefficient of diffusion attenuation	φ	potential difference
		φ_{ohm}	ohmic potential difference
η	polarization of electrode; viscosity	$\varphi_{m,e}$	potential difference between metal and electrolyte
		Ψ	electrostatic potential
$\eta_{(+)}$	polarization of positive electrode	ω	polarization resistance (constant) at low current densities; frequency
$\eta_{(-)}$	polarization of negative electrode	ω°	polarization resistance for $c_j = 1$
η_{conc}	concentration polarization		

Au commencement de l'année 1800 (la date d'une aussi grande découverte ne peut être passée sous silence), à la suite de quelques vues théoriques, l'illustre professeur imagina de former une longue colonne, en superposant successivement une rondelle de cuivre, une rondelle de zinc et une rondelle de drap mouillé, avec la scrupuleuse attention de ne jamais intervertir cet ordre. Qu'attendre à priori d'une telle combinaison? Eh bien! je n'hésite pas à le dire, cette masse en apparence inerte, cet assemblage bizarre, cette pile de tant de couples de métaux dissemblables séparés par un peu de liquide, est, quant à la singularité des effets, le plus merveilleux instrument que les hommes aient jamais inventé, sans en excepter le télescope et la machine à vapeur.

François Arago

("Alexandre Volta", Oeuvres complètes, V. 1, p. 219–220, Paris, Leipzig, 1854)

Introduction

Electrostatic devices based on electrostatic induction and accumulation of electric charge remained the only useful source of electric power until the end of the 18th century. These electrostatic generators produced high voltages across the coatings of "Leyden jars" (up to several tens of kilovolts), sufficient to experiment with spark discharges, but the total charge generated was virtually infinitesimal, from 10^{-6} to 10^{-4} coulomb.

The Italian physician Luigi Galvani carried out in 1786 his famous experiments in which two different metals applied to frog's nerve caused muscle contraction quite similar to that produced by discharge of Leyden jars. Galvani explained this effect in terms of "animal electricity". The first to come up with the correct interpretation was the Italian physicist Alessandro Volta who in 1794 pointed to the contact of two different metals with each other and the muscular tissue as the cause of this "galvanic" effect. In March 1800 Volta made public the "construction of an apparatus . . . of unfailing charge, of perpetual power". This device, which is known nowadays as the Volta pile, was the first chemical power source, or "galvanic battery". Better and more efficient types of chemical power sources were developed as time went on.

However primitive was the first chemical power source, its appearance marked the beginning of a new era in the history of electricity. Although the concepts of positive and negative charge were already formed, and fundamental laws of electrostatics discovered (Coulomb's law, for example), the phenomenon of continuous flow of electrical charges remained unknown. In other words, the concept of electric current had not yet evolved. The various effects produced by electric current were equally unknown.

The chemical effects of electric current were discovered virtually within a few months of Volta constructing his battery. Nicholson and Carlisle reported electrolysis of water in May of 1800. Metal electrodeposition processes were discovered in 1803, and in 1807 Davy isolated alkali metals by electrolysis of molten salts.

The magnetic field of an electric current was discovered by Oersted in 1819. The fundamental laws of electrodynamics and electromagnetism were formulated during the next several decades: Ampere's law of interaction between electric currents (1820); Ohm's law of proportionality between current and voltage (1827); the law of electromagnetic induction (Faraday, 1831); Joule's law of heat release in an electric circuit (1843), and many others. All these phenomena were discovered in experiments carried out with electric current supplied by chemical batteries.

The appearance of electrochemical cells provided an impetus to research into practical applications of electric current. The first prototype of the electric telegraph appeared in 1804. The first practical electric motor was designed by Jacobi in 1834, and in 1838 Jacobi experimented with a motor-driven boat on the Neva river not far from St. Petersburg. The same year he formulated the principles of electroforming.

Electrochemical batteries remained the only practical source of electric power throughout the first half of the nineteenth century, and substantial progress was achieved. This led to rapid development of the theory and practice of electrical engineering, and the seventh decade of that century saw the appearance of a revolutionary new power source: the electromagnetic generator. These generators soon surpassed their predecessors both in electrical and economic parameters. Electromagnetic generators brought about the development of electric grids, and electric energy was put to large-scale domestic and industrial use.

Although by the end of the 19th century electrochemical cells had ceased to be the only source of electric power, they were being perfected and used as autonomous power sources for the emerging communications facilities and in portable devices. It is interesting to note that battery-powered vehicles were developed by the end of the century, and at the time proved competitive with as yet primitive cars powered by internal combustion engines.

A new surge of interest in chemical batteries began in approximately 1920, in connection with momentous progress in radioelectronics. Primary cells and storage cells monopolized power supply in radio receivers for almost two decades. Another powerful impetus to development of electro-chemical cells was provided by the automobile industry which needed to organize large-scale production of starter batteries. The post-war progress in electronic circuitry and in aerospace technology required that not only the production output be increased but that cell performance be drastically improved.

Electrochemical cells are employed nowadays in almost every field of human activity. The yearly production of the cell manufacturing industry around the globe has reached billions of individual primary and storage cells. It is quite impressive that, should all the electrochemical cells now in use be switched on simultaneously, the total electric power thus obtained would be comparable to that of all electric power plants in the world taken together (about 10^9 kW). This comparison disregards, of course, the fact that power plants operate continuously while batteries deliver current for short periods with long intervals between them, so that the total energy delivered is small compared to energy produced by power plants.

Other types of devices can also be used as autonomous power sources. Thermal energy is directly converted into electrical energy in thermoelectric generators and thermoionic converters. The sun's radiation is converted in solar cells, and radiation energy of radioactive isotopes is converted in isotopic batteries. Nevertheless, electrochemical cells remain the most versatile and the most widely used autonomous power packs. The reason for this lies in their operational advantages: they are independent of external sources of heat or radiation, and are constantly ready to deliver power. In contrast to diesel engine/generator sets, also used for autonomous power supply, electrochemical cells are noiseless.

Another reason for widespread acceptance of electrochemical cells is the tremendously wide range of electric power that can be delivered. Electronic wristwatches are powered by miniature cells with power of about 10^{-5} W, while submarine storage batteries reach up to 10^7 W. The mass of a single "power unit" may vary from 0·1 gram to a hundred tons. It is striking that both huge and miniature cells operate with the same high efficiency which, furthermore, is maintained in the most varied operational conditions. No other type of electric power source could be said to be as flexible or as versatile.

The electrochemical and physicochemical processes occurring in electrochemical cells are fairly complicated. A clear understanding of these processes is essential for maintenance and operation of chemical power sources, as well as for their improvement or for development of new cell types.

BIBLIOGRAPHY

Monographs

1. G. W. Heise and N. C. Cahoon (eds.), "The Primary Battery", Vol. 1. Wiley, New York (1971).
2. N. C. Cahoon and G. W. Heise (eds.), "The Primary Battery", Vol. 2. Wiley, New York (1976).
3. K. V. Kordesch (ed.), "Batteries", Vol. 1, Manganese Dioxide. Dekker, New York (1974).

4. K. V. Kordesch (ed.), "Batteries", Vol. 2, Lead Acid Batteries and Electric Vehicles. Dekker, New York (1977).
5. G. W. Vinal, "Storage Batteries". McGraw-Hill, New York (1955).
6. S. Falk and A. J. Salkind, "Alkaline Storage Batteries". Wiley, New York (1969).
7. R. Jasinski, "High-energy Batteries". Plenum Press, New York (1969).
8. E. W. Justi and A. W. Winsel, "Kalte Verbrennung. Fuel Cells". Steiner, Wiesbaden (1962).
9. W. Vielstich, "Brennstoffelemente". Verlag Chemie, Weinheim (1965).
10. Y. Miyaka and A. Kozawa (eds.), "Rechargable Batteries in Japan". JEC Press, Cleveland (1977).
11. M. Breiter, "Electrochemical Processes in Fuel Cells". Springer, Berlin, Heidelberg, New York (1969).
12. J. Kořita, J. Dvořak and V. Bohackova, "Electrochemistry". Methuen, London (1970).
13. V. S. Bagotzky, and V. N. Flerov, "Recent Developments in Chemical Power Sources", Energiya Publishing House, Moscow (1963) (In Russian).
14. M. A. Dasoyan, "Chemical Power Sources". Energiya Publishing House, Leningrad (1969) (In Russian).
15. V. V. Romanov and Yu. M. Hashev, "Chemical Power Sources". Soviet Radio Publishing House, Moscow (1978) (In Russian).

Periodicals and Series

1. *Journal of Power Sources* (D. H. Collins, ed.). Elsevier Sequoia S.A., Lausanne.
2. *Energy Conversion*. Pergamon Press, Oxford.
3. *Journal of Applied Electrochemistry*.Chapman and Hall, London.
4. *Journal of the Electrochemical Society*. Princeton, N.J.
5. *Electrokhimiya* (Nauka Publishing House, Moscow)—English translation—Soviet Electrochemistry, Consultants Bureau, New York.
6. *Proceedings of International Power Sources Symposia in Brighton* (D. H. Collins, ed.).
   1962 Batteries. Pergamon Press, Oxford (1963).
   1964 Batteries 2. Pergamon Press, Oxford (1965).
   1966 Power Sources. Pergamon Press, Oxford (1967).
   1968 Power Sources 2. Pergamon Press, Oxford (1969).
   1970 Power Sources 3. Oriel Press, Newcastle upon Tyne (1971).
   1972 Power Sources 4. Oriel Press, Newcastle upon Tyne (1973).
   1974 Power Sources 5. Academic Press, London and New York (1975).
   1976 Power Sources 6. Academic Press, London and New York (1977).
   1978 Power Sources 7. Academic Press, London and New York (1979).
7. *Proceedings on Intersociety Energy Conversion Engineering Conferences* (published by the host organization).
   7 IECEC, San Diego (1972).
   8 IECEC, Philadelphia (1973).
   9 IECEC, San Francisco (1974).
   10 IECEC, Newark (1975).
   11 IECEC, State Line (1976).

  12 IECEC, Washington (1977).
  13 IECEC, San Diego (1978).
  14 IECEC, Boston (1979).
8. *Proceedings of United States Army Power Sources Symposia.* Red Bank, N.J., PSC Publication Committee 24–28 Power Sources Symposia (1970, 1972, 1974, 1976, 1978).
9. *International Electric Vehicle Symposia*
  4 EVS, Düsseldorf (F.R.G.) (1976), preprints.
  5 EVS, Philadelphia (1978), preprints.
10. *Progress in Batteries and Solar Cells* (A. Kozawa, K. V. Kordesch *et al.*, eds.) JEC Press Inc., Cleveland (USA), Vol. 1 (1978); Vol. 2 (1979).

Part One

General Aspects

Chapter One

Working Principles of Chemical Power Sources

1.1 Basic concepts

Chemical power sources are chemical-to-electrical energy converters. In the process of operation (discharge) of a chemical power source two reactants interact by way of a chemical reaction. The energy of this reaction is delivered as the energy of direct electric current.

Other devices are known for conversion of chemical energy into electrical energy, for example, the thermal power plant and diesel engine/generator sets. The energy of fuel combustion is transformed in these converters first into thermal energy (in furnaces or combustion chambers), then into mechanical energy (in turbines or cylinders of internal combustion engines), and finally into electrical energy (in electrical generators). In contrast to this multiple-stage process, electrochemical cells operate on a single-stage basis, avoiding intermediate transformation into other types of energy.

A chemical power source is thus a device in which the energy of a chemical reaction occurring within it is directly converted into electrical energy.

The numerous existing types of chemical power sources vary in size, structural features, and the nature of the chemical reactions. Correspondingly, they vary in performance and parameters. This variety reflects diversified conditions in which cells operate, each field of application imposing its specific requirements.

A chemical power source comprises one or several single cells, referred to as galvanic cells. Each such cell generates a comparatively low voltage, typically from 0·5 to 4 V for different classes of cells. If higher voltages are required, the necessary number of cells are connected in series, comprising a galvanic battery.

By their principles of functioning, galvanic cells (and batteries consisting of such cells) are subsumed under the following headings:

3

(a) **Primary cells** (single-discharge cells, sometimes also called galvanic cells). A primary cell contains a finite quantity of the reactants participating in the reaction; once this amount is consumed (on completion of discharge), a primary cell cannot be used again.

(b) **Storage cells** (multiple-cycle cells, also called accumulators, and secondary or reversible cells). On completion of cell discharge, a storage cell can be recharged by forcing the electric current through the cell in the reverse direction; this results in regeneration of the initial reactants from the reaction products. Therefore, electric energy supplied by an external power source is accumulated in the storage cell in the form of chemical energy; during the discharge phase this energy is delivered to a consumer. Most storage cells allow a large number of such charge–discharge cycles (hundreds or even thousands), and though they cannot be operated continuously, their total service life is considerable.

(c) **Fuel cells.** In the fuel cell mode of operation, reactants are continuously fed into the cell while the reaction products are continuously removed. Hence, fuel cells can discharge continuously for a considerable length of time.

The classification into primary and storage cells is not rigorous since in certain conditions some primary cells can be recharged; on the other hand, storage cells are sometimes used for a single discharge.

Two intermediate types of galvanic cells are also known. Compound cells ("semi-fuel cells") contain a store of only one of the reactants while the second is continuously fed into the cell during discharge. The discharge duration of such cells is determined by the stored amount of the first reactant; if this reactant can be regenerated by the charge current, a compound cell functions as a storage cell. In renewable cells (mechanically and/or chemically rechargeable cells) it is possible, on completion of discharge, to replace the consumed reactants by fresh amounts. Therefore, in contrast to fuel cells, cells of this type operate with reactant storage which is replenished not continuously but periodically. Renewable cells are often of the compound type, with the supply of only one reactant renewed periodically.

1.2. Chemical reactions in cells

(a) Current-producing reactions

Any chemical power source uses a chemical reaction between an oxidizer and a reducer. It is a well established fact that in reactions of this type the reducer being oxidized releases electrons, while the oxidizer being reduced gains electrons. One example of a redox reaction is the reaction between silver oxide (oxidizer) and metallic zinc (reducer):

$$Ag_2O + Zn \rightarrow 2Ag + ZnO \qquad (1.1)$$

in which electrons are transferred from zinc atoms to silver ions incorporated into the crystal lattice of silver oxide.

If the reaction (1.1) takes place in a flask in which silver oxide is thoroughly mixed with fine zinc powder, no electrical energy is produced despite all the intergrain transfer of electrons. This is because electron transfers in the reaction mixture are spatially random so that the reaction energy is liberated as heat, considerably raising the temperature of the reaction mixture. The same reaction proceeds in chemical power sources but in an ordered manner controlled by an electrochemical mechanism, and therefore produces electric current.

In the simplest case a galvanic cell consists of two electrodes made of different materials immersed into an electrolyte† (Fig. 1). The electrodes are conducting metal plates or grids covered by reactants (sometimes called active materials); the oxidizer is applied to one electrode and the reducer to the other. In the silver–zinc cell the electrodes are metal grids, one of which is covered with silver oxide and the other with zinc; a solution of KOH serves as electrolyte.

When the electrodes are lowered into the electrolyte, a potential difference develops between them; it is called the open circuit voltage (o.c.v.), and denoted by $\mathscr{E}$. The potential of the electrode in contact with the reducer is more negative since it has a stronger tendency to release electrons. In the case of the silver–zinc cell $\mathscr{E} = 1\cdot60$ V and the zinc electrode is negative.

If the electrodes are connected by a conducting external circuit, the presence of o.c.v. causes the electrons to flow from the negative electrode to the positive one. This is equivalent to the electric current I flowing in the reverse direction (by convention, the electric current is considered as a flow of positive charge). Simultaneously, reactions start on the surfaces of the electrodes immersed into the electrolyte: zinc is oxidized on the negative electrode

$$Zn + 2OH^- \rightarrow ZnO + H_2O + 2e \qquad (1.2)$$

† Both in this phrase and hereafter the term "electrolyte" denotes a liquid or solid phase with ionic conductivity (for instance, an acid solution, molten salt, etc.). A different interpretation is sometimes used: An electrolyte is a substance which in the normal state has no ionic conductivity but becomes an ionic conductor as a result of dissociation into ions after dissolving or melting (for instance, a solid salt which may form an "electrolyte solution" or an "electrolyte melt").

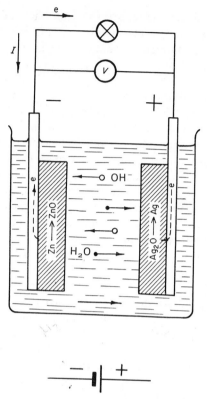

Fig. 1. Galvanic cell schematic. Shown at the bottom of the figure is the symbol representing a cell in drawing of electric circuits.

and silver oxide is reduced on the positive electrode

$$Ag_2O + H_2O + 2e \rightarrow 2Ag + 2OH^- \qquad (1.3)$$

Those chemical reactions which produce electrons or in which electrons participate are called the electrochemical or electrode reactions. The two electrode reactions are matched in the sense that their rates are always equal, that is the number of electrons released by one reaction per unit time is equal to the number of electrons consumed in the second reaction.

Electrode reactions sustain a continuous flow of electrons in the external circuit. The OH^- ions produced by reaction (1.3) in the vicinity of the positive electrode are transported through the electrolyte toward the negative electrode and enter the reaction (1.2). The electric circuit as a whole is thus closed. The same electric current then flows in each segment of the circuit, and charge is not accumulated either in electrodes or in other parts of the circuit. When the total amount of at least one reactant is consumed, the current flow stops.

The total reaction on both electrodes, referred to as the overall current-producing reaction, is the reaction (1.1). In contrast to the process in a flask, in chemical power sources the reaction proceeds as two spatially separated partial reactions. Electric current is generated because random transfer of electrons between reactants is replaced by a spatially ordered process: electrons supplied by the reducer particles first go to the negative electrode, then to the positive electrode through the external circuit, and only from this electrode to the oxidizer particles. The electrical energy generated by a chemical power source is the work of charge transfer in the external circuit.

Two basic principles built into the cell structure make possible the above-described electrochemical mechanism: (a) the reactants, namely the oxidizer and the reducer, are spatially separated, and this results in spatial separation of electrode reactions; (b) each reactant contacts the electrolyte which both allows the electrode reactions to take place and at the same time provides the closing segment of the electric circuit in which the current can flow.

The silver–zinc cell is one example of a storage cell: after discharge, it can be recharged by forcing through it an electric current in the reverse direction. In this process both electrode reactions (1.2) and (1.3) and the overall reaction (1.1) go from right to left: zinc oxide is being reduced on the negative electrode, while silver is being oxidized on the positive one.

Those electrode reactions which proceed readily in either of the two directions are called reversible. The unobstructed reversible functioning of electrodes is a prerequisite for making high-performance storage cells.

Electrochemical cells. A device comprising two electrodes in contact with an ion-conducting electrolyte is called an electrochemical cell. The cell o.c.v. is different from zero if the electrodes are of different materials. Two directions of current flow are then possible in a cell: the "natural" direction, when the current in the external circuit flows from the positive to the negative electrode, and the direction of forced flow, when an external voltage forces the current to flow in the reverse direction. The first case corresponds to the discharge of the cell (the galvanic cell mode), and the second corresponds to its charging and to electrolysis processes (the electrolyser mode).

The electrode through which the current flows from the external circuit into the electrolyte is termed the anode, and its counterpart with the opposite direction of current is termed the cathode. The oxidation electrode reactions (releasing electrons) always occur at the anode, and are thus termed anodic reactions. Correspondingly, the reduction electrode reactions are termed cathodic reactions.

When an electrochemical cell is being discharged, its negative electrode is the anode, and its positive electrode is the cathode (Fig. 2(a)). In charging of a cell, and also in the operation of an electrolyser, the situation is reversed (Fig. 2(b)). Therefore in cell nomenclature the terms anode and cathode are linked not to the electrode polarity (positive and negative) but to the current direction, and using them calls for certain caution.

Electrochemical systems. The combination of the reactants of the current-producing reaction, that is the oxidizer and the reducer, and the electrolyte is the so-called electrochemical system of a given chemical power source. The electrochemical system determines not only the nature of the current-producing reaction but also a number of performance characteristics of cells.

By convention, an electrochemical system is written in the notation

$$(-) \text{ reducer} \mid \text{electrolyte} \mid \text{oxidizer} (+)$$

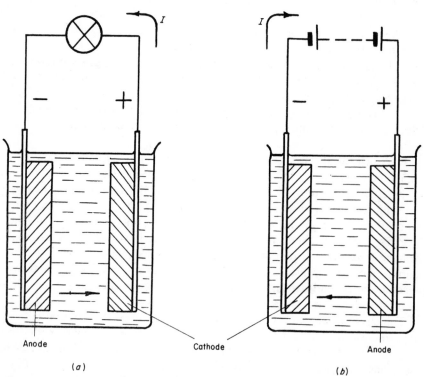

Fig. 2. Current direction (a) in a discharging cell, and (b) in a storage cell being charged or in an electrolyser.

Vertical bars denote interfaces of two conductors, for instance, electrode–electrolyte interfaces at which electrode reactions occur. In the case of the silver–zinc cell the electrochemical system is written as

$$(-)\, Zn\,|\,KOH\,|\,Ag_2O\,(+)$$

(the chemical formula of water which is part of aqueous electrolyte solutions, is not included in this notation). The reducer (negative electrode) must be written at the left-hand side. In naming an electrochemical system or a chemical power source, however, it is usual to place the oxidizer first; for instance, "silver–zinc battery" (it would be more correct to say "silver oxide–zinc battery").

Sometimes a cell contains two electrolyte solutions contacting through a porous diaphragm. In this case the interface between the two liquids is shown as a vertical dashed line. For a copper–zinc cell in which the copper electrode is immersed in a copper sulphate solution, and the zinc electrode in a zinc sulphate solution, the electrochemical system is shown as

$$(-)\, Zn\,|\,ZnSO_4 : CuSO_4\,|\,Cu\,(+)$$

When the reactants are liquids or gases, the current-producing reaction occurs on the surface of the current-collecting electrode whose material is not directly involved in the reaction. Sometimes such a non-consumable electrode is called an inert electrode. Usually, however, the "inert" electrode is responsible for a strong catalytic effect on the electrode reaction, so that the cell performance may be substantially improved by a suitable electrode material. Therefore, it is good practice to include into the notation for the electrochemical system the material of which the current-collecting electrode is made. For oxygen–hydrogen fuel cells we thus have

$$(-)\, Pt, H_2\,|\,KOH\,|\,O_2, Pt\,(+)$$

(b) Side reactions

Storage of chemical power sources and the main current-producing reaction during charging and discharging may be accompanied by chemical or electrochemical side reactions. A number of metals are corroded on contact with aqueous solutions, whereby water is decomposed and hydrogen is liberated. This process is especially intensive in the case of alkali metals. Likewise, strong oxidizers may give rise to oxygen evolution from water or may react with structural parts of cells, such as spacers, separators, and so on. Strong reducers are readily oxidized by the oxygen of air. These and some other processes result in unproductive consumption of reactants, in other words, in self-discharge of electrodes.

In addition to self-discharge of individual electrodes, a cell as a whole may be self-discharging owing to direct interaction of reactants or because of internal short-circuiting. For example, metal deposition may lead to formation of thin metal "bridges" between electrodes. The electrodes are then discharging through such bridges by way of normal current-producing reactions, but the current bypasses the external circuit and thus cannot be utilized, the reaction energy being released as heat. The variable-valency ions, such as the iron ions, produce similar effects. Alternately diffusing toward the positive and then the negative electrode, these ions are oxidized ($Fe^{2+} \rightarrow Fe^{3+}$) on the former, and then reduced on the latter ($Fe^{3+} \rightarrow Fe^{2+}$), each time consuming a fraction of the reactants. This "shuttle" effect is equivalent to a continuous—and unproductive—transfer of electrons from the negative to the positive electrode. The presence of variable-valency ions is therefore considered a detrimental factor in cell electrolytes.

Side reactions also take place in the case of overcharge of a storage cell. Current can only flow through an electrochemical cell if electrode reactions are taking place at the electrodes. Hence, a charge current forced through a completely charged battery, having depleted the available amount of reactants, involves new species in the electrode reactions. In storage cells with aqueous electrolyte solutions, anodic evolution of oxygen starts at the positive electrode

$$4OH^- \rightarrow O_2 + 2H_2O + 4e \qquad (1.4)$$

and cathodic evolution of hydrogen at the negative electrode

$$2H_2O + 2e \rightarrow H_2 + 2OH^- \qquad (1.5)$$

(the reactions are given for alkaline solutions). Oxygen and hydrogen evolution commences after the respective electrode is completely charged, and thus may not necessarily start simultaneously. The overall process on the two electrodes is the electrolysis of water. Similar processes of electrolytic decomposition of the solvent or electrolyte occur in storage cells with non-aqueous electrolytes. For certain storage cell types such post-charge processes are impossible (for example, for solid-electrolyte cells). In such cases the steady-state current drops off on completion of charging.

All the side reactions described above lower the efficiency of a cell by reducing the electric energy delivered in the course of cell discharge and by raising the energy required for charging the cell. The ratio of the main current-producing reaction and the unproductive side reactions depends on the cell design, temperature and some other factors.

1.3. Open-circuit voltage, on-load voltage and current density

Each electrode of the electrochemical cell brought into contact with the electrolyte develops a certain electrode potential E, also referred to as the redox potential of a given electrode reaction. The higher the reducing power of the reactant the more negative the electrode potential, and the higher the oxidizing power of the reactant the more positive the potential. The concept of the "electrode potential" (or the "half-cell" potential) is by no means a trivial one; we give it a detailed definition and analyse its physical meaning in Section 4.1. Nevertheless, this parameter will be used throughout the first three chapters of the book to characterize individual electrodes. The potentials of a number of electrode materials are listed in Tables A.1 and A.2 (see Appendix).

The open-circuit voltage (o.c.v.), $\mathscr{E}$, of a galvanic cell, is the potential difference between the positive and negative electrodes with no current flowing in the external circuit:

$$\mathscr{E} = E_{(+)} - E_{(-)} \qquad (1.6)$$

According to this definition, the o.c.v. is always positive. The voltage depends on the nature of both the electrodes and electrolyte (see Table A.3 of the Appendix) but not on the size or structural features of a cell. The value of the o.c.v. $\mathscr{E}$ is related to the thermodynamic value of the electromotive force (e.m.f.) $\mathscr{E}^{(T)}$; these two concepts will be more rigorously defined in Section 4.1. In various cell types the values of o.c.v. may be equal to, or smaller than, the values of the e.m.f.

When a galvanic cell delivers current, the potential difference between electrodes is changed. This potential difference is called the on-load (charge or discharge) voltage and denoted by U. The higher the current the lower the discharge voltage U_d and the higher the charge voltage U_c; both U_d and U_c tend to the value of the o.c.v. for very low current drains.

Two factors determine changes of voltage across the cell in on-load conditions: the ohmic potential difference $\varphi_{ohm} = IR_{ohm}$ determined by the internal cell resistance R_{ohm} (this ohmic voltage develops mostly across the electrolyte layer between the electrodes); and electrode polarization. The polarization accounts for changes in electrode potentials caused by current flow, from the initial (off-load) value E to a new value $E^{(i)}$. Quantitatively, the polarization (overpotential) is the magnitude of the shift in potential

$$\eta = |E^{(i)} - E| \qquad (1.7)$$

The cathode current shifts the potential toward more negative values, and the anode current toward more positive values (the causes of polarization will be elaborated in Section 4.3). As a result, the voltage across the cell electrodes diminishes in the discharge·phase and increases during charging; ohmic potential difference produces the same effect.

Summarizing and taking into account equations (1.6) and (1.7) we can express the on-load discharge voltage U_d as

$$U_d = E^{(i)}_{(+)} - E^{(i)}_{(-)} - I_d R_{ohm} = \mathscr{E} - \eta_{(+)} - \eta_{(-)} - I_d R_{ohm} \qquad (1.8)$$

and the charge voltage as

$$U_c = E^{(i)}_{(+)} - E^{(i)}_{(-)} + I_c R_{ohm} = \mathscr{E} + \eta_{(+)} + \eta_{(-)} + I_c R_{ohm} \qquad (1.9)$$

Both the ohmic and the polarization energy losses are related not only to the discharge or charge current but also to electrode dimensions, or rather to the area S of the surface in contact with the electrolyte. Losses are determined by the current density i defined as a quotient of current to surface area S ($i \equiv I/S$). As current density increases, polarization is enhanced (but in contrast to ohmic voltage, polarization is not proportional to current). Polarization is small for low current densities, and electrode potentials approach off-load values. Therefore it is preferable to achieve high current drain in batteries by increasing the surface area of the electrodes rather than by raising the current density. One possible method is to use thin electrodes because of their high surface-to-volume (surface-to-mass) ratio.

Porous electrodes are often used in cells for this purpose. The active surface of electrodes is then enlarged by electrolyte-filled inner pores. The true surface area Σ of such electrodes is many times larger than the apparent (projected) surface area S. As a result, the true current density i_σ is much smaller than i calculated from the apparent surface area S (apparent current density). Because of difficulties involved in measuring Σ, these data are not always available; as a result, the terms "electrode surface" area and "current density" in the literature dealing with cells usually refer to S and i.

1.4. Faraday's laws: specific consumption of reactants

In the general case a current-producing reaction can be written as

$$v_O O + v_R R + v_X X \rightarrow v_P P + v_Q Q + v_Y Y \qquad (1.10)$$

and its partial electrode reactions as

$$v_O O + v_X X' + ne \rightarrow v_P P + v_Y Y' \qquad (1.11)$$

$$v_R R + v_X X'' \rightarrow v_Q Q + v_Y Y'' + ne \qquad (1.12)$$

where O and R stand for the initial oxidizer and reducer (silver oxide and zinc in the silver–zinc cell), P and Q denote their reaction products (silver and zinc oxide), X and Y denote all the other participants of the reaction (hydroxyl ions, water molecules), v_j are the stoichiometric coefficients, and n is the number of electrons taking part in a single elementary step of the reaction† ($n = 2$ for the silver–zinc cell).

The laws established by Faraday in 1833–34 <u>state that the total charge that passes through an electrochemical cell is directly related to the mass of the reactants consumed in the reaction.</u> When v_O gram-moles of oxidizer and v_R gram-moles of reducer are consumed in reaction (1.10), the charge that has passed through the circuit is nF, where F is the Faraday number (or simply Faraday) equal to 96 491 (roughly 96 500) C/mol or 26·8 ampere-hours. For example, the total charge passing through a silver–zinc cell per one mole Ag_2O (231·74 g) and one mole Zn (63·37 g) is $2F$, that is 53·6 Ah. From this it is easy to calculate specific consumption of reactants per unit charge, for instance one ampere-hour: 4·32 g/Ah of Ag_2O and 1·22 g/Ah of Zn. (The mass of the reactant corresponding to one Faraday is called the equivalent mass.)

The specific consumption of reactants $g_Q^{(T)}$ thus calculated is a theoretical, or ideal, parameter. In actual cells specific consumption g_Q is always greater than the theoretical value. This stems from diminishing efficiency of the electrochemical reaction resulting from the accumulation of reaction products, or for other reasons, so that the reactants are not fully consumed (this is the so-called passivation of the electrodes). Moreover, the reactants may be wasted in side reactions, producing what look like deviations from Faraday's laws. The ratio $\lambda = g_Q^{(T)}/g_Q$ is called the utilization coefficient of a given reactant. This coefficient is strongly dependent on cell design and on the operation mode, and may vary for different reactants from 0·2 to 0·98.

Another practically important characteristic is the specific consumption of reactants per unit energy produced g_W. The energy released by a cell discharging into the external circuit is found as the product of the total transferred charge and the potential difference between the electrodes. Maximum energy W_{max} is given by

$$W_{max} = nF\mathscr{E}^{(T)} \qquad (1.13)$$

as $\mathscr{E}^{(T)}$ is the maximum possible value for the o.c.v. (unless stated otherwise, energy in equation (1.13) and further in the text is always referred to the transferred charge nF, i.e. to the reaction of molar quantities of reactants in reaction (1.10)). The maximum energy gives the theoretical (minimum)

---

† This n should not be confused with z_j, the number of elementary charges of the jth ion; for instance, $z = -2$ for the SO_4^{2-} ions.

consumption of reactants per unit energy $g_W^{(T)}$. For a silver–zinc battery ($\mathscr{E}^{(T)}$ = 1·60 V) these consumptions are 2·70 g/Wh for silver oxide and 0·76 g/Wh for zinc. When the real specific consumption is calculated it is necessary to take into account not only the reactant utilization coefficients but also the actual discharge voltage. Indeed, discharge at lowered voltage diminishes the amount of released energy and enhances the reactant consumption per unit energy.

The specific consumption of one reactant per ampere-hour is independent of the nature of the second reactant, and can therefore be specified for each individual electrode (see Tables A.1 and A.2). At the same time the consumption per watt-hour is a function of the voltage across the cell, and can be indicated for each electrode only if the nature of the second electrode is known (Table A.3).

As follows from Faraday's laws, the current passing through a cell is proportional to the reaction rates of both the overall current-producing reaction and the partial electrode reactions. Consequently, the value of the cell current may be regarded as a measure of the reaction rates given in electrical units.

1.5. Thermodynamics of cell reactions

(a) E.m.f. and thermodynamic functions

We know from the thermodynamics that the maximum possible work of a chemical reaction is equal to the decrease in free energy G of the reaction components, $W_{max} = -\Delta G$. Hence, taking into account equation (1.13),

$$\mathscr{E}^{(T)} = -\Delta G/nF \tag{1.14}$$

This is the fundamental thermodynamic equation relating the e.m.f. to the nature of a current-producing reaction.

The heat effect of the reaction $\bar{q}$ (i.e. heat liberated if the reaction takes place in a flask) is equal to the decrease in the total energy (enthalpy) H of the reaction, $\bar{q} = -\Delta H$. This quantity is not the same as the decrease in free energy. Therefore, even when a cell is operated in optimal conditions and its voltage is very close to e.m.f., not all of the reaction energy is transformed into electrical energy; some part of it, $\bar{q}_{entr}$ (the unavailable energy), is always released as heat:

$$\bar{q} = nF\mathscr{E}^{(T)} + \bar{q}_{entr} \tag{1.15}$$

The unavailable energy is determined by the change in the reaction entropy S, $\bar{q}_{entr} = -T\Delta S$ (T is the absolute temperature).

The heat effect of an electrochemical system is conveniently expressed in electrical units, that is referred to unit charge passed through the cell: $\bar{\mathscr{E}} = \bar{q}/nF$. By convention, $\bar{\mathscr{E}}$ (in volts) is called the thermoneutral voltage. The unavailable energy can then be expressed in the form $\bar{q}_{entr} = nF(\bar{\mathscr{E}} - \mathscr{E}^{(T)})$. If a cell is operated in off-optimum conditions, that is at a voltage below the e.m.f., the difference $\bar{q}_{\Delta U} = nF(\mathscr{E}^{(T)} - U_d)$ is also released as thermal energy. The total heat released in a discharging cell is found as the sum of these two components:

$$\bar{q}_{cell} = \bar{q}_{entr} + \bar{q}_{\Delta U} = nF(\bar{\mathscr{E}} - U_d) \tag{1.16}$$

For some reactions the maximum work is greater than the heat effect, i.e. $\mathscr{E}^{(T)} > \bar{\mathscr{E}}$. In this case $\bar{q}_{entr}$ is negative (ΔS is positive) so that a cell operated in optimal conditions, instead of liberating heat absorbs it from the ambient medium and transforms it into electric energy; heat absorption in the course of the reaction results in cooling of the cell. This cooling is very rarely observed, however, because of the effect of $\bar{q}_{\Delta U}$.

The Gibbs–Helmholtz equation, one of the most important equations of chemical thermodynamics, can be presented for electrochemical systems in the following form (in electrical units):

$$\mathscr{E}^{(T)} = \bar{\mathscr{E}} + T\frac{d\mathscr{E}^{(T)}}{dT} \tag{1.17}$$

Equation (1.17) clearly shows that for systems with $\mathscr{E}^{(T)} < \bar{\mathscr{E}}$ the slope $d\mathscr{E}^{(T)}/dT$ is negative, so that e.m.f. falls off as temperature increases. For systems with $\mathscr{E}^{(T)} > \bar{\mathscr{E}}$, the e.m.f. rises as temperature increases.

The quantities G, H and S are called the thermodynamic functions. These functions have been measured for a large number of materials and can be found in reference books.† The decrease in free energy in the course of reaction (1.10) can be calculated if G is known for all reacting components:

$$\begin{aligned}-\Delta G = &(v_O G_O + v_R G_R + v_X G_X) \\ &- (v_P G_P + v_Q G_Q + v_Y G_Y)\end{aligned} \tag{1.18}$$

If the jth component of reaction (1.10) is a liquid or a gas, its free energy G_j is a function of thermodynamic activity a_j (related to solution concentration or gas pressure):

$$G_j = G_j^\circ + RT \ln a_j \tag{1.19}$$

where R is the gas constant, $R = 8\cdot313$ J/K mole. Normally the tables list

---

† Absolute values of G and H cannot be determined; therefore, the relative values are used: the values of ΔG and ΔH for the reaction of formation of the specific compounds from the elements; G and H of the individual elements are assumed equal to zero.

the values of G_j° for the standard state in which $a_j = 1$. The e.m.f. calculated from equation (1.14) for the standard $-\Delta G^\circ$ is called the standard e.m.f. and denoted by $\mathscr{E}^\circ$.

The quantities $-\Delta H$ and ΔS are found from listed values of H and S by means of expressions similar to equation (1.18), and can be used to calculate $\mathscr{E}^{(T)}$, $d\mathscr{E}^{(T)}/dT$ and other thermodynamic characteristics.

Table A.3 lists $\mathscr{E}^\circ$ and $\bar{\mathscr{E}}$ for a number of practically important cell reactions. The following should be said about these data.

(1) Thermodynamic parameters are determined not only by the nature of a specific current-producing reaction but also by any modifications employed of the reactants. Changes are possible in $\mathscr{E}^{(T)}$ and $\bar{\mathscr{E}}$ when a reactant is modified (e.g., with respect to its crystal structure, degree of hydration, etc.). The changes correspond to the transition energy of the modification involved and as a rule do not exceed $0\cdot05$ V. When a reactant is a non-stoichiometric compound (mixed compounds, variable composition phases, etc.), wider ranges of variation are possible for thermodynamic parameters. The same is true for modifications of reaction products.

(2) As a rule, $\bar{q}_{entr}$ is very small at room temperature so that $\mathscr{E}^{(T)}$ nearly equals $\bar{\mathscr{E}}$. As temperature rises the difference increases, mostly as a result of changes in $\mathscr{E}^{(T)}$. Figure 3(a) illustrates temperature dependence of $\mathscr{E}^{(T)}$ and $\bar{\mathscr{E}}$ for the hydrogen–oxygen reaction in fuel cells. In this reaction $\bar{q}_{entr} > 0$ and $d\mathscr{E}^{(T)}/dT < 0$. The step on the curve for $\bar{\mathscr{E}}(T)$ at 100°C reflects the fact that at lower temperatures the reaction product is liquid water, and at higher temperatures it is water vapour. The heat effect is higher below 100°C since water condensation heat is added to the reaction heat (in applied thermodynamics it is customary to refer to the higher and lower calorific power of hydrogen, respectively).

The ratio of the electrical energy W obtained to the reaction heat effect $\bar{q}$ is a measure of efficiency; this ratio is sometimes used to characterize electrochemical cells (and especially fuel cells) and to compare their parameters with those of other power-producing devices. The limiting value of the thermodynamic efficiency of a cell is a function only of the thermodynamic parameters of the current-producing reaction:

$$\eta^{(T)} = W_{max}/\bar{q} = \mathscr{E}^{(T)}/\bar{\mathscr{E}} \tag{1.20}$$

According to the second law of thermodynamics, the theoretically possible efficiency of a heat engine is determined by the maximum and minimum temperatures, T_{max} and T_{min}:

$$\eta^{(T)} = 1 - T_{max}/T_{min} \tag{1.21}$$

Figure 3(b) shows thermodynamic efficiency as a function of temperature for the oxygen–hydrogen fuel cell and for an ideal heat engine (where it is assumed that $T_{min} = 300$ K). Obviously, heat engines are inferior to fuel cells in this respect up to about 1200 K. The thermodynamic efficiency of a heat engine is lower because only a small fraction of thermal energy can be converted into the highly ordered mechanical or electric energy.

Actual efficiency depends also on performance characteristics. In the case

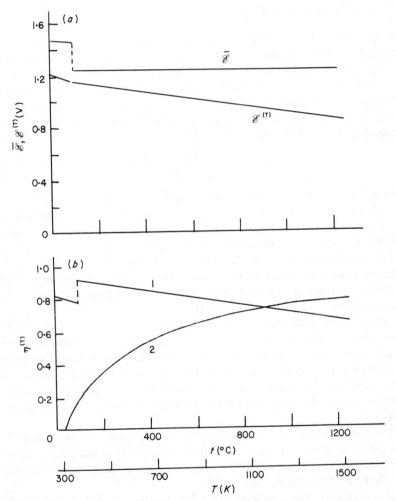

Fig. 3. (a) E.m.f. $\mathscr{E}^{(T)}$ and thermoneutral voltage $\bar{\mathscr{E}}$ as functions of temperature for the hydrogen–oxygen fuel cell; (b) Theoretical thermodynamic efficiency of the hydrogen–oxygen fuel cell (1) and of the ideal heat engine (2) as functions of temperature.

of cells, these are the voltage efficiency $\eta_U = U/\mathscr{E}^{(T)}$ and the reactant utilization coefficient λ. The net efficiency is thus

$$\eta^{(f)} = \lambda W/\bar{q} = \lambda\eta_U\eta^{(T)} \tag{1.22}$$

In the existing cell types $\eta^{(f)}$ reaches 60–80%. The corresponding figure for the best heat engines is only 45%, and reaches 55–60% only in magneto-hydrodynamic generators at $T_{max} \sim 2000$ K.

(b) The Nernst equation

The potential of an individual electrode is related to the thermodynamic parameters of the electrode reaction by an expression similar to equation (1.14) (a more rigorous analysis will be given in Section 4.1). In the case of dissolved or gaseous components both free energies G and electrode potentials are determined by the component activities; the corresponding relationship is called the Nernst equation. For the electrode process (1.11), equations (1.14), (1.18) and (1.19) yield

$$E = E^\circ + \frac{2\cdot3RT}{nF} \log \frac{a_O^{\nu O}a_X^{\nu X}}{a_P^{\nu P}a_Y^{\nu Y}} \tag{1.23}$$

where 2·3 is the numerical coefficient of conversion of natural to common logarithms; at $t = 25°C$ the factor $2\cdot3RT/F = 0\cdot059$ V. The first term E° in equation (1.23) is a constant specific for each electrode reaction, and is referred to as the standard electrode potential. It corresponds to the value of the electrode potential when the activity of all reactants is equal to unity. The numerator in the expression under the logarithm in equation (1.23) includes activities of the more oxidized form of the reactant and other components on the left-hand side of (1.11) or the right-hand side of (1.12); likewise, the denominator includes the activities of the more reduced form and other reaction products. Equation (1.23) takes into account the activities of only those components whose concentrations or partial pressures are changed (i.e. solutes and gases). The activity of constant-composition solids and that of water in dilute aqueous solutions is constant and is assumed equal to unity. The activity of water is reduced in concentrated solutions.

The electrode potential of any electrode material is mostly determined by E°; the concentration term in the Nernst equation is of secondary importance. Ten-fold variation of activity at room temperature (for $n = 2$) changes the electrode potential by only 0·03 V.

Unfortunately, the tabulated data† for activities of cell electrolytes are

---

† The activities of the positive and negative ions cannot be determined separately, consequently, tables list the mean ionic activities $a_\pm$ or the total activity of the solute, a_j.

available only for solutions of pure acids, alkalies and salts, but not for frequently used multicomponent high-concentration solutions. Consequently, activity values are used as a rule for solutions of pure acids, alkalies and some salts, and concentration values for other types of solutions. The activity coefficients (the activity-to-concentration ratios) are in the latter case assumed equal to unity. The activity of gases is also substituted by their partial pressures.‡ The results thus obtained are not very accurate but as a rule this is not significant because of the relatively small contribution of the concentration term.

As an illustration, here are the Nernst equations for three cell electrodes:

(a) **Copper electrode:** the reaction $Cu^{2+} + 2e \rightleftharpoons Cu$

$$E_{Cu} = E_{Cu}^{\circ} + \frac{2 \cdot 3RT}{2F} \log c_{Cu^{2+}}$$

(b) **Hydrogen electrode:** the reaction $2H^+ + 2e \rightleftharpoons H_2$

$$E_{H2} = E_{H2}^{\circ} + \frac{2 \cdot 3RT}{2F} \log \frac{a_{\pm}^2}{p_{H_2}}$$

(c) **Silver oxide electrode:** the reaction

$$Ag_2O + H_2O + 2e \rightleftharpoons 2Ag + 2OH^-$$

$$E_{Ag2O} = E_{Ag2O}^{\circ} - \frac{2 \cdot 3RT}{F} \log a_{\pm} + \frac{2 \cdot 3RT}{2F} \log a_{H2O}$$

If the electrolyte composition or that of solid reactants varies in the course of cell discharge, the off-load electrode potentials and the cell o.c.v. are accordingly changed (if, for example, we measure them while temporarily breaking the current). If the composition remains unaltered and only the ratio of solid reactants to reaction products is changed, both the electrode potentials and the o.c.v. remain at their initial values.

1.6. Analogues of chemical power sources

A number of devices and processes can be considered which are functionally analogous to chemical power sources or to processes in them.

‡ Gas pressures that have to be inserted into the Nernst equation must be expressed in the now obsolete units, normal atmospheres (atm), since the values of E° listed in the tables are calculated for the pressure of 1 atm which in the SI system corresponds to 101325 Pa (approximately 0·1 MPa).

The energy conversion in electrolysers proceeds in the direction opposite to that realized in galvanic cells, that is the electrical energy is converted into chemical energy. In principle, however, the structural design of electrolysers is identical to that of cells; the nature of electrode reactions also remains the same. Electrolysis units are distinguished by being large-scale and stationary. The total power of electrolysis plants reaches hundreds of thousands of kilowatts. This is an indication of the feasibility of developing large-scale electrochemical power plants for production of electrical energy.

Electrochemical cells can be employed to generate electric power by converting not only chemical but other types of energy as well, for instance, thermal energy. In the simplest case a symmetric cell design can be used in which two identical electrodes are kept at different temperatures. Because of the non-zero temperature coefficient of the electrode potential, an e.m.f. is produced between the two electrodes, and discharge current can be delivered into the external circuit. The same electrode reaction takes place at each electrode but in the opposite directions, and there is no overall current-producing reaction; electric power is generated at the expense of the thermal energy input required to maintain the temperature difference between the electrodes. A similar principle may prove feasible in some cases when one electrode of the symmetric circuit is irradiated with light. Numerous other classes of electrochemical devices are known for conversion of thermal or light energy into electric power.

Corrosion processes, especially those at the contact of two materials, also involve reactions similar to the reactions in a shorted galvanic cell. The same galvanic-cell principle can be used for cathodic protection of metals.

Instead of storage cells, electric capacitors can be used to store electric energy. A single capacitor can be charged to a much higher voltage than a storage cell (up to the breakdown voltage, typically of the order of several hundred volts) but the voltage across the capacitor electrodes is continuously changing in the process of charging or discharging. The specific energy that can be stored in a capacitor is much lower than that of a storage cell. The capacitor discharge current is determined, however, solely by the parameters of the external circuit, and very high peak power can be delivered by capacitor discharge (flash-bulbs are one example) while the storage cell discharge current is limited by its internal resistance.

The energy released in oxidation of food in living organisms is directly converted into the mechanical energy of muscular motion. The efficiency of this conversion is very high, from 60% to 70%, because biochemical reactions like electrochemical cell reactions are spatially ordered. As a result it becomes possible to convert chemical energy directly, without the intermediate stage of producing highly disordered thermal energy. It is interesting to note that a very close analogue of a chemical power source can be found in nature, for instance, the electric generator of the eel *Electrophorus electricus*. This organ can be regarded as a "battery" comprised of individual membrane cells. When excited, this battery generates voltages up to 1000 V, and can deliver pulse discharge currents up to 1 A. Similar organs are found in other species of "electric" fish. The mechanism by which biological galvanic cells function is very similar to that of man-made batteries.

Chapter Two

Cell Types

2.1. Electrochemical systems of cells

(a) Requirements imposed on electrochemical systems

As we mentioned in Chapter 1, an electrochemical system comprises two reactants, that is, an oxidizer and a reducer, and also an electrolyte.

The number of available oxidizers, both inorganic and organic (e.g. higher oxides of metals, oxygen-containing acids, and their salts, oxygen, halogens, organic nitrocompounds, etc.) and reducers (e.g., metals, their lower oxides, sulphites, hydrogen, hydrocarbons, etc.) is quite large.

As to electrolytes, these are materials with ionic conductivity, that is materials conducting electric current owing to mobility of positive and/or negative ions. Electrolytes are classified into (a) aqueous solutions of acids, alkalis or salts; (b) non-aqueous solutions with ionic conductivity, prepared by dissolving salts in organic or inorganic solvents (for example, $LiAlCl_4$ in propylene carbonate); (c) molten salts; and (d) the group of solid compounds with ionic lattices in which one of the ionic species has sufficiently high mobility.

The number of possible combinations of oxidizer, reducer and electrolyte (in other words, the number of electrochemical systems) is thus very large. Not all of them, and in fact rather few, are suitable for the development of acceptable cells. The number of electrochemical systems actually in use is very limited since the following requirements must be met.

(1) Both the oxidizer and the reducer must be thermodynamically active. The reactivity can be characterized by the electrode potential which appears when the electrode with the active compound is in contact with the electrolyte. The electromotive force is higher the stronger the reducing and oxidizing power of the reactants.

(2) The specific consumption of reactants must be low. The lower the total reactant consumption in the reaction the greater the electric charge and the higher the energy that can be obtained from a fixed amount of reactants (in other words, the better the cell characteristics). Therefore it is preferable to utilize reactants with low molecular mass or those transferring a high number n of electrons in the electrode reaction, that is reactants with low equivalent mass. Also the reactant utilization coefficient must be high.

(3) The reactants must readily enter electrochemical reactions, and these must proceed with sufficiently high rates. From the practical viewpoint this means that polarization-induced voltage loss must be acceptably low even for elevated discharge currents.

(4) The reactants must be stable in storage and in contact both with electrolytes and with the ambient medium; the rates of counter-productive side reactions must be very low.

(5) Both the reactants and the electrolyte must be technologically convenient. The reactants' properties must be such that electrodes of prescribed shape and size can be manufactured.

(6) Preferably, preparation of the reactants and the electrolyte must require only easily available and low-cost materials.

Some of the requirements itemized above are contradictory. In particular, it is far from easy to achieve the shelf-life stability of active oxidizers and reducers.

Table A.3 lists the electrochemical systems currently used for manufacturing electrochemical cells; several systems of purely historical value are also given.

Table A.3 demonstrates that electrochemical systems differ in reversibility. In some cases a current-producing process is practically irreversible, and a cell can be recharged only with substantial difficulties (for instance, mercury–zinc cells). In cases of high reversibility storage cells can serve over many charge–discharge cycles.

(b) Brief historical background

A considerable number of cell systems appeared in the first half of the 19th century. The following proved most important.

The Volta Pile

This battery (Fig. 4) employed disk-shaped electrodes made of silver (or copper) and zinc; each pair of disks in the pile was separated by broadcloth disks soaked in water. The pile was thus a battery with bipolar electrodes (see Section 2.2). The anode reaction was zinc oxidation (dissolution):

$$Zn \rightarrow Zn^{2+} + 2e$$

No special oxidizer was introduced into the cell, and this role was played by water molecules which were reduced at the silver electrode to gaseous hydrogen:

$$2H_2O + 2e \rightarrow H_2 + 2OH^-$$

As a result, the oxidizer was very weak in this system, and the o.c.v. of each cell in the pile was only about 0·4 V. The most significant shortcoming of this design was evolution of hydrogen gas in the course of discharge; accumulation of bubbles on the cathode surface resulted in cathode polarization and rapid decrease of voltage across the pile. Later the water-soaked separators were replaced by liquid electrolyte, although cell performance was not sufficiently improved. If high discharge current was needed, very large batteries had to be built. One pile manufactured in 1803 comprised 2100 individual cells.

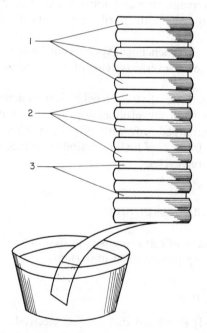

Fig. 4. Volta pile. 1, zinc disks; 2, silver disks; 3, water-soaked broadcloth disks.

Daniell, Jacobi and Meidinger copper–zinc cells

In 1836 Daniell developed a cell with oxidizer in the form of copper ions in an aqueous copper sulphate solution. The introduction of special oxidizers into chemical cells was a very important step forward. The overall current-producing reaction was

$$Zn + Cu^{2+} \rightarrow Zn^{2+} + Cu$$

In order to suppress direct chemical reaction between zinc metal and copper ions (which would result in depositing a spongy copper layer on the surface of zinc), Daniell used two electrolyte solutions in the cell; the zinc electrode was immersed in zinc sulphate solution, and the copper electrode in copper sulphate solution (Fig. 5). The two solutions were separated by a porous ceramic diaphragm which ensured that ionic current could flow within the cell. In Jacobi's version, developed also in 1836, the zinc electrode was immersed in an ammonium chloride solution. In 1859 Meidinger suggested separating the solutions by using the difference in their densities: the more dense solution of copper sulphate in the lower part of a cylindrical container, and a less dense solution of magnesium sulphate (into which the zinc electrode was immersed) in its upper part. The elimination of the

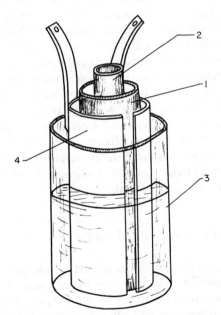

Fig. 5. Daniell copper-zinc cell. 1, ceramic cup; 2, copper cathode; 3, glass container; 4, zinc cathode.

porous diaphragm resulted in a reduction of the internal resistance of the cell.

These cells considerably surpassed Volta piles both in service life and in stability of operation, and soon became widespread. It is interesting to note that the unit of tension used in the forties and fifties of the last century was "one Daniell" equal to the o.c.v. of a copper–zinc cell (approximately 1·10 V).

Grove–Bunsen and Poggendorff–Grenet cells

These cells achieved greater discharge power due to using stronger oxidizers than those in Daniell cells, particularly solutions of nitric acid which was reduced on platinum (Grove, 1838) or graphite (Bunsen, 1841) current-collecting electrodes. The anode was again of zinc immersed in dilute sulphuric acid. The solutions in the anode and cathode compartments were separated by a porous diaphragm. Accelerated self-discharge caused by violent reaction between zinc and nitric acid which slowly diffused through the diaphragm constituted a serious disadvantage of these cells. Sodium bichromate dissolved in sulphuric acid proved a better oxidizer than nitric acid due to its less violent reaction with zinc (Poggendorff, 1842). In later devices the diaphragm was eliminated, so that the zinc electrode was immersed directly into the bichromate solution (Grenet, 1856). In order to minimize self-discharge losses, special means were built into the cells to remove the electrodes from the electrolyte when a cell was idle (Fig. 6).

It can be seen from these examples that cell parameters were improved by the utilization of better oxidizers. For a long time it was generally assumed that the oxidizers react with hydrogen evolving at the cathode, thereby removing the gas film and suppressing electrode polarization. The erroneous concept gave rise to the term "depolarizer" which meant a solid or dissolved oxidizer. From the standpoint of oxidizer's role, the term is misleading. Oxidizers do not facilitate the reaction of hydrogen reduction, nor do they lower hydrogen-induced polarization; in fact, oxidizers are themselves electrochemically reduced.

(c) Chemical power sources of today

Cell systems with dissolved oxidizers proved inconvenient for practical use in spite of all precautions and measures to the contrary. Direct zinc–acid solution interaction resulted not only in self-discharge but also in intensive evolution of gas and hence in difficulties for cell sealing.

Transition to solid oxidizers enabling complete and reliable spatial

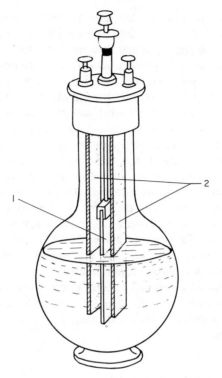

Fig. 6. Grenet cell. 1, removable zinc anode; 2, graphite cathodes.

separation of reactants was a substantial accomplishment achieved in the second part of the 19th century. A cell of this type, containing manganese dioxide mixed with graphite as oxidizer, was described by Leclanché in 1865. Ammonium chloride solution was used as electrolyte. These cells proved inexpensive and stable, and their service life reached into months and even years. Further improving Leclanché cells, Gassner and others later developed "dry" cells with immobilized "spillproof" electrolytes. In 1881 Lalande and Chaperon developed copper oxide–zinc cells with an alkaline electrolyte; other systems with solid oxidizers appeared later.

Solid reactants made it possible to manufacture smaller cells suitable to power portable units. Another important advantage consisted in the possibility of developing storage cells. The first version of lead storage cell appeared in 1859 (Planté). The first alkaline storage batteries were developed by the end of the century: the nickel–cadmium battery in 1899 (Jungner) and the nickel–iron battery in 1901 (Edison).

All these electrochemical systems comprising solid reactants and aqueous electrolytes proved so efficient that they are still in use, almost a century after they appeared. As in the past, commercially produced chemical cells are mostly based on these familiar systems. This long-term technological viability can be paralleled by only a few other inventions.

Aqueous alkaline solutions are used in a number of widespread electrochemical systems. Corrosion problems in alkaline electrolytes are not as severe as in acid solutions. Electrochemical processes in alkaline solutions are less inhibited and proceed at higher rates than in neutral solutions.

Oxidizers utilized in alkaline solutions are oxides of copper, nickel, mercury, silver and of some other metals; zinc is the most common reducing reactant. Cadmium and iron are also utilized as reducers (for example, in frequently used alkaline storage batteries). Of the three, zinc has maximum activity, but entails corrosion problems. Cadmium is not corroded in alkaline solutions and, in contrast to zinc, ensures a high degree of reversibility and therefore long cycle life. The disadvantages are the limited availability and high cost of cadmium. The iron electrode is comparatively inexpensive but has a number of shortcomings.

Among chemical cells with non-alkaline electrolytes, the most widespread are lead batteries with sulphuric acid for electrolyte (corrosion of lead in sulphuric acid is acceptably low) and manganese–zinc Leclanché-type cells with salt solutions.

(d) Future of cell development

Intensive research and development effort aimed at creating electrochemical cells with better parameters and wider operational capabilities has been under way in many countries since the middle of this century.

Electrochemical systems with aqueous electrolytes and solid reactants were very thoroughly investigated earlier, and now the possibilities of further improvement of cells based on such systems prove limited. It became clear in recent years that a breakthrough in cell technology could be achieved only by giving up either one or both of the two "compulsory" components of today's electrochemical cells: aqueous electrolytes and solid reactants.

Wider utilization of liquid and gaseous reactants (these were used in pioneer electrochemical cells) considerably extends the scope of possible current-producing reactions. In a number of such systems the consumption of reactant per unit of delivered energy is much lower than in solid-reactant systems.

Utilization of non-aqueous electrolytes makes it possible to operate with extremely active oxidizers and reducers, including alkali metals, which are unstable in aqueous solutions. Electrochemical cells with non-aqueous electrolytes may be operative at elevated temperatures; in the case of molten or solid electrolytes, for example, it is the 300–1000°C temperature range.

Certain progress has already been reported in the investigation of these "novel" electrochemical systems, and several models of new cell types are already marketed.

2.2. Cell designs

(a) Solid-reactant electrodes

The requirements of cell electrodes with solid reactants are many-fold. The electrodes must contain the necessary amount of reactants; the electrode–electrolyte interface should facilitate electrode reactions; each element of the electrode surface must deliver current to the current-collector. As a rule, electrodes are thin plates 0·2–15 mm thick (for this reason, cell electrodes are frequently referred to as plates). An electrode current collector (lug) connected to the cell terminal is usually fixed at the upper edge of the electrode.

The simplest shape of a metal electrode (zinc plate, for example) is a thin metal sheet formed by rolling or casting. It is a reactant and a conductor at the same time. Such electrodes are not very efficient since true current density is higher at a smooth surface than at a porous one. Another disadvantage is the impossibility of consuming the reactant completely: anodic dissolution (oxidation) of metal results in thinning of the electrode and increase of its electrical resistance. Furthermore, a non-uniform reaction over the electrode surface may lead to "pitting" of some areas thus disrupting current collection from the neighbouring, not yet completely consumed, areas. Smooth metal electrodes are nevertheless employed in some cases.

The so-called surface metal electrodes, used in some types of storage cells, are made from comparatively thick (6–15 mm) sheet metal whose surface layer undergoes preliminary loosening by multiple cycles of oxidation and reduction (the surface formation process). The porous layer thus formed provides the reactant for the current-producing reaction during charging and discharging, while the dense metal substrate serves as current collector. As the storage cell is cycled, the loose porous layer partially falls off but the dense substrate is at the same time being further loosened. As a result, the amount of reactant in the active layer of the electrode is maintained constant, resulting in longer service life of surface electrodes. The shortcoming lies in a large amount of metal not participating in the current-producing reaction, that is in a low coefficient of utilization of the metal reactant. The electrode characteristics are often improved by special processing of plates which increases their surface; this is illustrated in Fig. 57 by a surface electrode of a lead battery.

An electrode of the most widespread design, the so-called pasted electrode, comprises a metal (or graphite) current collector and the so-called active mass (active layer). This latter consists of powdered reactant and various additives ensuring electrical conductance, strength and other properties. The active layer is applied by pressing, rolling-on, pasting, or by other methods, to one or both sides of a frame current collector (this may be a mesh or grid, a thin metal sheet, or a rod). The layer thus obtained is porous; the depth of transformation in the active layer may be quite large because the current-collecting function is carried by the grid.

In box-type electrodes (iron-clad, armoured, pocket and tubular designs) the active mass is normally compressed into tablets and placed into perforated thin-wall elongated boxes of rectangular or circular cross-section (see, for instance, Fig. 63); the active mass contacts electrolyte through the perforation holes. The boxes, made of plastic or metal, mechanically contain the powdered active mass. If made of metal, they also serve as current collectors. Usually the box cross-section is small; a large number of them are fixed in a special frame forming a flat plate of the required size.

(b) Primary and storage cells with solid reactants

A cell in its simplest form (sometimes called the "Box" assembly) consists of an electrolyte-filled container and vertically aranged electrodes (Fig. 60). Several electrodes of each polarity are normally installed with a view to increasing the effective surface. Electrodes of identical polarity form a single electrode group. All electrode collectors of the group connect to a common collector bus which in turn connects to the cell terminal (also called the pole). All electrodes of identical polarity are connected in parallel, so that the total cell current is the sum of currents of individual electrodes. The two individual electrode groups form a common electrode group in which positive electrodes alternate with negative ones.

A cell container may be of rectangular or circular cross-section (rectangular or cylindrical cells). The container material may be plastic, metal, glass, etc. The positive and negative terminals are mounted on the container cap (or cover), and the current collectors and electrode groups are sometimes fixed to these terminals. The cover normally carries a plugged hole for filling the cell with a liquid electrolyte. A plug may be provided with a vent or safety valve to release gases liberated inside the cell. If the cell cover is of metal, at least one terminal must be electrically insulated from it. Likewise, in cells with metal container at least one electrode group and its current collector must be carefully insulated from the container.

The alternating electrodes of opposite polarity are spaced by electrolyte-filled gaps, with gap widths varying from 0·2 to 10 mm in different cell types. An accidental contact between electrodes of opposite polarity (this would be equivalent to internal shorting) is prevented by placing separators in the gaps. The separator design must be such that the ionic current in the electrolyte is not impeded; in other words, electrodes should not be screened too much.

There are cases when the cell container is made of the metal serving as the negative electrode reactant. In many types of manganese–zinc cells, for example, a cylindrical zinc container is at the same time the negative electrode (Fig. 50). The positive electrode, compressed into a cylinder of smaller diameter, is placed inside the cell.

The liquid–electrolyte cells described above are not as a rule leak-proof, especially when the cell is turned upside down. In some applications, however, it is required that cells be completely sealed, usable in arbitrary spatial orientations, and able to withstand mechanical loads (i.e. be vibration-proof and shock-proof).

"Dry" manganese–zinc cells are made leakproof by adding either starch or other additives to the electrolyte (an aqueous solution of zinc chloride and ammonium chloride) and thus gelling it (forming immobilized electrolyte). Immobilization leaves electric conductivity of the initial electrolyte almost unchanged.

The danger of leakage through the filling hole is absent in sealed-cell design, since electrolyte is put into the cell in the course of manufacturing, before the hole-free cover is hermetically sealed onto the container. Complete sealing of cells is admissible only if gassing within a cell is not likely to swell or even rupture the cell container, or if special means are provided for chemical removal of liberated gas (see Section 6.5).

One variety of sealed cells is the button (or disk) cell (Fig. 64). The reactants of one electrode in such cells are inside, and in direct contact with, the metal container while those of the second electrode contact the convex cover. A cell is sealed by rolling in the edges of the container and cover separated and insulated by a rubber or plastic gasket; the container and the cover serve as cell terminals. The reliability of this design is improved by utilization of immobilized electrolytes or by containing the electrolyte in a porous material (such as filter paper). Unlike cylindrical cells the height of button cells is less than their diameter.

In order to increase geometric surface, it is sometimes possible to use coiled-type electrodes made of one or two very thin reactant-carrying foils. A separator is required if both electrodes are made of foil.

(c) Cells with liquid or gaseous reactants

Electrode reactions involving liquid-phase or gas-phase reactants occur on the surface of solid electrodes (current collectors). These electrodes catalyse the reactions but, not entering into them chemically, are not consumed.

Utilization of liquid or gaseous reactants for one or both electrodes creates difficulties related to the necessity of spatial separation of the oxidizer and reducer. The same obstacles are encountered if the reactants are soluble in the electrolyte.

Several techniques are available for overcoming these difficulties:

(1) **The use of solid electrolytes.** A solid electrolyte serves as a partition between the compartments with the reactants (and current collectors) of the corresponding electrodes. In order to prevent interpenetration of the reactants, the electrolyte has to be non-porous. For instance, this principle was realized recently in the sulphur–sodium storage cell (see Section 16.3).

(2) **The use of semipermeable membranes.** Ideally, a semipermeable membrane immersed into the electrolyte between the electrodes lets through only the current-carrying ions and blocks the reactants dissolved in the electrolyte. Unfortunately, no ideal membranes have been developed so far. In practice the available membranes slow down the interpenetration of dissolved reactants between the electrode compartments but do not eliminate it. The copper–zinc cell described in Section 2.1 is an electrochemical cell in which this principle was used for the first time.

(3) **The use of diffusion electrodes.** This method is the most frequently employed in electrochemical cells with liquid or gaseous reactants. A diffusion electrode (for instance a gas-diffusion electrode) is a porous metal plate simultaneously serving as a current collector (Fig. 7). The plate separates the liquid electrolyte on one side of the electrode from a liquid or gaseous reactant on the other side. The pores are filled partially with the electrolyte and partially with a reactant or its solution in the electrolyte. In the course of discharge, the reaction takes place on the pore walls. The reactants diffuse through the pores and compensate for the amount consumed in the reaction. The supply rate must be such that the reactant is completely consumed within the porous electrode and thus barred from entering the interelectrode layer of electrolyte and reaching the second electrode. The reactant is fed into the electrode either by diffusion or by a

pressure gradient. The term "diffusion electrode" does not, therefore, accurately represent the nature of the electrode.

In fuel cells, diffusion electrodes are used both for the oxidizer and for the reducer. The reactants are fed continuously into the electrode chambers and from them into the porous electrodes. As a rule, the reaction products are transferred from the diffusion electrode into the electrolyte. Therefore continuous operation of a fuel cell requires that reaction products be removed continuously or intermittently. This is achieved by circulating the electrolyte through special chambers where the reaction products are removed, or by other means.

(4) **The use of semipassive electrodes.** The requirement of complete separation of the reactants is not always compulsory, for example, in Grenet's bichromate–zinc cell (see Section 2.1). Although the oxidizer solution in this cell is in direct contact with the zinc electrode surface, the rate of their direct interaction is rather low since the zinc electrode rapidly develops a thin-film coating which slows down the reaction. At the same time, the main reaction of zinc dissolution proceeds unimpeded. This state of the zinc electrode is sometimes said to be semipassive.

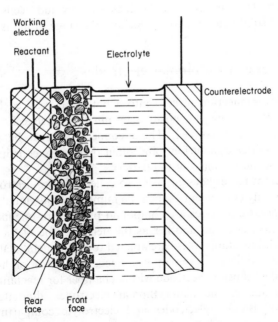

Fig. 7. Schematic of the diffusion electrode.

(d) Cells with continuous supply of solid reactants

Numerous attempts have been made to develop fuel cells or compound cells with continuous supply of not only liquid or gaseous but solid reactants as well (such as metals, oxides and so on). Several methods were suggested. Metals may be used as thin wires or ribbons wound into reels and continuously fed into the electrolyte. Another, and more intricate technique consists in dissolving a metal in mercury forming an amalgam. The amalgam is fed into the cell where the dissolved metal is then anodically dissolved and is transferred into the electrolyte in ionic form. The released mercury is to be returned into the reactor where it will dissolve more metal.

A method utilizing powder or suspension electrodes would be more versatile, being applicable not only for metal but for other solid reactants as well. Powder from a special bin is continuously fed into the reaction zone where it comes in contact with the current collector and electrolyte. Mobility of powder particles in the course of a reaction on a static powder electrode is quite low. In a suspension electrode a fine-powdered reactant is suspended in liquid electrolyte and moves in the high-speed flow (produced by a pump) onto the current collector surface. The electrode reaction occurs during the short time of contact on this surface.

Despite a considerable research effort and advanced development of several models, numerous difficulties still preclude cells with continuous supply of solid reactants from reaching the stage of industrial production.

(e) Batteries with bipolar electrodes

The series connection of cells into a battery is realized by fixing intercell connectors between the positive terminal of each cell and the negative terminal of the next cell. These connectors must be designed for the maximum discharge current, and often their mass constitutes a substantial fraction of the net mass of the battery.

A different battery design, using so-called bipolar electrodes, is possible— and especially convenient—when the number of individual cells is high (Fig. 8(a)). Bipolar electrodes are compound plates with one face operating as a positive electrode and another as the negative one of the next cell. In the case of solid reactants, these are applied to a thin metal plate: reducer to one face, and oxidizer to the opposite face. In cases of liquid or gaseous reactants, a diffusion electrode with a chamber for reactant supply is placed on each side of the plate. Bipolar electrodes alternate with layers of electrolyte. Bipolar electrodes and electrolyte compartments are tightly sealed around the edges, preventing the leakage of electrolyte from the cell

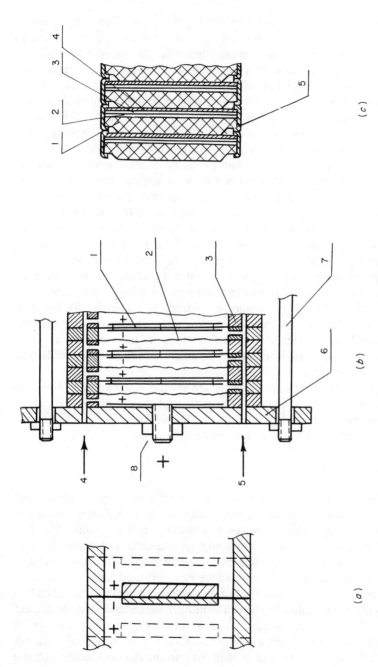

Fig. 8. Batteries with bipolar electrodes. (*a*) Cross-section of a bipolar electrode. (*b*) Filter press battery. 1, bipolar electrode; 2, diaphragms; 3, sealing gaskets; 4, oxidizer input; 5, reducer input; 6, end plate; 7, tightening bolts; 8, positive current collector. (*c*) Flat cell battery. 1, positive electrode; 2, diaphragm with electrolyte; 3, negative electrode; 4, current-conducting layer; 5, plastic seal cups.

and ensuring leak-proof separation of neighbouring electrolyte compartments by the electrode plate. Separating plates simultaneously function as walls of cell containers and as intercell connectors (the current between two contiguous cells passes through a thin wall with negligible resistance). This results in substantial reduction of mass and volume of a battery.

There are several types of battery construction involving bipolar electrodes. The first is the filter press type (Fig. 8(b)) mostly used with gaseous and liquid reactants. In this system the whole group of electrodes and separators for electrolyte compartments is tightened by means of end plates and long bolts; sealing is achieved by using elastic gaskets which are compressed when the group is tightened. After sealing, compartments are filled with electrolyte through special small-diameter channels in gaskets or close to the edges of the electrodes (recirculation of the electrolyte is also possible). Other channels are used to supply reactants to the operating battery. Although the construction described for the filter press makes it possible to replace faulty parts, the process of simultaneous sealing of all cells by tightening the electrode groups is difficult and requires high precision in machining of the parts. In the block-type construction, either each cell or group of cells is sealed by a plastic coating, for example, by injection moulding. Finally, each cell (bipolar electrode and electrolyte compartment) of a flat-cell battery (Fig. 8(c)) is sealed along its perimeter by a plastic jacket. This design is used, for example, in dry manganese–zinc cells with immobilized electrolyte.

The bipolar electrode constructions are mostly employed at present to produce batteries comprising primary or fuel cells; storage batteries utilize other principles, presumably because the removal of gas liberated in the course of the charging phase poses serious difficulties.

(f) Reserve cells

It may be expedient to improve cell characteristics by selecting electrochemical systems with very active reactants which, however, cannot be kept in contact with electrolytes over a suitable period of shelf-life. This specific problem is solved by designing reserve cells which can be kept inactive when in storage and are activated directly before the discharge phase. As a rule, an activated cell cannot be inactivated later.

In one version, no electrolyte is introduced into the manufactured cells; a liquid electrolyte is added to the cell container immediately before the cell operation. A cell is filled either manually through a vent in the cover, or automatically. In the latter case the battery carries a special ampoule containing the electrolyte which is pumped into the electrode compartments when required.

In another version the cells referred to as thermal reserve cells contain solid electrolyte and can be stored for a long time; the activation procedures consist in rapid heating of the cell to the electrolyte melting point (for instance, by means of an explosive cartridge).

Chapter Three

Performance

3.1. Electrical characteristics

(a) O.c.v. and discharge voltage

The o.c.v. of a cell depends on the selected electrochemical system, but is somewhat affected by the electrolyte concentration, gas pressure, degree of discharge of the cell, temperature and other factors. O.c.v. is a fairly reproducible quantity, provided the above parameters are fixed. The spread in o.c.v. of some cell types (such as standard cells, see Section 13.6) does not exceed $10^{-4}\%$, justifying their use as reference sources.

The discharge voltage of a cell is a function of the current passing through it (equation (1.8)). Plotted in Fig. 9 is a typical curve of discharge voltage U_d as a function of discharge current I_d, referred to as the current–voltage curve. Very often (although not always) these curves are S-shaped. As a result of non-linearity of current–voltage curves, the formal effective internal resistance of a cell

$$R_{eff} \equiv -\frac{dU_d}{dI_d} = R_{ohm} + R_{pol(+)} + R_{pol(-)} \tag{3.1}$$

is not constant but changes with changing current owing to variable polarization resistance of the electrodes, $R_{pol} \equiv d\eta/dI$.

The functional dependence of U_d on I_d is sometimes represented by a simplified linear equation:

$$U_d = \mathscr{E} - I_d R_c \tag{3.2}$$

where the apparent internal resistance R_c is assumed constant. This is rather a crude approximation, especially in cases of S-shaped current–voltage curves. The approximation is improved if $\mathscr{E}$ in equation (3.2) is replaced by

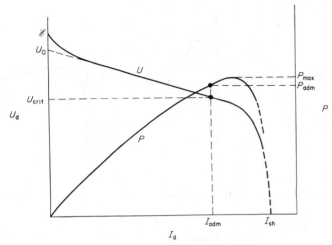

Fig. 9. Typical current–voltage curve, and cell power as a function of load current.

the intercept U_0 of the central (usually linear) segment of the current–voltage curve extrapolated to zero discharge current.

The discharge voltage is strongly dependent on structural and technological features of a cell, on temperature, and on numerous other factors. The spread in discharge voltage is considerably greater than that in the o.c.v.

When a cell is being discharged at a constant current and with other essential conditions fixed, a trend of gradual decrease of voltage is normally observed. Typical curves of U_d as function of time τ, or discharge curves, are plotted in Fig. 10. The degree of voltage fall-off is different in different cell types, varying from 5–10% of the initial level U_{init} for some cells to 50% in other systems. The decrease in voltage may be caused by (a) decrease in the o.c.v. in the course of discharge as a result of altered ratio of the amount of reactants to that of reaction products, and (b) increased polarization of electrodes and increased ohmic resistance. The drop in voltage is often especially steep at the beginning of discharge (and particularly in freshly charged storage cells). By convention, a more stable voltage which sets on after a fraction (3–10%) of cell capacity is delivered, is sometimes considered as the initial discharge voltage. The initial phase of discharge is sometimes accompanied by an increase in cell voltage, while in other cases the cell voltage first sharply drops and then rises again (the initial "dip" in voltage).

The decrease of voltage at the end of discharge may be either steep or gradual. Because of this decrease it is necessary to terminate the discharge at a certain cut-off voltage U_{fin} even though the reactants are not yet com-

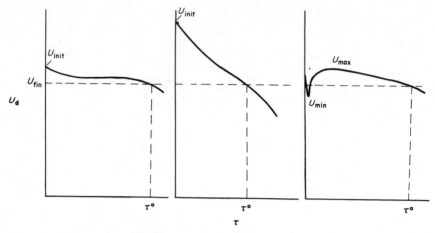

Fig. 10. Typical cell discharge curves.

pletely consumed. The choice of cut-off voltage is determined by specifics of the current-consuming circuit. The depth of discharge of a cell with a falling discharge characteristic has to be limited if the permissible voltage range is narrow.

Summarizing, the following characteristic values of discharge voltage must be distinguished: initial voltage U_{init}, maximum U_{max}, minimum U_{min} and cut-off voltage U_{fin}. All these voltages depend on the discharge mode, that is discharge current, temperature and so on. Lower cut-off voltage is permissible as a rule for high-current drain than for low-current drain of a cell.

Another characteristic convenient for energy estimates is the mean voltage $\bar{U}$ in given discharge conditions; $\bar{U}$ is defined as the mean integral or mean arithmetic value of discharge voltage over the discharge period τ° from U_{init} to U_{fin}:

$$\bar{U} = (1/\tau^\circ) \int_0^{\tau^\circ} U \, d\tau \approx (1/N) \sum_{i=1}^{N} U_i \qquad (3.3)$$

where N is the number of voltage measurements spaced by identical time intervals (at least 5–6 readings over τ°).

(b) Discharge current and discharge power

The discharge current I_d of a cell depends on the external circuit resistance R_{ex} and is given by Ohm's law as $I_d = U_d/R_{ex}$. Calculation of discharge current for a given R_{ex} requires that the U_d versus I_d current–voltage curve be available. Power delivered by a cell being discharged is found as

$$P = U_d I_d = U_d{}^2/R_{ex} = I_d{}^2 R_{ex}$$

Neither the discharge current nor the power delivered are characteristics of a cell since both are determined by an arbitrarily chosen resistance R_{ex} of the external circuit. The current drain may be low if R_{ex} is high, or it may be high if R_{ex} is small. The power delivered first rises as R_{ex} diminishes and current increases and then falls off since the effect of voltage decrease becomes predominant over that of increasing current. By using the simplified expression (3.2) for voltage, we obtain

$$I_d = \frac{\mathscr{E}}{R_{ex} + R_c}; \qquad P = \frac{\mathscr{E}^2 R_{ex}}{(R_{ex} + R_c)^2} \qquad (3.4)$$

The maximum power is therefore $P_{max} = \mathscr{E}^2/4R_c$, and it is delivered when $R_{ex} = R_c$ and $U = 0.5\mathscr{E}$. With actual current–voltage characteristics, the cell power has a maximum too (Fig. 9) which may be reached, however, at different values of R_{ex} and U.

The following systems of discharge may be distinguished, depending on the specifics of the load circuit: (a) constant external circuit resistance, $R_{ex} = $ const (I_d falling off as voltage decreases); (b) constant current discharge, $I_d = $ const; (c) constant power discharge, $P = $ const (load current increases as cell voltage decreases); and (d) variable-load discharge, with prescribed load variation program. Discharge may be continuous or intermittent, with breaks of uneven duration.

Each cell type is characterized, by convention, by the rated mode, that is fixed values of discharge current or load resistance, cut-off voltage, temperature and some other parameters of discharge. The round-off mean voltage in this discharge mode is referred to as the rated voltage of a given cell type. Documentation supplied with cells may indicate, along with the rated mode parameters, parameters of other modes which are closer to real conditions of cell operation.

The maximum admissible discharge current I_{adm} and the related maximum admissible discharge power P_{adm} give an important characteristic of any cell type. These performance characteristics correspond to a critical lower bound on cell voltage U_{crit}; certain factors (such as cell overheating) make it impossible or undesirable to operate at cell voltages below U_{crit}. To a certain extent, the choice of U_{crit} is arbitrary; consequently, it is normal to indicate only approximate values of I_{adm} and P_{adm}. In particular, short-duration discharges (pulse loads) admit significantly higher currents and power of discharge than prolonged discharges.

A cell may also be characterized by the short-circuit current I_{sh} measured when the discharge voltage drops to zero. Although this parameter is of no special interest to a consumer, it is sometimes indicative of the internal conditions or quality of the cell.

(c) Cell capacity and stored energy

The electric charge Q_d that has passed through the external circuit over the cell discharge interval τ is given by $\int_0^\tau I_d \, d\tau$ or simply $I_d\tau$ if $I_d = $ const. In the literature dealing with cells, this charge is expressed not only in coulombs $(1\text{ C} = 1\text{ As})$ but also in ampere-hours $(1\text{ Ah} = 3600\text{ C})$ or in Faradays $(1F = 96\,500\text{ C/mol})$.

The electrical energy delivered by a discharging cell is found as

$$W = \int\limits_0^\tau U_d I_d \, d\tau \approx \bar{U} Q_d \tag{3.5}$$

This energy is measured not only in joules $(1\text{ J} = 1\text{ Ws})$ but also in watt-hours $(1\text{ Wh} = 3600\text{ J})$.

The maximum amount of electricity delivered by a completely discharged cell is referred to as the Ah capacity, C. Correspondingly, the maximum energy thus delivered is called the Wh capacity. Both the Ah capacity and Wh capacity are obviously determined by the amounts of reactants in a cell at the start of discharge. The greater this amount the longer the current-producing reaction is sustained and the higher both the Ah capacity and the Wh capacity.

The rated capacity C_0 and the rated Wh capacity refer to the rated discharge mode, and are guaranteed by the manufacturer. Sometimes these parameters are included in the cell specification. Technological factors result in a spread in the actual Ah capacity and Wh capacity which normally exceed the guaranteed values by 5–15%.

As the discharge current increases, both the discharge voltage and the reactant utilization coefficient diminish, thereby diminishing the effective cell capacity. This effect is clearly demonstrated if the discharge curves are plotted as functions of the delivered charge.

A number of attempts to find a relationship between the actual cell capacity and discharge current were reported in the literature. An empirical formula

$$C = k/I_d^\alpha \quad (\text{or } I_d^{(1+\alpha)}\tau^\circ = k) \tag{3.6}$$

found by Peukert in 1897 for lead batteries, is employed more often than other approximations (k and α are constants). This equation provides a good fit to various experimental data, although the empirical constants are determined not only by the cell type but also by cell temperature and other factors and have to be found for each specific situation. The values of α vary from 0·2 to 0·7, and are smaller the lower U_{fin}. An "ideal" cell would have $\alpha = 0$ (C independent of I_d). Similar and more complicated empirical relationships are also known, and can be used for interpolation of experimental data within the respective ranges for which their validity has been established. They cannot, however, be used for extrapolation to other discharge modes.

The cell capacity is sometimes higher if the discharge is intermittent than if it were continuous; from the capacity standpoint, therefore, an intermittent discharge is equivalent to a continuous discharge at a lowered current drain. This effect is caused by the "recovery" of cell electrodes during the breaks in the intermittent discharge mode. It may happen that, on the contrary, the discharge capacity is reduced when the discharge duration is prolonged (by smaller I_d or longer breaks). This is a result of self-discharge, significant for long-duration discharge modes. The maximum capacity is realized, therefore, for intermediate times of discharge. If the ambient temperature is low, the cell heating may result in an apparent increase of capacity with increasing current.

If one considers the power supply of a load circuit, the Ah capacity is less important than the Wh capacity. Practically, however, it is easier to monitor the discharge current (by means of an ammeter) than power and to determine the expected discharge duration from the discharge Ah capacity and not from the Wh capacity. Therefore both the cell documentation and the cell specification indicate only the Ah capacity. Evidently, it is assumed that the mean voltage of the cell is known.

(d) Electrical characteristics of storage batteries

When a storage cell is being charged, the charge voltage is higher the higher the charge current (equation (1.9)). The charge voltage also increases with time as the charge stored rises (Fig. 11). A battery is charged until either a certain final charge voltage $U_{c.fin}$ is reached (recognized at the completion of charging by a sufficiently steep rise in the battery voltage, as in curve 2) or until the battery accepts a prescribed amount of charge defined as the charge capacity Q_c°.

Both the Ah capacity and the Wh capacity of a battery are referred to one complete discharge of this battery after it has been charged to full capacity. It can happen that a battery is charged following an incomplete

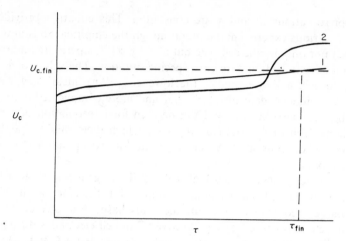

Fig. 11. Typical charging curves without (1) and with a hike (2) at the end of charging.

discharge; the depth of discharge (d.o.d.) $\theta_d = Q_d/C°$ of a thus-cycled battery is then less than unity (or $C/C°$). The ratio of the residual capacity C_{res} in a given state (for example, after a preliminary partial discharge) to the rated capacity $\theta_c = C_{res}/C°$ is a characteristic of the charge state of the battery ($\theta_c = 1 - \theta_d$).

The discharge Ah capacity of the ideal storage battery is equal to its charge Ah capacity. Charging, however, is frequently accompanied by side processes. For instance, side reactions (1.4) and (1.5) in aqueous solutions often start before charging is completed, and for some time proceed concurrently with the main charging reaction. The so-called ampere-hour efficiency of capacity (or current) utilization of a storage cell

$$\mu_Q \equiv C/Q_c \tag{3.7}$$

is less than unity because part of the current is partially consumed by side reactions. Another important characteristic of a storage cell is its watt-hour efficiency

$$\mu_W \equiv \bar{U}_d C/\bar{U}_c Q_c = \mu_Q \bar{U}_d/\bar{U}_c \tag{3.8}$$

The mean discharge voltage being always lower than the charge voltage, the Wh efficiency is lower than the Ah efficiency, and is below unity even for an ideal cell.

This is illustrated in Fig. 12 by plotting Q_d/Q_c and W_d/W_c as functions of Q_c (or $\theta_c = Q_c/C$) for alkaline nickel–cadmium storage batteries. These curves demonstrate that the degree of utilization of the charge current and charge energy diminishes as θ_c increases.

The ratio of the charge and discharge voltages (either of mean values or of

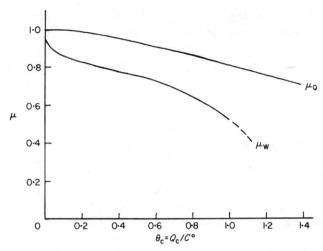

Fig. 12. Ah and Wh efficiency of nickel–cadmium storage cells as functions of the state of charging.

the values at a specific phase of operation) is an important factor in operation of storage batteries. Not only is the Wh efficiency of storage batteries decreased if these voltages differ greatly, but it also becomes difficult to use a battery in a buffer mode (see Section 7.2), that is with the load switched on to the battery not only during discharge but during charging as well. Excessive variation of battery voltage in the course of charge–discharge cycling impairs the operation parameters of the load circuitry and may even lead to circuit failure.

(e) Electrical characteristics of fuel cells

When a fuel cell delivers current, the reactants are constantly fed into the cell and the reaction products constantly removed. The system is therefore in a stationary state, the cell voltage at a constant discharge current either being constant or varying slightly owing to temperature fluctuations and other factors. Secondary factors (wear, etc.) become significant only after very long discharge times, resulting in gradual decrease of cell voltage.

A fuel cell power unit comprises an electrochemical generator (a battery of fuel cells with all the necessary auxiliary devices) and a store of reactants in tanks or pressure bottles. The mass of the generator is determined by the maximum prescribed discharge power P_{pre} and can be written as γP_{max}, where γ is the fuel cell generator mass per unit delivered power. The required amount of stored reactants is determined by the prescribed operation time τ_{pre} (the time to replenishment of the consumed reactants) or by the prescribed amount of the total delivered electrical energy W_{pre}. If the reactant consumption per unit of electrical energy (taking into account the container mass) is denoted by g_W, then the total mass of the power unit is

$$M = \gamma P_{pre} + g_W W_{pre} \qquad (3.9)$$

These two parameters, γ and g_W, are the most important electrical characteristics of a fuel cell battery. They are functions of the conditions in which a battery is operated. As the current drain is increased the power delivered by the generator is stepped up, thus reducing γ and, owing to a reduced discharge voltage, the specific reactant consumption g_W is increased. This emphasizes the necessity of selecting an operating point on the current–voltage curve (to select current and voltage) with a view to optimizing the net characteristics of the power unit in a given field of application.

3.2. Operational characteristics

Cell characteristics strongly depend on the operating temperature. As a rule, both the discharge voltage, the maximum admissible power, and the reactant utilization coefficient are lower at lower temperatures. On the other hand, increased temperatures are conducive to side reactions (such as corrosion processes) and thus reduce the cell efficiency. Each cell type must operate, therefore, in a specific temperature range in which the characteristics are within the prescribed limits.

Another group of important cell characteristics includes a number of lifetime parameters. In the case of primary cells, factors of paramount importance are the shelf-life, that is the maximum admissible interval of storage between manufacturing and the beginning of cell discharge, and the service life, that is the total interval including time of discharge (this time may be considerable for light-drain and intermittent discharge modes). It is required that during these intervals self-discharge should not reduce the cell capacity by more than a rated fraction (for example, by more than 20 % of the initial or rated capacity); equally, the discharge voltage must also be maintained within a prescribed range.

In the case of reserve cells, it is necessary to distinguish between the shelf-life in a non-activated state (for example, in cells not yet filled with the electrolyte) and that after activation (filling with the electrolyte).

Normally the discharge capacity of a storage battery (its Wh and Ah capacities) is reduced in the course of cycling; sometimes the capacity increases over the first several cycles, passes through a maximum and then drops. The practical cycle life of a battery is a measure of the admissible number of charge–discharge cycles not resulting in a capacity or voltage decrease below a given limit. This number is a function of the depth of cycling: the admissible number of cycles increases if only a fraction of the battery capacity is discharged during each cycle. The technical documentation lists the cycle life in the rated mode, with charge and discharge phases involving either the full capacity or its prescribed fraction.

It is also of interest, from the economic standpoint, to know the total energy delivered by a storage battery during its whole cycle life. As a rule, this characteristic depends only slightly on the depth of cycling, since a decrease in the depth of discharge is balanced out by the increased number of cycles. The total delivered energy of some cell types reaches a maximum for a certain depth of discharge.

The service life of a storage battery gives the admissible duration of its operation (after the battery is filled with electrolyte). It is expedient sometimes to indicate the admissible shelf-life of a charged and of a discharged battery, as well as the degree of the discharge capacity loss after storage of a charged battery.

The main characteristic of a fuel cell is the maximum duration of continuous or intermittent operation.

The characteristic times given above are limited by side processes such as corrosion, ageing of materials, and so on, reducing the cell capacity ("self-discharge" processes) and other parameters of batteries. Each characteristic time is strongly dependent on the conditions of storage and operation (temperature, humidity).

The last group of characteristics is related to handling of cells. These are the mechanical strength (for example, with respect to vibrations and shocks accompanying transportation), the possibility of operating in any spatial orientation, reliability, maintenance difficulties, "fool-proofness" (the term was introduced by Edison in 1911 as one of the characteristics of storage batteries) and numerous others. An important property of storage cells is a low sensitivity to overcharging and to other accidental deviations from the maintenance procedures given in manuals.

3.3. Comparative characteristics

It is frequently necessary to compare electric or other characteristics of cells differing in size, design or electrochemical system. The easiest method is to use the normalized (reduced) parameters.

The current-producing electrochemical reactions occur on the surface of electrodes in contact with the electrolyte. The current density is a measure of the relative reaction rate. Therefore the curve plotting voltage as a function of current density is a useful characteristic of a cell, reflecting only its specific properties and independent of its size.

Not only the charge or discharge current but also such cell characteristics as power, capacity or energy can be referred to unit surface area of the electrodes; they are given then in mW/cm^2, mAh/cm^2 or mWh/cm^2, respectively. The effective internal resistance can likewise be referred to unit surface area, but the relevant unit of measurement is then ohm cm^2 (in contrast to current, resistance diminishes as the surface area S increases).

It is more convenient to a user (for whom the surface area of the electrodes is, as a rule, of only minor importance) to refer the basic parameters to unit mass or unit volume of a cell, or to its rated capacity.

Discharge modes are conveniently compared by using the normalized current j defined as the ratio of current (in amperes) to the rated capacity (in ampere-hours), $j = I/C^\circ$ (the values of j are commonly used without mentioning the units h^{-1}). Its inverse value, that is the normalized operation time in hours, θ, is also used. If, for example, a cell with the rated capacity of 35 Ah is discharged at a 7 A drain, the normalized current is 0·2, and the normalized time of operation 5 h. The actual time of complete discharge τ° may differ from the normalized time of discharge θ because in the general case the actual capacity is not equal to the rated one: the greater the normalized current the lower the actual capacity. The relative capacity C/C° is plotted in Fig. 13 as a function of normalized current in a number of cell types.

Sometimes the current is referred not to the rated capacity C° but to the actual capacity C in a specific mode (i.e. to different capacities realized for different discharge currents). The normalized time thus defined coincides with the actual discharge duration. The cell current is then expressed as a fraction of the actual capacity, that is $I = (1/\tau^\circ)C$. If the actual operation time in the above example is four hours (28 Ah capacity), the normalized current is 0·25C. One may also refer to a four-hour discharge (or charge) mode. This system is practicable only if the actual capacity is a sufficiently reproducible and well known function of current. The quantitative characteristics cited henceforth in this book invariably refer to the rated capacity.

In order to characterize the cell power by a parameter independent of the cell size, it is sometimes useful to operate with the normalized internal resistance $\bar{r}_{eff}$ defined as the product of the effective internal resistance by the rated capacity, $\bar{r}_{eff} = R_{eff}C^\circ$. As follows from this definition and from equation (3.1), $\bar{r}_{eff}$ is equal to the derivative of voltage with respect to normalized current (taken with the negative sign).

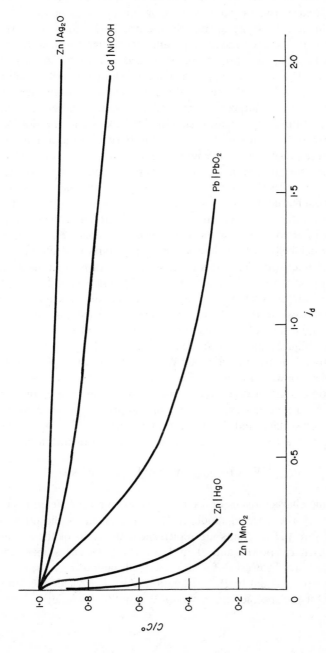

Fig. 13. Relative capacity $C/C°$ as a function of normalized discharge current j_d for silver–zinc, nickel–cadmium and lead storage cells, and for mercury–zinc and manganese–zinc primary cells.

A widespread approach consists in using the specific energy and specific power (per unit mass) $(w_m; p_m)$ or unit volume $(w_v; p_v)$ of the cell. The density of most cells being within the range 1.5–2.5 g/cm^3, the specific volume parameters exceed the mass parameters by just this factor.

In each cell type, the specific energy is a diminishing function of the specific power. The w versus p plots give a clear illustration of the electrical parameters of a given type of cell. For a detailed comparison, these curves must be plotted for a number of temperatures since the specific energy is strongly temperature-dependent. It should also be kept in mind that the cell specific energy is a function of its state, and in particular the conditions and duration of the preceding storage and cycling.

The volume-specific and mass-specific energy is considerably reduced in small-scale cells with their inherently low absolute energies. This effect is rooted in an increased fraction of the cell volume and mass accounted for by structural parts (the container, current collectors, etc.) and the correspondingly reduced fraction taken by the active reactants. Each cell system has its own critical size below which the specific energy starts to drop off.

Figure A.1 (see Appendix) plots w_m as a function of p_m for cells normally discharged by high normalized current $(j > 0.2)$, and Fig. A.2 plots similar w_v–p_v curves for cells with typically low discharge currents $(j < 0.05)$. The logarithmic scale is used for convenience. Dashed lines connect the points with equal time of complete discharge $(\tau^\circ = w/p)$.

These two figures demonstrate that different cell types have specific energies roughly in the range 5–500 Wh/kg; in typical applications they provide specific power of from 10^{-2} to 500 W/kg.

The maximum admissible specific power of a cell $p_{adm.m}$ or $p_{adm.v}$ is determined by the electrodes: the maximum admissible power density $p_{adm.s}$ at the electrode surface, and the structural parameters s_m or s_v indicating the electrode surface area per unit mass or unit volume. For instance,

$$p_{adm.m}[W/kg] = p_{adm.s}[W/cm^2] \times s_m[cm^2/kg]$$

The specific characteristics of a fuel cell power unit are determined to a large extent by the selected mode of operation, such as the duration of discharge before the reactants are replenished, the ratio of the average to maximum discharge power, and so on. If a fuel cell battery is operated with constant discharge power P for a prescribed period τ_{pre}, the delivered electrical energy is equal to $P\tau_{pre}$. Equation (3.9) then yields the following expressions for the specific parameters of the unit as a whole:

$$w_m = \frac{\tau_{pre}}{\gamma + g_w\tau_{pre}}; \qquad p_m = \frac{1}{\gamma + g_w\tau_{pre}} \qquad (3.10)$$

These formulas show how the specific parameters depend on τ_{pre}. The longer the discharge, the higher w_m; in the limit, it tends to the specific energy of the reactants, $1/g_w$, which manifests that the contribution of the electrochemical generator into the total mass of the power unit becomes negligible. Short duration of discharge results in sharply reduced w_m. Consequently, the advantages of fuel cell units in comparison with other cell types appear only for long operation times. As an example, Fig. 14 gives the net mass of a power unit per kilowatt as a function of a fixed discharge duration for oxygen–hydrogen fuel cells (assuming $\gamma = 50$ kg/kW and $g_w = 4$ kg/kWh) and for nickel–zinc cells ($w_m = 60$ Wh/kg). The net mass of the fuel cell unit is smaller for operation times above 4 hours.

In using all the above-mentioned comparative characteristics the fact should not be overlooked that electrical parameters do not cover all the maintenance requirements of chemical cells. Such characteristics as storage-ability, cycle life, service life, mechanical properties, as well as economics estimates are at least as important as electrical parameters. Hence, a meaningful choice of a chemical power source is impossible without a thorough analysis of all these characteristics.

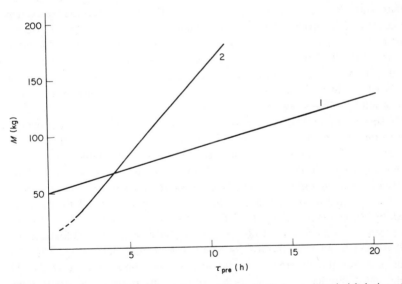

Fig. 14. Net mass M of a 1 kW power unit comprising fuel cells (curve 1) and nickel–zinc cells (curve 2), as a function of prescribed discharge duration τ_{pre}.

Chapter Four

Electrochemical Aspects of Cell Operation

4.1. Refined concept of electrode potential

(a) The Galvani potential

By definition, in electrostatics the difference $\Delta\Psi_{B,A} \equiv \Psi_B - \Psi_A$ between electrostatic potentials at two points A and B is the work done by external forces which transfer a positive unit charge from A to B in the field of electrostatic forces of the surrounding charges. A conductor has identical potential in all its points if the electric current is zero; otherwise the electric field should be displacing the available free charge carriers. When a current is flowing through a conductor, the potential difference across it is governed by Ohm's law.

If two non-identical conductors, for instance, two metals, are brought into contact, a certain potential difference (or jump) determined by the properties of the two conductors sets in across the contact interface. It results from the difference in the chemical forces acting on free charge carriers (electrons) in the surface layers on both sides of the interface. Some of these electrons are therefore transferred from one metal to the other, making it negatively charged, while the first metal is depleted of electrons and thus becomes positively charged. A double electric layer is formed (Fig. 15) in which, in the equilibrium state, the electric field inside the layer balances out the difference in chemical forces applied to electrons. The potential difference $\varphi_{2,1} = \Psi_2 - \Psi_1$ between any point in the first conductor and any point in the second is said to be the Galvani potential of the interface. The sign of the Galvani potential is determined by the order in which the conductors are considered; hence, $\varphi_{1,2} = -\varphi_{2,1}$.

One essential feature must be emphasized. Although the Galvani potential between two non-identical conductors is a physically real quantity, it

52

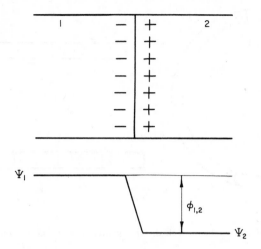

Fig. 15. Double layer at the contact of two metals, and the potential distribution.

cannot in principle be measured. Indeed, we must discard the method of determining it by measuring the work done in transferring a unit charge, since in transferring the charge (which is inseparable from its "material" carrier) the work is done not only against the electric forces but also against the chemical ones (the work vanishes in the equilibrium state since the two forces cancel out); secondly, any measuring instrument connected to the circuit (a voltmeter, a potentiometer, etc.) forms at least one additional interface whose Galvani potential is added to the one meant to be measured. Neither can the Galvani potential be calculated theoretically, at the present moment at any rate, since the complexity of the sum of chemical forces affecting electrons and other charged particles is quantitatively unmanageable.

It should also be added that the Galvani potential is not restricted to metal–metal interfaces; it also appears at the interfaces between metals and electrolytes ($\varphi_{m,e}$), and in fact between any two conductors.

(b) Conducting circuits

Several conductors connected in series are said to form a conducting circuit, which may be either closed or open (Fig. 16). A circuit is said to be correctly open if its terminal segments are constituted by identical conductors; otherwise it is an incorrectly open circuit. The simplest example of the latter case is that of two non-identical conductors in contact with each other.

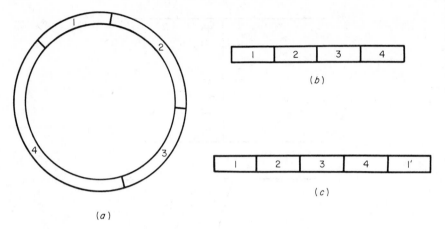

Fig. 16. Conductor circuits: (*a*) closed; (*b*) incorrectly open, and (*c*) correctly open.

The potential difference between the terminal segments of a circuit is the algebraic sum of Galvani potentials at each interface.

As in the case of the Galvani potential of an individual interface for an incorrectly open circuit, this sum can be neither measured nor calculated. It is measurable only in correctly open circuits, that is when the terminal segments are of the same chemical nature, and is referred to as the circuit o.c.v. (sometimes e.m.f.):

$$\mathscr{E} \equiv \Delta\Psi_{1',1} = \varphi_{2,1} + \varphi_{3,2} + \varphi_{4,3} + \varphi_{1',4} \qquad (4.1)$$

The components of this sum remain non-measurable.

The o.c.v. of a circuit comprising only metals or other electron conductors always equals zero (Volta's law). Hence,

$$\varphi_{4,1} = -\varphi_{1',4} = \varphi_{4,3} + \varphi_{3,2} + \varphi_{2,1} \qquad (4.2)$$

Therefore, metallic conductors 2 and 3 included in the circuit between conductors 1 and 4 have no effect on the potential difference between the terminal segments. An o.c.v. may be generated in a purely metallic circuit only by external factors, such as temperature gradient, variable magnetic field, and so on.

A circuit is called galvanic if it includes at least one ionic conductor. In the general case, the o.c.v. of this circuit is distinct from zero, because in contrast to purely metallic circuits in which charge carriers (namely, electrons) are identical in each segment, currents in galvanic circuits are carried by electrons in some of the segments and by ions in the remaining ones. An interface between the two is the place of "relay transfer" of current by an electrode reaction. A device realizing a galvanic circuit is called an electrochemical cell (galvanic cell, or electrolyser cell).

It should be kept in mind that whenever the o.c.v. of a galvanic circuit is discussed, what is meant is the potential difference in a circuit with identical terminal segments; for example, in the case of silver–zinc cells the circuits may be of the types

$$(-) \, Ag \, | \, Zn \, | \, KOH \, | \, Ag_2O \, | \, Ag \, (+)$$
$$(-) \, Zn \, | \, KOH \, | \, Ag_2O \, | \, Zn \, (+)$$
$$(-) \, Cu \, | \, Zn \, | \, KOH \, | \, Ag_2O \, | \, Cu \, (+)$$

and so on. It is not difficult to demonstrate on the basis of Volta's law that the nature of the terminal material does not effect the o.c.v., which is determined by only the metals contacting the electrolyte, so that the o.c.v. is the same in all the systems listed above. Consequently, it is customary to drop the terminal lead material in the notation of electrochemical systems.

(a) Localization of the o.c.v. (e.m.f.)

The o.c.v. of a galvanic cell is the algebraic sum of at least three Galvani potentials, namely, two at the electrode–solution interfaces and one at the metal–metal contact. It took quite a long time to resolve the problem of which of these interfaces contributes decisively to the o.c.v. Volta's concept was that the metal–metal interface is of predominant importance, with the solution being but a separating intermediary phase; the potential jump between the metals is observed in the circuit precisely because there are no potential jumps at the metal–solution interfaces. The existence of the so-called Volta potential, that is the difference between potentials of two points on the surface of two metals brought into contact in vacuum, was an argument in favour of this "contact theory of the e.m.f.". The German electrochemist Nernst, who at the end of the 19th century found a relationship between the o.c.v. and the concentration of the current-producing species, took into account only the potential jumps at the electrode–solution interfaces, that is at the places where the electrode reactions are located (the "chemical theory of the e.m.f."). This contradiction, called the Volta problem in electrochemistry, found its solution in the twenties of this century in papers published by A. Frumkin and his school.[1]† According to the concepts accepted now all the interfaces are equally important; none of them can be singled out in advance as predominant. The metal–metal interface also participates in the current-producing reaction since it is crossed by the flow of electrons from the anode to cathode.

† Literature cited will appear at the end of each chapter.

(d) Electrode potentials

An electrochemical cell involves two electrodes and it is naturally desirable to try to express the o.c.v. as the difference of two quantities, each of them characterizing a single electrode. At first, the electrode–electrolyte Galvani potentials were used as such characteristics: $\mathcal{E} = \varphi_{2,e} - \varphi_{1,e}$. We know, however, that this relationship neglects the metal–metal Galvani potential. Furthermore, it is practically useless since the required quantities cannot be measured.

A characteristic suitable for the stated purpose is the electrode potential E_i (which should not be confused with the potential difference between the electrode and solution!). The electrode potential is the o.c.v. of a correctly open circuit consisting of the electrode in question and a selected reference electrode. One reference cell is the normal hydrogen electrode (an electrode made of platinized platinum and partially immersed in a sulphuric acid solution with unit activity of hydrogen ions (approximately $3 \cdot 4$ M H_2SO_4); the electrode is placed in a stream of pure hydrogen at a pressure of $0 \cdot 1$ MPa, and the reaction occurring on it is $2H^+ + 2e \rightleftharpoons H_2$). In the case of the copper–zinc cell, the electrode potentials correspond to the o.c.v. of the circuits

$$Cu \mid Pt, H_2 \mid H_2SO_4 : CuSO_4 \mid Cu \qquad \text{for } E_{Cu}$$

and

$$Cu \mid Pt, H_2 \mid H_2SO_4 : ZnSO_4 \mid Zn \mid Cu \qquad \text{for } E_{Zn}$$

The electrode potential is the algebraic sum of the Galvani potentials of the circuit. For example, the copper electrode potential is

$$E_{Cu} = \varphi_{Cu,e} + \varphi_{e,Pt(H_2)} + \varphi_{Pt,Cu} = \varphi_{Cu,e} - \varphi_{Pt(H_2),e} + \varphi_{Pt,Cu} \qquad (4.3)$$

As follows from equation (4.3), the electrode potential relative to a selected reference electrode characterizes the metal–solution Galvani potential to the accuracy of a constant term, $E_{Cu} = \varphi_{Cu,e} + \text{const}$. If the Galvani potential is changed by $\Delta\varphi_{Cu,e}$ as a result of a change in a relevant factor (such as a change in solution concentration, or the appearance of a cell current), then the electrode potential assumes the same increment $\Delta E_{Cu} = \Delta\varphi_{Cu,e}$. Consequently, the measured electrode potential is a fairly good indication of the properties of a given electrode. Wherever the term "potential of an electrode" is used henceforth it designates the electrode potential with respect to a specified reference electrode. It is clear from the above that in the copper–zinc cell $\mathcal{E} = E_{Cu} - E_{Zn}$. The o.c.v. can thus be expressed as the difference of two measurable quantities each characterizing one electrode. The value of φ of the reference electrode in the expression for

$\mathscr{E}$ cancels out, so that its specifics do not affect the result. It is essential, however, that the potentials be referred to the same reference electrode.

By convention, the o.c.v. is defined in electrochemistry as the difference between the potentials of the right-hand (in the notation of a galvanic circuit) and left-hand electrodes:

$$\mathscr{E} = E_{rh} - E_{lh} \qquad (4.4)$$

In the cell notation, the reducer (negative electrode) is written on the left, and the oxidizer on the right; consequently, equation (4.4) is identical to equation (1.6), and the $\mathscr{E}$ of a cell is always positive. If a single electrode is considered, the reference electrode is written on the left; the sign of the electrode potential is then plus or minus, depending on whether the electrode is positively or negatively charged relative to the reference.

(e) Equilibrium and non-equilibrium electrode potentials

Electrode reactions involve the transfer of charges (electrons or ions) across the electrode–electrolyte interface. We have already shown above that the Galvani potential at the interface balances out the chemical and electrical forces applied to the charges. This means that the electrode reaction is at equilibrium, and does not proceed in either direction. The corresponding electrode potential is said to be the equilibrium potential (or reversible potential), defined by equation (1.14) as the change in free energy for the overall reaction proceeding both at the investigated and the reference electrodes; in the case of the copper electrode, for example, this is the reaction $Cu^{2+} + H_2 \rightleftharpoons Cu + 2H^+$.

The equilibrium at the electrode surface is essentially dynamic. Each second a certain number of charge carriers crosses the interface in one direction, and the same number of carriers crosses it in the opposite direction. Consequently, although the net reaction rate is zero, some exchange of charge occurs between the electrode and the electrolyte owing to uninterrupted cathodic and anodic partial reactions. The rate of this process, expressed in electrical units, is referred to as the exchange current (or current density) and denoted by I° or i°, respectively.

The range of the exchange current is very wide in different electrodes. The greater the exchange current, that is the easier the charge transfer, the more easily the equilibrium potential sets in and the higher its stability with respect to external perturbations (for instance, equilibrium perturbations due to the passage of current through the electrode).

The stability of the electrode potential is low for electrodes with low exchange current. It is also possible that the equilibrium potential corresponding to a given reaction does not set in even at zero external current because other concurrent processes distort the potential. A typical example is the oxygen electrode (on a platinum or silver current collector, for example): the actual potential of this electrode is more negative by 0·15–0·3 V than the value calculated thermodynamically from the free energy. This zero-current potential is said to be non-equilibrium with respect to the reaction in question; it is also called the steady-state potential provided it is reproducible and shows no significant fluctuations. The non-equilibrium potential produced by two or more concurrent reactions is termed the mixed potential. In the general case the non-equilibrium potential is not governed by the Nernst equation.

The electromotive force is the equilibrium potential difference of the galvanic circuit, that is it is defined by equation (4.4) if equilibrium values of electrode potentials are substituted into this equation. The value of the e.m.f. can be calculated from equation (1.14) by using the value of ΔG of the current-producing reaction. The potential difference across a galvanic circuit is non-equilibrium, that is, not described by equation (1.14), if the potential of at least one of the electrodes is non-equilibrium. The corresponding value is referred to as the open circuit voltage (o.c.v.) to distinguish it from the equilibrium e.m.f. This term is also used when the stoichiometry of the current-producing reaction is not clear, or when the identification of the reactant modifications is not unambiguous. In contemporary cell nomenclature it is customary to apply the term o.c.v. to the actual zero-current potential difference across a cell (equilibrium as well as nonequilibrium), and to use the term e.m.f. only for the quantity calculated on the basis of the thermodynamic data (see Section 1.5). The e.m.f. of the oxygen–hydrogen cell at room temperature is 1·23 V while its o.c.v. is 0·9–1·1 V. The lead storage battery is an example of a cell with coinciding e.m.f. and o.c.v. The value of the o.c.v., $\mathscr{E}$, cannot as a rule exceed the e.m.f., $\mathscr{E}^{(T)}$, provided the stoichiometry of the reaction is taken correctly into account in calculating the e.m.f.; however, sometimes it may exceed the standard value of the e.m.f., $\mathscr{E}^{\circ}$, because of the effect of reactants concentration or that of the electrolyte.

(f) Scales of electrode potential

In addition to the normal hydrogen electrode, other reference systems are used, such as the calomel, mercury oxide and other types of electrodes. This gives different scales on which electrode potentials differ by the potential differences between reference electrodes.

A point of practical importance is the dependence of the electrode

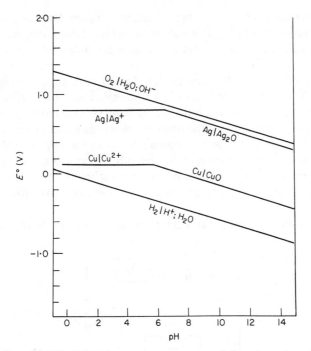

Fig. 17. Potential ($E°$) of the hydrogen, oxygen, copper and silver electrodes as function of pH.

potential on the pH of a solution. A plot of such dependence is called the Pourbaix diagram[2] (see, for example, Fig. 17). In the case of the silver electrode, neither hydrogen nor hydroxyl ions participate in the electrode reaction in solution with pH < 6. Hence, the potential of the electrode is independent of the solution pH. The situation is different in the case of the silver oxide electrode in alkaline solution in which one hydroxyl ion reacts per electron: according to the Nernst equation, the potential diminishes by 0·059 V when the pH is increased by unity. The same dependence on pH is found for the hydrogen electrode potential. Electrodes of this type are sometimes measured on a special scale with reference to the potential of the reversible hydrogen electrode (r.h.e.) in the same solution (the corresponding symbol is E_r). On this scale, the silver oxide electrode potential is independent of the pH.

(g) The measurement of electrode potentials

The measurement of zero-current electrode potentials is realized as that of the o.c.v. in a circuit including the electrode in question and a convenient reference electrode. The o.c.v. measurement requires that polarization-inducing currents be suppressed in the circuit. This is achieved by employing instruments with very low current consumption (valve voltmeters, potentiometric circuits, etc.).

The non-zero current electrode potentials also have to be measured in some cases. The appropriate means is a three-electrode circuit (Fig. 18) in which the current passes through the electrode investigated and an auxiliary one. No current flows through the reference electrode so that its potential is not distorted. In order to eliminate ohmic losses in the reference electrode circuit resulting from the current in the other electrodes, the reference electrode compartment is connected to the surface of the investigated electrode by a thin glass capillary filled with the electrolyte (the Luggin capillary).

If the electrode in question and the reference electrode are immersed in

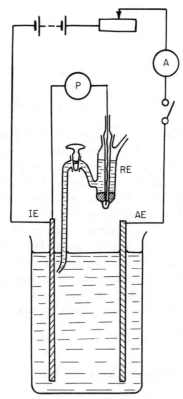

Fig. 18. Simplified potentiometric circuit. IE, investigated electrode; RE, reference electrode; AE, auxiliary electrode; P, potentiometer (high-resistance voltmeter); A, ammeter.

different electrolytes, an additional potential jump, called the diffusion potential, is formed at the boundary of the solutions. The diffusion potential is lower the smaller the difference in solution compositions. Therefore it is advisable to use the alkaline mercury oxide reference electrode in alkaline solutions and the calomel electrode in solutions of chlorides. The diffusion potential can also be lowered by introducing an intermediate concentrated solution of KCl. This potential can be measured (at least approximately) if formed at the interface of two phases of similar nature (such as two aqueous electrolyte solutions), and in this respect differs from other Galvani potentials.

4.2. Electrolytes: passage of current and transfer of ions and reactants

(a) Ion migration and electrolyte conductivity

By definition, an electrolyte is a phase with mobile ions. When an electric field of strength E is applied to the electrolyte, the ions begin to move (migrate) along the field direction. The rate of migration of each ion is proportional to E, that is $v_j = u_j E$ for the jth ion species; the coefficient u_j is the migration rate at $E = 1$ V/cm, called the mobility of a given ion species. The number of gram-ions transferred per unit time across a unit cross-section area is called the flux of ions, J. For migration in a solution with the ion concentration c_j the flux of the ions j is

$$J_{j,\,\text{migr}} = c_j u_j E \tag{4.5}$$

The charge transferred by this flux is the partial current density of this ion species:

$$i_j = |z_j| F J_{j,\text{migr}} = |z_j| F c_j u_j E \tag{4.6}$$

(equation (4.6) operates with the magnitude of z_j since the current density is assumed essentially positive). The total current density is the sum of partial current densities, $i = \Sigma_j\, i_j$, because the directions of migration of ions with opposite signs are opposite, and hence the direction of the respective currents is common. The ratio of the partial to total current density,

$$t_j = \frac{|z_j| c_j u_j}{\Sigma_j\, |z_j| c_j u_j} \tag{4.7}$$

is said to be the transference number of a given ion species. Obviously, $t_j \geqslant 1$ and $\Sigma_j\, t_j = 1$. For fixed ions, $u_j = 0$ and consequently $t_j = 0$. Electrolytes with mobile ions of one sign only are said to have unipolar conductivity.

The (specific) conductivity κ of the electrolyte is defined as the current density at $E = 1$ V/cm,

$$\kappa = F \sum_j |z_j| c_j u_j \qquad (4.8)$$

and measured in $\text{ohm}^{-1} \text{cm}^{-1}$. The inverse quantity $\rho \equiv 1/\kappa$ (ohm cm) is called the resistivity of the electrolyte. In a constant cross-section electrolyte layer, the ohmic potential drop between any two points along the current flow direction is given by

$$\varphi_{\text{ohm}} = i\rho l \qquad (4.9)$$

where l is the distance between the points.

With the exception of the boundary zones, containing the electric double layers, both the bulk of the electrolyte as a whole and any of its parts are electrically neutral, that is the algebraic sum of charges of all the ions involved is zero: $\Sigma_j z_j c_j = 0$.

(b) Classification of electrolytes

In addition to the ionic conductivity κ, other important properties of electrolytes must be considered in cell technology: their chemical and electrochemical stability, solubility of basic components, melting and boiling points, phase transition point, and so on. The nature of the electrolyte strongly affects the character and overall rate of the electrode reactions, and hence the polarization losses. Therefore the maximum conductivity may not be the decisive criterion in the electrolyte selection.

Various electrolytes are utilized in cells:

Aqueous solutions.[3] The electrolytes utilized in electrochemical cells are still for the most part aqueous solutions of acids, alkalis and salts. The solutes dissociate into sufficiently mobile hydrated ions of both signs, with the possibility of varying their concentration over a wide range; the solution conductivity is varied accordingly. As shown by equation (4.8), the conductivity κ of very dilute solutions is directly proportional to concentration. As concentration increases, the u_j diminish owing to the interionic interaction, increased viscosity and other factors. Furthermore, the degree of dissociation of the electrolyte may also be reduced. As a result, the increase of κ slows down, and often κ passes through a maximum. As an example, Fig. 19 plots κ as a function of concentration for two most important electrolytes used in cells: solutions of potassium hydroxide (KOH)[4] and sulphuric acid (H_2SO_4).[5] Conductivity strongly depends on temperature, mostly because of the effect on mobility u_j.

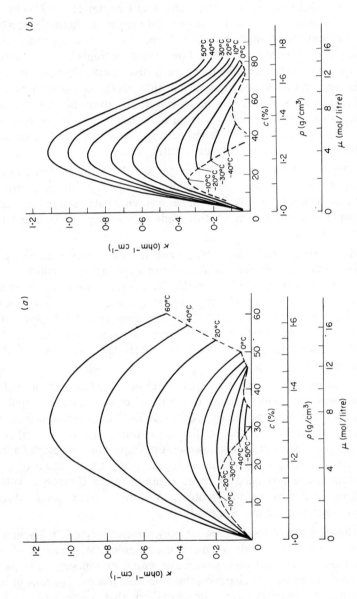

Fig. 19. Electrical conductivity of (*a*) potassium hydroxide and (*b*) sulphuric acid solutions as a function of concentration and temperature.

Non-aqueous solutions.[6,7] Non-aqueous solutions are suitable for cells if the dissolved salts undergo dissociation. This is the case for solvents with not too low a dielectric permittivity. Table 4 of Chapter 15 (p. 311) lists a number of solvents and some of their physical properties. Benzene, acetone and similar solvents cannot be utilized because of low solubility and negligible dissociation of salts. Most of the organic solvents listed in Table 4 are, in contrast to water, aprotic, that is their dissociation does not produce hydrogen ions. In particular, this means that alkali metals immersed in these solvents remain stable, with no hydrogen evolution that would result from metal dissolution. Liquid ammonia is an example of a non-aqueous but not aprotic solvent which behaves very similarly to water but remains liquid at temperature much lower than water does.

Molten salts.[8,9] Mobile ions of both signs are also formed when a salt is melted. In contrast to the case of solutions, a melt contains a strictly constant concentration of ions. The range of conductivity variation in melts is correspondingly more narrow; it depends only on temperature and the nature of the melt.

Solid electrolytes.[10] These are ionic crystals in which most often only one ion species is mobile. Therefore, unlike solutions and melts, the conductivity of solid electrolytes is, as a rule, unipolar. The dependence of conductivity on temperature is usually very strong, appearing sometimes as sharp jumps in plots of conductivity as a function of temperature (Fig. 93, Chapter 16). These jumps are caused by the fact that not all phase modifications of a given material have high ionic conductivity, and the few that have are stable in specific temperature ranges.

Matrix electrolytes. These are liquid electrolytes (solutions or melts) filling the pores of a solid insulator, called the electrolyte carrier. Capillary forces restrict the liquid's fluidity. Although the liquid's conductivity is somewhat lowered by the matrix, the basic properties are nevertheless retained. Matrix electrolytes lack, however, one important property of a free liquid, namely stirring by liquid flows (convection). One specific type of matrix electrolyte is produced by immobilization of the electrolyte with a gelling agent, whereby its macromolecules form a fairly stable three-dimensional network serving as a matrix.

Ion-exchange electrolytes (ionites, or polymer electrolytes).[11] These are specific solid electrolytes containing a macromolecular skeleton of an organic (or sometimes inorganic) compound with fixed ionogenic groups of one sign, such as negatively charged sulphonic acid groups. The ions of the opposite sign, or counterions, are not localized, that is they are mobile. Electrolytes of this type have unipolar conductivity; on being soaked in water, they swell, and the counterion mobility increases. Counterions may be substituted by a different species of the same sign by exposure to a selected solution. Electrolytes with positive counterions are referred to as cationites (cation-exchange electrolytes), and those with negative counterions as anionites. If an ion-exchange electrolyte is impregnated with a solution of an acid, alkali or salt, its conductivity is no longer unipolar; ions of both signs can now penetrate into the swelled network, turning it into a sort of matrix electrolyte. Ion-exchange electrolytes are mostly applied as thin membranes.

The solid, matrix, immobilized, and so forth, electrolytes facilitate the development of sealed cells, and enable the design of compact models. On the other hand, liquid electrolytes also offer certain advantages by fulfilling buffer functions: levelling of temperature inside the cell, facilitation of heat transfer, compensation of volume changes of the reactants and products, acceleration of concentration levelling, and so on. The utilization of liquid electrolytes, in contrast to solid ones, does not entail difficulties of ensuring sufficiently tight contact between the electrodes and electrolyte.

Another important factor is the purity of the electrolytes. Even minute admixtures of foreign compounds may drastically reduce the cell performance. Water is an unwanted ingredient in non-aqueous solutions. Alkaline electrolytes readily absorb carbon dioxide from the air (are carbonated) and therefore have to be protected.

(c) Mass transfer in electrolytes

The cell electrolyte is not only the link closing the electric circuit but is also a source of ions for the electrode reaction. It has already been mentioned (see Section 1.2) that the OH^- ions formed at the positive electrode of the silver–zinc cell are then reacted at the negative plate. Consequently, one OH^- ion must be transferred from the cathode to the anode for each reacted electron in steady-state conditions; hence, the total flux is given by $J_{OH^-} \equiv i/F$ ($|z_j| = 1$). The K^+ ions do not participate in the reaction, and their transfer is unnecessary ($J_{K^+} = 0$). Therefore, from the point of view of mass conservation, the transference numbers of the OH^- and K^+ ions must equal 1 and 0, respectively. In migration, however, they are 0·79 and 0·21, that is a fraction of the current is inevitably carried by the K^+ ions moving in the electric field. The material balance in actual conditions is maintained by the reactants supplied not only by migration but also through diffusion and convection.

Let us consider first the silver–zinc cell with a KOH solution contained in a matrix which impedes mixing. Formation of the OH^- ions at the cathode in a current-delivering cell is only partially balanced out by their migration away from the surface. On the other hand, the cations K^+ are transported to the cathode surface by migration. As a result, the electrolyte concentration at the cathode surface increases; at the same time, and for similar reasons, it diminishes at the anode. A concentration gradient is thereby built up, with both the OH^- and K^+ ions transported by diffusion toward the anode. The concentration gradient eventually established is such that the migration transfer of potassium is completely balanced out by diffusion which enhances the OH^- transfer to the required rate, so that a steady-state situation is achieved.

According to Fick's first law, the diffusion flux of each species is proportional to the concentration gradient:

$$J_{j, \text{diff}} = D_j \frac{dc_j}{dx} \tag{4.10}$$

where D_j is the diffusion coefficient (for simplification, the analysis is limited to a single dimension). The diffusion coefficient D_j and mobility u_j are quantities of similar nature; in dilute solutions they are related by Einstein's equation

$$D_j = \frac{RT}{|z_j|F} u_j \tag{4.11}$$

The migration and diffusion fluxes of the K^+ ions have opposite directions, and must cancel out in the steady-state discharge, that is

$$J_{+, \text{migr}} - J_{+, \text{diff}} = c_+ u_+ E - D_+ \frac{dc_+}{dx}$$

$$= u_+ \left(c_+ E - \frac{RT}{F} \frac{dc_+}{dx} \right) = 0 \tag{4.12}$$

In the case of OH^- ions these components add up and must correspond to the stoichiometrically stipulated flux

$$J_{-, \text{migr}} + J_{-, \text{diff}} = c_- u_- E + D_- \frac{dc_-}{dx}$$

$$= u_- \left(c_- E + \frac{RT}{F} \frac{dc_-}{dx} \right) = \frac{i}{F} \tag{4.13}$$

By virtue of electroneutrality, $c_+ = c_-$ throughout the KOH solution. Equations (4.12) and (4.13) then yield that

$$J_{-, \text{diff}} = J_{-, \text{migr}} = \frac{i}{2F} \tag{4.14}$$

so that one-half of the OH^- transfer is driven by electric field and another half by diffusion. The necessary concentration gradient is

$$\frac{dc}{dx} = \frac{i}{\alpha F D_-} \qquad (4.15)$$

where $\alpha = 2$. The coefficient α is distinct from 2 if an electrolyte is not univalent. If the electrolyte contains an excessive concentration of a foreign salt (i.e., not participating in the reaction), the migration transfer of the reactant diminishes and α is below 2 (the limiting value is $\alpha = 1$); the concentration gradient is therefore increased.

The steady-state concentration gradient is assumed constant throughout the whole diffusion zone (a dependence of the ion diffusion coefficient on concentration may result in a slight variation of the concentration gradient within the diffusion layer). Consequently, for a given gradient calculated by equation (4.15) (i.e. for a given current density i) the total concentration difference is proportional to the length of the diffusion zone x, that is

$$\Delta c = (dc/dx)x$$

The concentration gradient in the above example of a matrix electrolyte is established across the whole interelectrode distance L; the total concentration difference is the greater the longer this distance (the thicker the matrix).

The distribution is changed drastically if a liquid electrolyte is used instead of a matrix one. The convective flows of the fluid transporting ions of both signs constitute an additional transfer mechanism. Convection results not only from stirring but also from slight temperature gradients and differences in solution density (natural convection). This process produces a levelling-off of the concentration over the whole volume of the solution, with the exception of a thin layer close to the electrode surface which remains only slightly affected by convection flows. The thickness δ of this layer is the smaller the more intensive the convection, and varies from 0·01 to 0·1 mm. Both the concentration gradient and the diffusion transfer of the reactant operate within this layer (hence, the term "diffusion layer"). The diffusion layer being quite thin, the concentration difference Δc is much lower than in the case of no convection ($\Delta c = c_b - c_s$, where c_s is the concentration close to the electrode surface, and c_b is the bulk concentration averaged by stirring). The concentration difference may nevertheless become appreciable at high current densities.

In the general case, the flux of the jth species towards (or away from) the electrode surface equals, according to equations (1.11) or (1.12), $v_j i/nF$ regardless of whether an ion or a neutral molecule is transported. Under conditions of diffusion transfer the corresponding concentration gradient of a given species is

$$\frac{dc}{dx} = \frac{v_j i}{\alpha n F D_j} \qquad (4.15')$$

One characteristic feature of the diffusion transfer is that it reaches a certain upper bound which would be realized if the concentration c_s of the reactant species at the electrode surface were reduced to zero. This maximum transfer corresponds to the so-called limiting diffusion current

$$i_l = \alpha \frac{n}{v_j} F D_j \frac{c_{b,j}}{\delta} \qquad (4.16)$$

The limiting current in a solution containing several reactant species is determined by the species with the least value of the maximum possible diffusion flux.

There are therefore three components in the total flux of ions in the electrolyte: migration flux J_{migr}, diffusion flux J_{diff} and convective flux J_{conv}. The component predominant in liquid electrolytes (excepting the diffuse layer) is, as a rule, J_{conv}, and in matrix electrolytes it is usually J_{diff}. No concentration changes are possible in a pure molten salt, and $J_{diff} = 0$ (this does not cover the case of molten mixtures). Only migration transfer is possible in solid electrolytes, so that stoichiometric transfer can only be ensured if the non-reacting ions are not transferred. This means strictly unipolar conductivity by an ion formed at one electrode and reacted at the other.

4.3. Polarization of the electrodes

Polarization of the electrodes, that is shifting of the electrode potential in a current-delivering cell, is one of the main factors causing energy losses in operating cells. The quantitative characteristic of polarization η is a function of the relative rates of electrode reactions, that is of current density i. A graph of this function is called the polarization curve.

Several factors bring about electrode polarization. An analysis of the more important of them is given below.

(a) Concentration polarization

According to the Nernst equation, the changes in concentration at the electrode surface of a current-delivering cell cause corresponding changes in the electrode potentials. This violates the equilibrium between the electrode and the initial (or bulk) concentration of the reacting species. The greater the concentration changes the greater the shift in the potential, that is the higher the concentration polarization.

Equation (1.23) gives the following expression for the potential shift caused by a change of the reactant (j) concentration at the electrode surface from the initial value $c_{o,j}$ to $c_{s,j}$ (for instance, a decrease of the OH^- concentration at the anode of the silver–zinc cell):

$$
\eta_{conc} = \frac{2 \cdot 3RTv_j}{nF} \left| \log \frac{c_{s,j}}{c_{b,j}} \right|
$$

$$
= \frac{2 \cdot 3RTv_j}{nF} \left| \log \left(1 - \frac{i}{i_l} \right) \right| \tag{4.17}
$$

This shift becomes significant ($\eta_{conc} > 3$ mV) when the current exceeds 10% of the limiting diffusion current. Some estimate of i_l is therefore required in order to evaluate the effects of concentration polarization.

The range of concentrations of cell electrolytes is, as a rule, from 2 to 10 mole/litre. The diffusion coefficients of most ions and molecules in aqueous solutions are about 10^{-5} cm^2/s. It is then easily calculated, by using these values and the diffusion layer thickness in a liquid electrolyte as given in Section 4.2, that the limiting diffusion current is in the range of 1 to 10 A/cm^2. The concentration polarization is thus only slight if a liquid electrolyte is used and current densities do not exceed $0 \cdot 1$ A/cm^2.

Concentration polarization produces a much larger effect in matrix (including immobilized) electrolytes where convection is suppressed. The effective diffusion coefficient in such electrolytes is reduced by a factor of 3 to 8. For an interelectrode spacing of 1 mm, the limiting diffusion current is of the order of 10^{-2}–10^{-1} A/cm^2, so that the effect of concentration polarization is already appreciable at very low currents.

Concentration polarization is also appreciable if the concentration of one of the components of the reaction is very low. The effect is especially well pronounced in electrode reactions involving gaseous components (the case of hydrogen, oxygen, or other gas electrodes) with inherently low solubility in liquid electrolytes.

When current is interrupted, diffusion equalizes the concentration differences in the bulk of the electrolyte and at the electrode surfaces. The characteristic time of this levelling-off is of the order of 1 s; during this interval the potential of a disconnected electrode returns to its initial value.

Concentration polarization is often observed in solid cell reactants such as oxides, for example. The reduction of some oxides requires a partial or total removal of oxygen ions from the lattice, or proton penetration into it. These processes proceed uninhibited in subsurface layers of individual crystallites. This change in the layer composition triggers the outward diffusion of oxygen ions or the inward diffusion of protons. The diffusion in a solid phase is much slower than that in a solution so that considerable modification of surface layer composition in the reactants and resultant concentration polarization is possible even at low currents. A pronounced feature of concentration polarization in solid phases is its non-steady-state character. Low-rate diffusion processes result in a slow continuous variation of the potential which returns very slowly (for hours) to its steady-state zero-current value after the current is switched off.

(b) Activation polarization

A current across an interface may disrupt the equilibrium between the chemical and electric forces acting on free charge carriers of the surface layer. If an electrode reaction involving electric current across the interface is inhibited, that is occurs at a low rate and has high activation energy (referred to as rate-determining reaction), certain excess energy is required to intensify the reaction; hence, a unidirectional force must be applied to the transferred charge carriers. This is achieved by changing the potential difference across the interface. The more retarded the electrode reaction, the greater is the potential change required to sustain any given current. This effect is referred to as activation polarization or transfer polarization (or reaction overvoltage).

The activation polarization is proportional to current density provided the latter is small compared with the exchange current i°:

$$\eta_a = \omega i \qquad (4.18)$$

That is the polarization resistance r_{pol} equals the constant ω. Polarization in the range of high current densities is often a linear function of the current density logarithm (the Tafel equation),

$$\eta_a = a + b \log i \qquad (4.19)$$

where a and b are constants. The constants ω and a which mainly determine the polarization are related to the exchange current: the higher i°, the smaller ω and a and the lower the polarization. The constant b, which chacterizes the dependence of polarization on current, is close to 0.12 V for many electrodes. This means that a ten-fold increase of current density raises the polarization by approximately 120 mV. Sometimes η_a is found to be a more complicated function of i.

The polarization is an undesirable process for the main electrode reaction since it leads to energy losses. It is advantageous, however, with respect to counterproductive side reactions (such as metal corrosion liberating hydrogen). The higher the polarization produced by these reactions the lower their rates and the smaller their detrimental effects.

In any given conditions, the rate of an electrochemical reaction is temperature-dependent; that is the reaction is characterized by an activation energy. A decrease in temperature increases the electrode polarization at a given current density and hence, the energy losses in a cell.

(c) Crystallization polarization

The electrodeposition of metals which takes place, for example, when oxide electrodes are reduced to a metal, mostly passes through two stages: nucleation and crystal growth. Crystals continue growing during the whole crystallization process, which produces a certain (normally slight) polarization since the penetration of a neutralized ion into the lattice may be inhibited. On the other hand, nucleation at the start of the process is often very much inhibited, so that the corresponding polarization is quite high. Once the first nuclei are formed, however, polarization quickly drops to a level typical for the crystal growth stage. As a result, sharp voltage minima are observed on the discharge curve and sharp maxima on the charging curve at the beginning of the respective processes (see Figs. 10 and 56).

(d) Passivation of electrodes

It is observed sometimes in a current-delivering cell that the reaction rate (electric current) diminishes gradually or suddenly although the active reactants are yet far from being exhausted. If the current is maintained constant by external devices, the electrode polarization is increased; in some cases the potential is pushed so high that a competing electrochemical reaction is initiated (for instance, oxygen evolution instead of the anodic dissolution of the metal). The cell voltage is correspondingly changed. The phenomenon has been termed electrode passivation.

Passivation is caused by a number of factors. In porous electrodes it may occur owing to restructuring of the active mass, deteriorating reactant supply, and so on. Passivation is observed most often in the case of the anodic dissolution of metal electrodes. In the past this was explained by the formation of thick, compact, oxide layers on the metal surface, thus screening and insulating the plate. It was later demonstrated[12] that more subtle effects may be at work, in which even minute quantities of adsorbed oxygen (or of other components) so modify the electrochemical properties that the reaction rate is drastically reduced and passivation sets in. These modifications are normally accumulated in a gradual manner but become apparent in a step-wise fashion, after a definite threshold state is reached.

Passivation of the main reactants of the current-producing reaction results in a reduced utilization coefficient and constitutes a negative factor. The tendency to passivation increases particularly when the temperature is lowered, constituting one of the main reasons for poor performance of cells at low temperatures. Conversely, passivation of structural materials (such as stainless steel) enhances their chemical stability and longevity.

4.4. Levelling-off effects. Distributed-parameter systems

An electric current flowing in a galvanic cell is usually distributed uniformly over the whole surface of the electrodes. In other words, the current density is the same at any point of the surface. Non-uniform distribution is nevertheless possible if different segments of the surface are not equally accessible to the current. A system in which this is the case is called a distributed-parameter system. The factors determining the actual distribution of current density in systems of this type are polarization relationships, ohmic potential differences and diffusion processes. In electroplating nomenclature, a system with uniformly distributed current is referred to as a bath with high throwing power.

Let us consider the simplest cell with one positive electrode and two negative ones placed at distances of l_1 and l_2 from the positive electrode (Fig. 20(a)). For the sake of simplification assume the electrode polarization to be proportional to current density, $\eta_j = \omega i_j$ (ω is the total polarization impedance of the two electrodes). The sum of ohmic losses and polarization voltage in the two halves of the cell connected in parallel must be the same since both generate the same voltage. The ohmic losses in the electrolyte are $\varphi_{ohm,j} = \rho l_j i_j$ (ρ denotes the electrolyte resistivity). Therefore, $\omega i_1 + \rho l_1 i_1 = \omega i_2 + \rho l_2 i_2$ or

$$\frac{i_1}{i_2} = \frac{\rho l_2 + \omega}{\rho l_1 + \omega} \tag{4.20}$$

If the polarization impedance is low ($\omega \ll \rho l_j$) the non-uniformity of the current density distribution is maximum ($i_1/i_2 = l_2/l_1$). As ω increases, the system tends to a uniform distribution, with the current density equalizing in both halves of the cell when $\omega \gg \rho l_j$. An increased polarization is

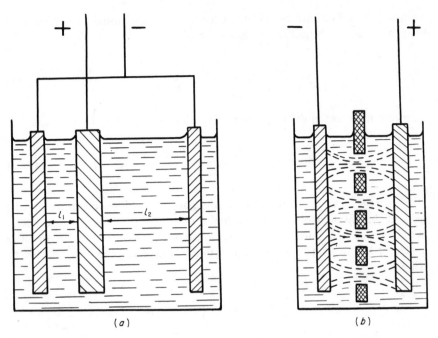

Fig. 20. Non-uniform distribution of current in an electrochemical cell.

therefore a factor producing the levelling-off of current densities in distributed-parameter systems.

Similar effects are observed if a perforated separator is positioned between the electrodes of a galvanic cell (Fig. 20(b)). Let α denote the relative area of the holes. If the electrode polarization is low, the current only flows along the shortest paths between the electrodes, with α giving the fraction of the electrode surface participating in the process; the remaining part of the surface is shadowed. The effective current density at the working electrode areas is correspondingly increased and produces a premature local depletion of the reactants. With a substantial electrode polarization the reactants are consumed more uniformly because of the bending of current paths and the resulting decrease in shadowing.

Let us consider another example of current distribution in a cell with long electrode plates where ohmic loss of voltage develops across the current collector (Fig. 21(a)). Let b denote the collector width, d its thickness, h its height, and ρ the resistivity of the collector material. Assume the electrode to be active on both sides (i.e. $S = 2bh$) and the current to be collected at the upper edge uniformly over the whole width of the plate (the collector cross-section is bd). The current density at the height x is denoted by i_x. The total current I_x at this height is obviously given by

$$I_x = 2b \int_0^x i_x \, dx \tag{4.21}$$

The ohmic potential difference across the segment from x to $x + dx$ is

$$d\varphi_{ohm} = \pm \frac{\rho \, dx}{bd} I_x = \pm \frac{2\rho \, dx}{d} \int_0^x i_x \, dx \tag{4.22}$$

(the signs + and − correspond to the cathode and anode, respectively; φ_{ohm} at the anode in the chosen frame of reference decreases as x increases, which yields the minus sign). The total potential difference across the collector length from x to the upper edge of the plate ($x = h$) connected to a bus is

$$\varphi_{ohm,\,x} = \pm \frac{2\rho}{d} \int_x^h dx \int_0^x i_x \, dx \tag{4.23}$$

The current density distribution along the plate is, as in the preceding example, non-uniform because the sum of ohmic voltage (in the current collectors of both electrodes and in the electrolyte) and polarization on both electrodes must be the same at any x. The current density is greater than the average value in the upper segments of the plates where the ohmic losses are smaller, and it is lower in the bottom segments. A quantitative calculation requires the knowledge of polarization as a function of current density; the higher the polarization the more uniform is the density distribution.

The most general form of a differential equation describing the potential and current distributions in a system with ohmic losses is derived by differentiating equation (4.22). Taking into account that $2/d = \sigma_v$, that is the ratio of the working surface area to the volume of the current collector, and that $d\eta = \pm d\varphi_{ohm}$ because of the constancy of the sum of $|\varphi_{ohm}|$ and polarization η, we obtain

$$\frac{d^2\eta}{dx^2} = \rho\sigma_v i_x(\eta) \tag{4.24}$$

In order to analyse a case where ohmic potential drops in the electrodes and in the electrolyte are negligibly small but the concentration gradients are produced, consider a cylindrical pore (Fig. 21(b)). Let Π be the pore

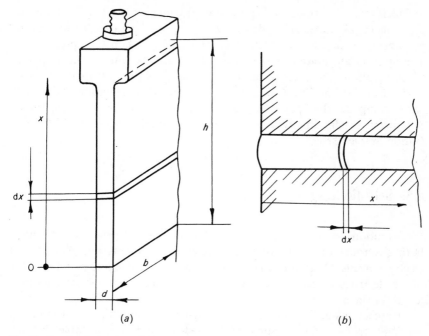

Fig. 21. On the derivation of the non-uniform current distribution equation (a) in a plane electrode and (b) in a cylindrical pore.

perimeter, S its cross-section. Evidently, the ratio Π/S again equals σ_v. The concentration gradient of the jth reactant species in an electrolyte layer at a distance x from the pore bottom is related to the current flowing through this layer by the expression

$$I_x = \pm (n/v_j)FSD_j(\mathrm{d}c_j/\mathrm{d}x)_x \qquad (4.25)$$

(the sign is determined by whether the jth species is produced or consumed in the reaction). On the other hand, the total current in the layer at x is equal (in analogy to equation (4.21)) to

$$I_x = \Pi \int_0^x i_x \,\mathrm{d}x \qquad (4.26)$$

By differentiating equations (4.25) and (4.26), we derive the general form of the differential equation in a system with variable concentration:

$$\frac{\mathrm{d}^2 c_j}{\mathrm{d}x^2} = \pm \frac{v_j \sigma_v}{nFD_j} i_x(c) \qquad (4.27)$$

Differential equations (4.24) and (4.25) can be solved if we know the dependence of the current density on polarization at constant reactant concentration or on the reactant concentration at constant polarization. The appropriate boundary conditions must also be known. In particular, these equations are used to calculate current distribution in porous electrodes (see Section 5.4). Calculations based on these equations are usually quite complicated. They become even more unwieldy if a distributed-parameter system involves both ohmic potential gradients and concentration gradients.

4.5. Self-discharge

By definition, self-discharge is a complex of chemical processes on one or both electrodes of a cell which consumes the reactants but generates no electric current. Quantitatively it is measured as an absolute or relative loss of the electrode or cell capacity resulting from this process, over a specified length of time.

The reactants utilized in cells are highly-activity ones, that is strong oxidizers and strong reducers. Some of these compounds are unstable and may undergo spontaneous decomposition. For instance, hydrogen peroxide H_2O_2, a strong oxidizer, easily decomposes liberating oxygen,

$$2H_2O_2 \rightarrow 2H_2O + O_2$$

and hydrazine, N_2H_4, a strong reducer, liberates hydrogen,

$$N_2H_4 \rightarrow N_2 + 2H_2$$

These processes are accelerated by a catalytic effect of some materials, including electrodes and even container walls.

Other reactants do not decompose spontaneously but react with electrolytes. One of the most relevant examples in cell technology is the reaction between zinc electrodes and aqueous electrolyte solutions. The zinc electrode potential being more negative than that of the hydrogen electrode in the same solution (by approximately 0·7 V in acidic and 0·4 V in alkaline solutions), a cathodic hydrogen evolution reaction is possible on the zinc surface. Electrons necessary for this reaction are supplied by the concurrent conjugate reaction of anodic zinc dissolution. The two reactions, proceeding on the same surface of the zinc electrode without spatial separation, are in fact "shorted" and deliver no electrical energy into the external circuit. The effect is therefore a gradual counterproductive consumption of the reactant, namely zinc self-dissolution.

The situation may be similar at the positive electrode: a strong oxidizer with more positive potential than that of the oxygen electrode may produce anodic evolution of oxygen, being itself reduced to a lower oxidation state.

It is therefore apparent that electrodes in an aqueous solution remain thermodynamically stable only within the range of potentials between those of the hydrogen and oxygen electrodes, that is from 0 to 1·23 V (relative to the reversible hydrogen electrode in the same solution). In principle, any electrochemical system utilizing an aqueous electrolyte and with o.c.v. above 1·23 V is unstable since the processes of hydrogen and/or oxygen evolution (in other words, an "internal" electrolysis of water consuming the corresponding amounts of reactants) are possible on the electrodes (or at least on one of them). In practice, however, such systems can often be considered stable when the side processes occur at a small reaction rate (i.e. high polarization) and so do not affect the process significantly over a specified length of time. The life-time of such cells may vary, depending on the rate of side reactions, from several minutes to several years.

Two cases are possible if $\mathscr{E} < 1·23$ V: (a) both electrode potentials are outside of the "danger zones", so that the system is absolutely (thermodynamically) stable; and (b) one of the electrodes is in the instability range and in principle may react with the aqueous solution.

The relative positions of the hydrogen and oxygen electrodes and the electrodes of a number of electrochemical systems with aqueous solutions are shown on the potential axis in Fig. 22. An example of a thermodynamically stable system is the system

$$(-)\ Cd(Hg)\,|\,CdSO_4(sat.)\,|\,HgSO_4\,|\,Hg\ (+)$$

used in the so-called standard cells (Section 13.6). Cases were recorded when such cells were kept for fifty years without any appreciable deterioration. The lead storage battery is an example of a system with high o.c.v., in which both electrodes are thermodynamically unstable and internal water electrolysis is possible. The hydrogen and oxygen evolution reactions are, however, very slow in this case, so that a charged lead battery is a comparatively stable system, enabling storage for several months without significant deterioration of parameters.

The situation is similar for non-aqueous electrolytes. Each electrolyte is characterized by a certain thermodynamic decomposition voltage, that is by the electrode potential values bounding the regions in which the reactants can react with the solvent. This voltage is higher in non-aqueous than in aqueous solutions; neverthelsss, thermodynamically, such active reducers as lithium may cause the reduction of the solvent. The decisive factor for the stability of a system of this type is again the rate of the reaction in question.

Electrodes not reacting with the electrolyte are sometimes self-discharged

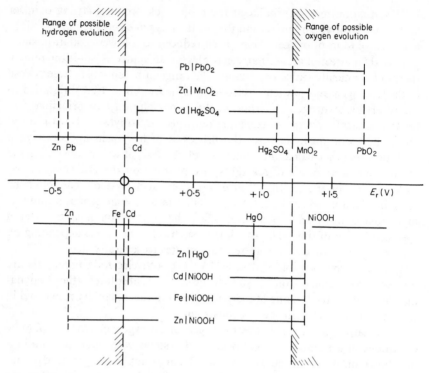

Fig. 22. The electrode potentials of a number of electrodes with respect to the potential of the hydrogen electrode in the same solution.

by reactions with other materials, such as the collector material or the additives to the electrodes. In the lead storage battery, for example, lead dioxide (PbO_2) may react at the positive electrode with the current-collecting lead grid producing divalent lead compounds (a reaction similar to the overall current-producing reaction in the battery but delivering no electrical energy). The self-discharge side reactions can also be illustrated by the oxidizer reacting with organic additives to the positive electrode (binders or hydrophobic agents) or with the separator material provided it comes into a direct contact with the electrode. Another example is oxidation of the reducer by the oxygen of the air.

In the general case the rates of the side reactions are not necessarily the same on both electrodes, since in this respect the electrodes behave independently. Consequently, the reactants in the electrodes may be unequally consumed. This may hinder further cycling of storage cells because one of the electrodes may be constantly undercharged.

4.6. Electrocatalysis

A high rate of the current-producing reaction is one of the most important targets when cell electrodes are developed. The higher the reaction rate at a given polarization, the lower the polarization at a given reaction rate, that is at a given discharge or charge current. The reaction rate is effectively increased by increasing the true active surface which is achieved by manufacturing electrodes from high-dispersion materials.

In the case of solid reactants the specific reaction rate (reaction rate per unit of true area) is determined first of all by the nature of the reactants. An effective electrode is realized if all the external factors retarding the reaction are suppressed, i.e., if the electrical conductance of the active mass is sufficiently high, the mass structure is optimized, the optimizing additives are selected, and so on. The situation is different in the case of liquid or gaseous reactants where the reaction rate is mostly determined by the nature of the "inert" electrode on whose surface the current-producing reaction really proceeds. The effect may be extremely large. For instance, the rate of cathodic hydrogen evolution per unit of true surface measured for the same polarization, is greater by a factor of 10^{10} on platinum than it is on lead.

The obvious step is then to select an electrode material with optimal catalytic properties for each electrochemical reaction involving liquid or gaseous reactants. This is the subject of electrocatalysis, a new branch of electrochemistry rapidly progressing after 1960 as a result of the fuel cell research effort.[13]

High activity is not the only requirement in a catalyst for an electrochemical reaction. According to the specifics of electrochemical reactions, it must be of the electron conductivity type. It must be corrosion-resistant in the reaction conditions, generally in concentrated acid or alkali solutions which, especially at elevated temperatures, attack a great number of materials. A catalyst must also be stable on contact with the reactants, that is strong oxidizers and reducers. The catalytic activity should not be lowered (poisoned) by admixtures or minor impurities. The catalyst's longevity affects the service life of a cell. Such economic aspects as the cost and availability of the catalyst are also important.

Platinum was empirically demonstrated to be an efficient and sufficiently versatile catalyst for a large number of reactions. Its activity and corrosion resistance are high. But being scarce and prohibitively costly, platinum cannot be used extensively as a cell catalyst. Two goals are therefore apparent; (a) to increase the efficiency of the platinum catalyst, thereby drastically reducing the platinum content in electrodes; (b) to find new, more abundant materials to replace platinum catalysts.

The efficiency of platinum utilization depends first of all on its dispersion.

The catalytic reaction takes place exclusively on the platinum grain surface, so that the core of the grain is mere ballast. Various methods are known for producing high-dispersion platinum black. As a rule, the basic compound is chloroplatinous acid, H_2PtCl_6; high-dispersion metallic platinum is deposited from it either electrolytically or chemically, by applying such reducers as hydrogen, hydrazine N_2H_4, sodium borohydride $NaBH_4$, and some others. The maximum possible true surface of high-dispersion platinum, with the grains comprising not more than eight to ten platinum atoms, each of which is therefore on the grain surface, is $270\ m^2/g$. It proved feasible to obtain platinum black deposits with a true surface of 20–50 m^2/g, with about 8–20% of platinum atoms occupying surface sites and hence being catalytically active. Platinum black becomes unstable when the specific surface area exceeds $30\ m^2/g$: the smallest particles recrystallize, merge and grow, so that the true surface decreases with time. High-dispersion platinum is often deposited onto some substrates, such as high-dispersion carbon-based materials. This measure increases stability of very small particles, and the true surface of stable platinum deposits on carbon reaches into the 90 to 100 m^2/g range.

Such platinum-on-carbon deposits are widely used in fuel cell research, especially in acid electrolyte cells (Chapter 17). The platinum consumption was initially quite high, up to 20–40 mg/cm^2 of platinum on the electrode surface, but was diminished almost ten-fold as a result of improved efficiency. Even smaller platinum consumption, 0·05–0·2 mg/cm^2, was reported for some electrodes. Evidently, the required amount of platinum depends on the specifics of the electrochemical reaction: it is smaller the easier it is for the reactant to enter the reaction. On the other hand, the reaction rate and the electrode efficiency are higher the greater the amount of catalyst. A compromise between technical and economic aspects has therefore to be reached in each individual case.

A shortcoming inherent to platinum catalysts (especially if they are applied in minute quantities) is their ease of poisoning, a drastic reduction of activity caused by irreversible adsorption of a number of so-called catalyst poisons, among them sulphides and arsenides, mercury, carbon monoxide and a number of other compounds. The poisons reach the surface either with the basic reactants or after their formation in the solution reacting with the structural materials (for example, leaching of sulphides from rubber spacers and seals). Complete elimination of these components is therefore a condition for the application of platinum catalysts.

Combination catalysts including several metals of the platinum group (such as platinum with added iridium, rhodium or ruthenium) are also used. Sometimes this results in a sharp increase of catalytic activity; platinum–ruthenium catalysts, for example, step up the electrochemical oxidation of methanol in acid solutions by two to three orders of magnitude.[14,15] The poisoning-resistance of the catalysts is also sometimes raised by these additives.

A considerable number of non-platinum catalysts are successfully used in alkaline solutions. The reaction of cathodic reduction of oxygen in alkaline solutions, an important one in chemical power sources, proceeds at a high rate on a high-dispersion silver catalyst, and at a somewhat lower rate on activated carbon. In the latter case the reaction rate can be increased by doping carbon with a number of promotor additives. Spinel-type oxides, such as cobalt spinel Co_3O_4 or cobalt–aluminium spinel $CoAl_2O_4$, proved to be effective promotors in carbon. Some complex organic metal-containing compounds, phthalocyanines and porphyrins (Fig. 23) can also be employed for this purpose.[16]

The anodic oxidation of hydrogen is effectively catalysed by the skeletal-type high-dispersion nickel catalyst (Raney nickel). In order to prepare this catalyst, a nickel–aluminium alloy (approximately 50% Al) is roughly ground and then treated with an alkaline solution which dissolves the aluminium and leaves the highly dispersed nickel skeleton. The time stability of this catalyst can be improved by such dopants as titanium which inhibit recrystallization and structure modifications. A shortcoming of the nickel catalyst consists in surface oxidation which starts when the hydrogen electrode polarization exceeds 0·15–0·2 V and reduces the catalyst's activity; oxidation becomes irreversible if polarization exceeds 0·3 V. A skeletal nickel catalyst can also be used for anodic oxidation of methanol in alkaline solutions.

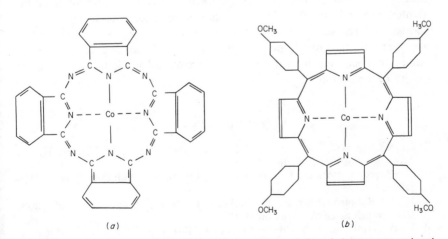

(a) (b)

Fig. 23. Structural formulas of (a) cobalt phthalocyanine and (b) cobalt tetrametoxyphenyl porphyrin.

The selection of non-platinum catalysts for acidic electrolytes is a much more difficult problem because of corrosion instability of many catalytically active materials (nickel, silver, etc.). The reaction of the cathodic reduction of oxygen on activated carbon proceeds in acidic solutions but at a low rate. It has been demonstrated recently[17] that this reaction rate can be considerably enhanced by introducing the above-mentioned admixtures of organic complexes. Unfortunately, these compounds are not sufficiently resistant to concentrated electrolytes over long periods of time. This resistance can sometimes be improved by a preliminary thermal treatment of the catalyst.[18]

A refractory compound, tungsten carbide, has a fairly high catalytic activity and sufficient stability in the reaction of hydrogen oxidation in acidic solutions.[19] Tungsten carbide can be produced as a high-dispersion powder deposited on graphite current collectors. This catalyst is very specific, being active with respect to the hydrogen reaction but not affecting other reactions, such as methanol oxidation, for instance. This constitutes an important advantage since tungsten carbide does not adsorb the compounds it cannot catalyse; hence, it is "poison-proof", with the ensuing relaxation of purity requirements. Tungsten carbide is irreversibly oxidized and becomes inactive at polarizations above 0·3 V.

Non-platinum catalysts for oxidation of organic reactants in acidic solutions have not been developed yet. This is an obstacle in the way of fuel cells with organic reducers. The development of such catalysts is an important problem facing the electrocatalysis specialists. Much attention must also be paid to increasing activity and stability of catalysts for oxygen reduction both in alkaline and acidic media. In recent years application-oriented research in electrocatalysis has been closely connected to the theorists' efforts aimed at determining the mechanisms of electrocatalytic processes, the nature of the catalytic activity of various materials, and ultimately at coming up with a theory capable of predicting the catalytic activity of a material.

REFERENCES

1. A. N. Frumkin, "Potentials of Zero Charge." Nauka Publishing House, Moscow (1979) (in Russian).
2. M. Pourbaix, "Atlas d'Equilibres Electro-Chimiques." Gauthier-Villars, Paris (1963).
3. R. A. Robinson and R. H. Stokes, "Electrolyte Solutions" (2nd edn.). Butterworth Scientific, London (1959).
4. L. E. Krutilova, I. N. Maximova, V. E. Razuvayev and M. U. Shcherba, Electric conductivity of alkali metals hydroxides at low and moderate temperature, *Ukr. Khim. Z.* **41**, No. 9, 925–928 (1976) (in Russian).

5. Gmelins Handbuch der anorganischen Chemie, 8. Auf. System N9, Teil 8, pp. 698–707. Verlag Chemie, Weinheim (1960).
6. Waddington, T. C. (ed.), "Non-aqueous Solvent Systems". Academic Press, London and New York (1965).
7. G. J. Janz and R. P. T. Tomkins, "Nonaqueous Electrolytes Handbook", Vols. 1 and 2. Academic Press, New York and London (1972, 1973).
8. G. Morand and J. Hladik, "Electrochimie des Sels Fondus", t. 1–2. Masson, Paris (1969).
9. H. Bloom and J. O'M. Bockris, Molten electrolytes, in "Modern Aspects of Electrochemistry" (Bockris, J. O'M., ed.), No. 2. Butterworth Scientific, London (1959).
10. Geller, S. (ed), "Solid Electrolytes." Springer, Berlin (1977).
11. R. A. Huggins, Ionically conducting solid-state membranes, in "Advances in Electrochemistry and Electrochemical Engineering" (Gerischer, H. and Tobias, C. W., eds.), Vol. 10, pp. 323–390. Wiley, New York (1977).
12. B. N. Kabanov, "Electrochemistry of Metals and Adsorption." Nauka Publishing House, Moscow (1966) (in Russian).
13. Sandstede, G. (ed.), "From Electrocatalysis to Fuel Cells." University of Washington Press, Seattle (1972).
14. A. N. Frumkin, Electrochemical oxidation of alcohols, in "Batteries-2" (Collins, D. H., ed.), pp. 537–543. Pergamon Press, Oxford (1965).
15. H. Binder, A. Köhling and G. Sanstede, Effect of alloying components on the catalytic activity of platinum in the case of carbonaceous fuels, in "From Electrocatalysis to Fuel Cells" (Sanstede, G., ed.), pp. 43–58. University of Washington Press, Seattle (1972).
16. R. Jasinski, Cobalt phthalocyanine as a fuel cell cathode, J. Electrochem. Soc. 112, No. 5, 526–528 (1965).
17. H. Jahnke and M. Schönborn, zur kathodischen Reduction von Sauerstoff an Phtalocyanin-Kohle-Katalysatoren, in "Proceedings, 3rd International Symposium on Fuel Cells", Brussels, 1969. Pp. 60–65.
18. V. S. Bagotzky, M. R. Tarasevich, K. A. Radyushkina, O. A. Levina and S. I. Andrusyova, Electrocatalysis of the oxygen reduction process on metal chelates in acid electrolyte, J. Power Sources 2, 233–240 (1977/78).
19. H. Böhm, New non-noble metal anode catalysts for acid fuel cells, Nature (London) 227, No. 5257, 483–484 (1970).

Chapter Five

Real Electrodes; Porous Systems

5.1. Properties of porous and disperse systems

One of the methods of intensifying electrode reactions in chemical power sources consists in using porous electrodes, as well as other parts made of porous and disperse materials: separators, matrices, pastes and so on. Some specific features are common to all these components.

The following parameters characterize porous and disperse systems:

Geometric parameters: size and shape of structural elements (particles, pores), their number and the characteristic distribution function. By convention, the systems are classified, according to the structural elements size, into low-disperse (cloths, grids, etc., characteristic size from 0·1 to several mm), medium-disperse (characteristic size from 100 to 0·1 μm), fine-disperse (characteristic size from 0·1 to 0·001 μm), and ultrahigh disperse systems (molecular-size structural elements).

Chemical parameters: the state of the surface of individual structural elements, such as the degree of oxidation, wettability by various liquids, and so on; chemical stability in a given medium at prescribed temperatures.

Mechanical parameters: strength characteristics with respect to applied stress and to crumbling.

(a) Powders, fibres[1]

Usually porous and disperse systems are prepared on the basis of powdered components. A powder is characterized by a primary and a secondary structure. The primary structural units are represented by the smallest compact grains, that is microcrystallites, of a given material. The largest primary particles (above 10–20 μm in size) may be easily separable from one

84

another. Smaller grains with larger surface-to-volume ratio tend to form secondary, or coagulation, structures termed agglomerates, allowing a limited mobility of primary particles. The shapes of agglomerates may be most varied: from comparatively compact spherical units to very loose branched units. The properties of porous and disperse materials obtained from powders depend strongly on the shape of secondary particles. For instance, branched agglomerates produce strong particle-to-particle cohesion, so that the materials with this structure have high mechanical strength.

Another important characteristic of powder is its bulk density, that is the mass-to-volume ratio obtained without especially compacting the poured powder. The bulk density of powders with loose branched structure is smaller by a factor of 5 to 20 than the true density of the material. Compression slightly deforms the secondary structure, increasing density.

The particle size δ is distributed within a certain range. This distribution is normally plotted by differential distribution curves (Fig. 24(a)), with $F_v(\delta) \equiv dV/d\delta$ (or $F_m(\delta) \equiv dM/d\delta$) characterizing the fraction of particles with diameter δ in volume (or mass). The integral distribution curve, where the ordinate $V(\delta) = \int_0^\delta F(\delta)\,d\delta$ is the total volume or mass of particles with

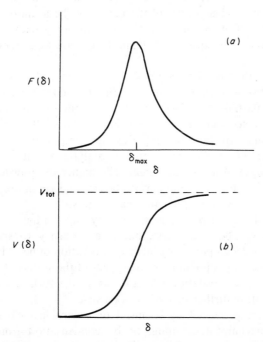

Fig. 24. Schematized curves of particle size distribution: (a) differential; (b) integral.

particle sizes less than or equal to δ, is also used. The term "linear size" is very ill defined if particles are not spherical, and so is used with reservation.

Various methods exist in granulometry (recording of particle-size distributions). Depending on the method used, the distribution of primary (δ_I) or secondary (δ_{II}) particles is found. A powder may be monodisperse, that is have particles of nearly the same size and a single sharp peak on the differential distribution curve, or polydisperse. The size δ_{max} corresponding to the maximum on the differential distribution curve is referred to as the most probable size.

The total, or true surface area of a powder Σ is a function of the size δ_I of primary particles. The surface area σ_m per unit mass of monodisperse primary particles is

$$\sigma_m = \frac{\Sigma}{\rho V} = \frac{\mu}{\rho \delta_I} \qquad (5.1)$$

where ρ is the true density of the material, μ is a numerical factor equal to 6 for cubic particles and to 3 for spherical ones. The parameter σ_m can be measured, for example, by determining adsorption of a compound which is adsorbed by the total surface and forms a monomolecular adsorption layer, and in particular by the low-temperature adsorption of argon or krypton (the BET techniques, an acronym of the names of the authors, Brunauer, Emmet and Teller). If the σ_m of a polydisperse powder is known, the mean size $\bar{\delta}_1$ can be found from equation (5.1); in the general case, $\bar{\delta}_1$ is not equal to $\bar{\delta}_{1,max}$.

As a rule, both the chemical composition and physical structure of the surface layer are very different from those in the bulk of a particle. The surface may be partially oxidized, or it may contain adsorbed foreign compounds and defects in the lattice. These factors affect interparticle cohesion, wettability by liquids, and also such important properties as contact resistance, degree of passivation, catalytic activity, as well as stability and strength of a porous system. The higher the powder's dispersion, the greater is the role of the surface layers, and the more important therefore is their composition and other parameters.

The reactivity of high-disperse powders is very high. In the case of metals this is seen as pyrophoric activity, that is a tendency to spontaneous oxidation in the air accompanied by intensive liberation of heat. Pyrophoric activity is suppressed by a preliminary treatment of the surface, for instance by a limited careful pre-oxidation which forms on the surface a thin dense oxide layer preventing further oxidation of particles.

The powders are produced by chemical or electrochemical deposition techniques, by mechanical crushing, or by condensation from vapour. Skeletal nickel powders prepared by the Raney method (see Section 4.6) are used nowadays as cell catalysts. Skeletal powders of other metals can also be obtained by this method.

The tendency now is to replace powders in porous materials by fibres. As a rule, the technology of fibre production is more complicated than that of powders, but fibres yield higher-strength materials.

(b) Porous bodies[2]

Two basic techniques are used to manufacture porous elements: (a) sintering of powders (e.g., in metal ceramics technology), and (b) leaching or thermal decomposition of granular components specially introduced into the mixture (so-called pore formers). The pores formed may either be closed, that is isolated from other pores and from the outer surface of the body, or open. Porous elements for use in cells are for the most part open-pore systems.

The most important geometric characteristics of porous systems are: (a) total porosity θ_p, that is the ratio of the total pore volume to the specimen volume; (b) curves of differential and integral distributions of pore radii, similar to distributions of particle sizes in powders; (c) true surface area per unit mass, σ_m (m^2/g), that per unit volume, σ_v (cm^2/cm^3), or per unit of the outer (geometric) surface area also called the surface roughness, σ_s (cm^2/cm^2); (d) the average pore radius $\bar{r}$ calculated from the true surface area by means of an equation of the type of (5.1), or the most probable radius r_{max} corresponding to the maximum on the distribution curve.

Pores may be of various shapes (see Fig. 25): from chains of spherical pores (obtained by removing spherical particles of a pore former) to gaps between regularly or irregularly spaced solid particles. A geometric model of a parallel array of cylindrical pores (Fig. 25(e)) is a convenient method of mathematical description of porous bodies; this model is very rarely realized, however, in real conditions. A better approximation is obtained if branching and corrugation of the pores are taken into account.

Second- and higher-order structures are often encountered in porous systems. Thus, porous carbon electrodes consist of individual granules with micropores. A secondary structure appears in a finished electrode: it is formed by a set of large-diameter pores between individual granules.

Wettability of a porous system by liquids, including liquid electrolytes, is an important chemical characteristic. Wettability determines penetration of aqueous or nonaqueous solutions, melts or liquid reactants into the internal pores; pores are not filled with electrolyte if wettability is poor.

Wettability depends on the conditions on the surface of a porous body and on liquid's properties (its surface tension). Highly wettable systems are

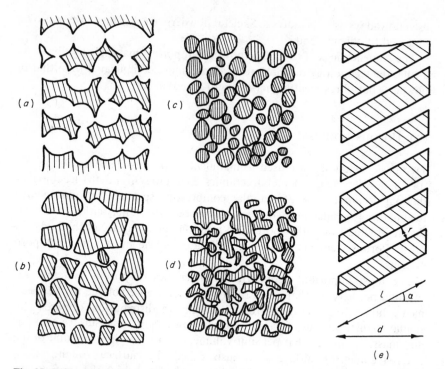

Fig. 25. Different pore shapes: (*a*) a string of spherical cavities, (*b*) cylindrical pores with corrugation and branching, (*c*) space between spherical particles, (*d*) real porous structure, (*e*) a simplified model of a porous structure with parallel arrangement of cylindrical pores.

termed lyophilic, or hydrophilic in the case of aqueous solutions. Unwettable systems are called lyophobic, or hydrophobic. In hydrophilic–hydrophobic porous electrodes which are essential in some cell types, a proportion of pores is wetted and filled with an aqueous electrolyte while the remaining pores are empty and serve to transfer a gaseous reactant into the electrolyte. This situation is achieved by employing non-uniform porous bodies consisting of two sorts of particles, one wetted and another unwetted by the electrolyte.

(c) Functional parameters of porous systems

Conductance of the liquid phase. The ionic current often flows through the electrolyte confined within the pores of a separator, matrix or electrode. It is therefore necessary to calculate the electrolyte resistance in the pores for known resistivity ρ of the free liquid electrolyte and known structure of the (insulating) porous element.

The problem in the general case is extremely complicated. Let us consider a simple model of a separator of thickness d with identical cylindrical pores of radius r, forming a parallel array and totally wetted by the electrolyte (Fig. 25(e)). Let l be the pore length, and N the total number of pores (all quantities are normalized to a unit area (1 cm^2) of the separator surface). The ratio $\beta = l/d$ ($\beta \equiv \cosec \alpha \geqslant 1$) characterizes the slope of the pores and is termed the pore tortuosity coefficient. The total pore volume is $N\pi r^2 l$, and the porosity is

$$\theta_p = \frac{N\pi r^2 l}{d} = N\pi r^2 \beta \tag{5.2}$$

The electrolyte resistance in the pores is

$$R_{pore} = \frac{\rho l}{N\pi r^2} \tag{5.3}$$

The ratio of R_{pore} to the resistance of an electrolyte layer of the same thickness d, that is to $R_e = \rho d$, is

$$\varepsilon \equiv R_{pore}/R_e = \beta/N\pi r^2 \tag{5.4}$$

It is referred to as the coefficient of resistance enhancement or of the conductivity attenuation (sometimes also the relative resistance). By definition, $\varepsilon \geqslant 1$. A comparison of equations (5.2) and (5.4) yields

$$\varepsilon = \beta^2/\theta_p \tag{5.5}$$

The factor ε increases as porosity decreases and tortuosity rises. With the porosity and tortuosity factor constant, ε is independent of the pore radius, that is a change in r is cancelled out by the opposite change in N if the total porosity is conserved.

The resistance enhancement factor is independent of the electrolyte resistivity in the separator pores. It will be useful for practical calculations to introduce the effective resistivity ρ_{eff} or effective conductivity κ_{eff} of the electrolyte in a porous body:

$$\rho_{eff} = \varepsilon\rho, \qquad \kappa_{eff} = \kappa/\varepsilon \tag{5.6}$$

which makes it possible to ignore the specific structure of a separator and calculate its resistance by using only its external geometric characteristics.

Actual porous systems deviate very strongly from the model as stated. Nevertheless, the corollary of the resistance being always an increasing function of diminishing porosity holds true. An empirical relationship (Archie's law) states that in actual porous systems

$$\varepsilon \sim \theta_p^{-m} \tag{5.7}$$

where $m = 1.8$–3.5. This demonstrates that the tortuosity factor β and porosity θ_p are correlated in random porous systems ($\beta^2 \sim \theta_p^{(1-m)}$), in contrast to the above ideal pore model where $\beta = \text{const.}$ and $m = 1$.

Diffusion. Likewise, the diffusion of dissolved components through a porous element is slower the lower its porosity and the higher the tortuosity factor. The diffusion attenuation factor ε_D coincides, as a rule, with the corresponding factor for conductivity. In analogy with equation (5.6), the effective diffusion coefficient D_{eff} can be found in terms of ε: $D_{eff} = D/\varepsilon$.

If pore diameters are small (for instance, exceeding the size of diffusing particles by a factor of not more than 3 to 5), the filtering effect of a pore and the effect of particle-pore wall interaction may lead to selective behaviour when the pore permeability (effective ε) is different for different particles.

Liquid flow. An external force Δp is required to produce the flow of liquid through a porous body. The bulk flow rate in the above model of a separator with cylindrical pores can be found by means of Poiseuille's law:

$$V = \frac{N\pi r^4}{8l\eta}\Delta p = \frac{\theta_p r^2}{8\beta^2 d\eta}\Delta p \tag{5.8}$$

where η is the liquid's viscosity. Equation (5.8) shows that in contrast to electrical conductivity, determined by the porosity and tortuosity factors, the flow also depends on the absolute pore size. A ten-fold decrease in pore radius with other conditions unchanged results in a hundred-fold drop in the flow rate. This factor is of extreme importance in cells, since in some cell types the flow rates across separators must be high while in other types they must be suppressed as much as possible.

In systems with small-diameter pores electro-osmosis effects are possible: displacement of the liquid when the charge or discharge current is switched on. These effects are usually negligible in cells because of high conductivity of typical electrolytes.

Electron conductance in the solid phase.[3] Electronic conductivity of a porous solid depends both on the intrinsic conductivity of the material and on contact resistance between grains, fibres and other structural units. Contact resistances, in their turn, depend on the method of preparation of the porous solid: simple pouring of powder, powder compression, subsequent sintering and so on. Intergrain contacts are best after sintering; those formed by simple pouring of powder are easily broken by oxidation of the surface or by a separating liquid film.

Electronic conductivity strongly depends on the volume fraction occupied by the material; the larger this fraction, the higher the number of contacts connected in parallel in the current flow. With small volume fractions (for instance, a small amount of metal powder in a paste), the connectivity of individual conducting particles breaks down and conductivity vanishes. Conductivity also depends on the shape of particles: a porous solid formed of fibrous particles has a higher conductivity, other factors being identical, than that formed of spherical particles. For a constant volume fraction and identical particle shape, conductivity may depend upon the particle size: the larger the size, the smaller the number of in-series contacts in the current paths, but the smaller the number of parallel circuits.

Pastes and immobilized electrolytes. Various pastes obtained by mixing fine-grain powders (reactants and other components) with liquids are used in cell technology. A property of pastes which is of considerable importance in a number of technological operations is its consistency. The consistency is determined both by the ordinary (kinematic) viscosity and by a specific structural viscosity which retards strain (flow, for example) at low external stress. Gelled (immobilized) electrolytes obtained from aqueous solutions by means of such gelling agents as starch, carboxymethyl cellulose, and so on possess a comparatively high structural viscosity. The consistency of pastes also depends on the nature of powder wetting—on its "moisture capacity".

Another important characteristic of pastes and immobilized electrolytes is their stability when liquifaction occurs (as a result of some chemical or physicochemical processes).

5.2. Active mass

The active mass of cell electrodes contains, in addition to the main reactants or catalysts, a number of additives affecting the performance of the electrodes. The electrode properties vary in a wide range even if the active mass formulation or preparation technology are only slightly varied. The "technological secrets" of high-quality cells can be said to be anchored in the active mass field.

The active mass is almost always porous, being composed of powdered components. This porous structure not only increases the electrode–electrolyte contact area but also compensates for the possible volume changes in the reaction components; this is a preventive measure against electrode deformation. The density of the active mass may serve as a measure of its porosity. Complicated physicochemical processes are constantly at work in the active mass of an operating cell.

High efficiency of reactant utilization is only achieved if the current-producing reaction proceeds at all points within the layer of the active mass. This situation is realized if the electron and ionic current flow and the supply of the reactants and removal of the products are provided for at each point (Fig. 26). In this section the electron current distribution in the active mass is analysed; the ionic current distribution and the reactant supply through the electrolyte within the electrode pores are treated in Section 5.4.

(a) Active mass additives

Conducting additives. It is quite typical for powdered reactants (oxides or salts) and some catalysts to have low electron conductivity and high contact resistance at grain contact points. To remedy this, high-conductivity additives are introduced into the active mass, such as fine-grain graphite, carbon black or chemically stable metals. These additives form, within the active layer, a conducting skeleton for current transfer to the collector. Two

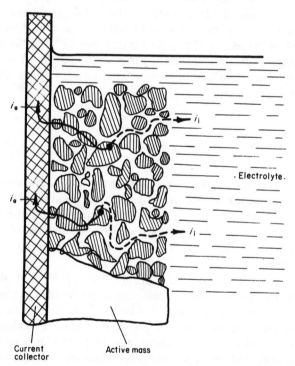

Fig. 26. Electronic (i_e) and ionic (i_i) currents in a porous electrode.

characteristics are not to be confused: the skeleton conductance which is a function of the volume concentration of the additive, and the conductance of the active mass volume from the reaction zone to the nearest point of the skeleton. The more dispersed the reactant and the conducting additive and the better they are mixed, the shorter the current path in the active mass and the higher the effective conductance.

On the other hand, the conductances of both the skeleton and the reactant depend on the contact resistance between neighbouring particles. If it is high, an increase in dispersion and the ensuing increase in the number of contacts may sometimes suppress conductivity. It is possible, therefore, to optimize the degree of dispersion of powders yielding a maximum conductance of the active mass. The length of the current path to the nearest conducting particle is typically 10^{-4} to 10^{-2} cm, and the path length to the nearest segment of the current collector is from 10^{-2} to 1 cm.

Binders. Various binders are introduced into the active mass to ensure mechanical strength and to prevent crumbling: polyvinyl alcohol, carboxymethyl cellulose, and so on. Binders serve to form a strong and elastic skeleton within the active mass layer. As a rule, this skeleton does not intersect the conducting skeleton; individual particles of reactants are not enveloped completely by the organic films. Hence, only a slight effect due to binders, at least for concentrations which are not too high, is observed in the conductivity of the active layer.

Hydrophobic and hydrophilic additives. In order to impart to the active mass partially hydrophobic properties, special organic additives are sometimes introduced, such as polyethylene, emulsified polytetrafluoroethylene (PTFE, or Teflon), and some others. These additives do not envelop the hydrophilic particles of the powder but form independent hydrophobic units within the porous structure, thereby interrupting the continuity of a liquid film on the surface of wettable grains. If, on the other hand, the basic components of the active layer are not sufficiently hydrophilic, then compounds (surfactants) improving wetting are sometimes added to the active mass.

Other types of additives. A number of compounds affecting the current-producing reaction are often added to the active mass of cell electrodes. As a rule, these additives serve to modify the structure of the reaction products, for example, to stimulate formation of small-grain deposits not shadowing the electrode segments they cover. Storage battery electrodes give an example in which the structure of the deposit is of great significance for the subsequent charge stages. The mechanical properties of the active mass are sometimes improved by adding plastifiers. All these effects are cell-specific and are therefore discussed in sections dealing with specific cell systems.

(b) Technological aspects concerning production of active mass and electrodes

The properties of cell electrodes are determined not only by the composition (formula) of the active mass but also by careful observance of the prescribed technology of its production. The following stages can normally be singled out in the electrode manufacturing process.

Monitoring of purity of raw materials. Foreign substances decisively affect the properties of the active mass. The active mass components must therefore comply with certain impurity content regulations. Furthermore, the process must eliminate all sources of spurious contamination of the electrodes.

Dispersion. Powders with required degree of dispersion are obtained by grinding in ball, vortex or other mills, by deposition from a solution, or by other techniques. Sometimes the powders are additionally fractionated, for instance, by sieves. The stage of drying the powder obtained by wet techniques may be of the utmost importance, since this may result in intensive oxidation of grain surfaces.

Mixing. A rather critical operation because the ultimate structure of the active mass, the current distribution in it, and a number of other parameters affecting the reactant utilization coefficient are determined by correct proportioning and thorough mixing of the blend.

Compression. The goal here is to improve the particle-to-particle contacts in the active mass. The techniques used to achieve this are pressing, rolling and extrusion.

Application of the active mass to the current collector. Many varieties of this operation are used, depending on the electrode design. The active mass prepared as paste may be applied to the current collector (a grid or a mesh) by pressing, pasting or other methods. The active layer thus formed is sometimes subjected to thermal treatment, and its strength is improved by wrapping or gluing-over with a film, paper or cloth. In the case of box-type electrodes the active mass is inserted into perforated boxes, such as tubes, pockets and so on. Prior to the application of the active mass, it is typical to subject the current collector surface to degreasing and roughening operations.

Forming. After manufacturing, the storage battery electrodes are usually "formed" by one or several charge–discharge cycles in a specific mode. This improves the active mass structure, increases the true reactant surface and removes undesirable impurities or sludge. Electrodes may be formed either prior to or after their assembly into a battery.

Additional technological operations specific to each cell type will be found in Part Two of the book.

5.3. Secondary transformation in electrodes

In cells with solid reactants, phase transitions take place on the electrodes, with the reactant phase disappearing and new solid phases forming, provided the reaction products are insoluble. A new solid phase may result either from a purely solid-phase reaction or from a reaction involving an intermediate formation of soluble compounds (reaction via solution).

Phase transitions in the active mass cause a number of bulk effects: swelling, strain, changes in mechanical strength and electric conductance (as a result of modified contacts between particles). It is therefore necessary to select the composition and structure of the active mass keeping these changes within acceptable limits and preventing deterioration of its characteristics.

The active mass of cycled storage cells sometimes retains some "memory" of its initial state. Some of the changes generated in the course of charging or discharging are not damped out by cycling but are gradually accumulated. The changes relevant here are those in the primary structure (i.e. in the size distribution in the mass) and in the secondary structure (the character of agglomeration of the particles). The memory is totally erased only if the reactant is completely dissolved at the discharge phase, to be deposited anew on the current collector during the subsequent charging phase (this is the case in the chlorine–zinc storage cell, see Section 14.4).

As both the reactants and the admixtures possess a high degree of dispersion and very large total surface area, a number of ageing processes are possible in the active mass; processes by which the structure and properties vary in time, independently of the charge–discharge cycles. One is recrystallization whereby the smallest grains merge into larger crystallites and the true surface area Σ decreases. The crystalline phase of the reactant or product may undergo modification. The surface of the conducting admixture particles may oxidize, thereby raising the contact resistance. Organic additives (binders, hydrophobic agents) may oxidize and partially decompose. And finally, the amount of liquid electrolyte in the pores of the active mass may either increase or diminish.

Another process, effective after a considerable number of cycles, appears in the form of crumbling (shedding) of the active mass or of some of its components. The reason for crumbling lies in loosening of the reactant's structure during one of the phases of a cycle, so that the particles are in poor cohesion with others and with the current collector. The particles crumbling off the plates accumulate at the lower part of the container as sludge. The process is accelerated by mechanical factors such as gas liberation in the course of overcharging, or vibrational loads.

The ageing and crumbling processes usually occur uniformly over the whole electrode surface. Other processes are, however, possible, with non-uniform distribution across the surface. One example is the gradual sliding of the active mass off the upper parts of the electrode and its accumulation in the lower part. Sliding is sometimes observed in tall storage cells, and cannot be treated as a consequence of crumbling of the active mass. Concentration changes in the electrolyte may constitute one of the causes: the more concentrated and denser solution formed during charging (or discharging) is accumulated in the bottom part of the container and displaces the lower-density liquid upward, causing electrolyte stratification. As a result, the conditions for the electrode processes become spatially non-uniform: dissolution of the active mass is favoured in the upper part and its deposition in the lower part of the cell.

The process of "shape changing" in the active mass is more involved. In contrast to sliding, it is not related to gravitational forces and may proceed in any direction. The nature of the process is quite complicated; it is probably most pronounced in the zinc electrode in alkaline solutions (e.g., in silver–zinc and nickel–zinc storage batteries) and will be discussed in detail in Section 12.1.

The sliding and shape changing of the active mass produces mass surplus in some areas. This does not necessarily lead to an increase in the electrode thickness: the alternative is a diminished porosity resulting in a smaller contact area with the electrolyte, and in deteriorated diffusion and conductance conditions. The net effects are the worsening of conditions in which the active mass operates and a reduction of the reactant utilization coefficient. In the limit, even complete local passivation of the active mass may occur.

If the active mass is distributed unevenly over the electrode surface owing to some structural features or technological irregularities, the resulting uneven distribution of current may either be smoothed out in the process of cycling or, the reverse, sharply amplified.

All these phenomena gradually lower the electrode and cell capacity, increase polarization and reduce the discharge voltage. They also constitute the main cause of limitation of the cycle life, service life and shelf-life of chemical power sources.

5.4. Macrokinetics of processes in porous electrodes

(a) General

The true internal surface area Σ of a porous electrode is much greater than the apparent geometric surface area S (in other words, the surface roughness factor $\sigma_s \equiv \Sigma/S$ is high), which enables one to obtain high currents with comparatively low polarization.

Let us denote the apparent (geometric) current density (i.e. calculated per unit projected area) by I. We here use I for the current density, replacing the familiar i to distinguish it more clearly from the true current density i_σ. In the case of ideally uniform current distribution, I should be greater by a factor of σ_s than the current density i_0 on a smooth electrode operating at the same polarization and with the same reactant concentrations. This is very rarely encountered in practice, however. More often than not, I is less than the maximum possible value $I_{max} = i_0\sigma_s$. The ratio

$$h = I/I_{max} = I/i_0\sigma_s \qquad (h \leqslant 1) \qquad (5.9)$$

is called the efficiency of the porous electrode.

A porous electrode is a system with distributed parameters; the efficiency is lowered because different points of the electrode are not equally accessible to the electrode reaction. The local (true) current density i_σ is a function of depth into the porous electrode because (a) ohmic potential differences are possible in the electrolyte filling the pores, and (b) the supply and removal of the reactants and products is slowed down in the pores, the concentration levelling-off is also retarded so that concentration gradients appear. Normally i_σ is a maximum at the outer surface where the feed conditions are most favourable and ohmic losses are minimum; i_σ diminishes as we move deeper into the electrode.

Several systems of porous electrodes are in use. The following characteristics are important for their classification.

Liquid and gas–liquid (liquid–gas) electrodes. All pores in a liquid electrode are filled with a liquid electrolyte (a melt or a solution). The electrodes with pores partly filled with a gas are referred to as gas–liquid (if the gas participates in the reaction) or liquid–gas (if the gas is the reaction product). The terms "two-phase" and, correspondingly, "three-phase" electrode are also used, thus assuming the solid to be a single phase even if it is formed from different materials, such as a metal and a water-repellent solid.

Variability of the electrode structure. In non-consumable electrodes (not participating in the current-producing reaction chemically, as in fuel cells), the porous structure of the electrode is either constant or slowly varying owing to secondary ageing processes. The structure of a consumable electrode participating in the reaction (the case of primary cells and storage batteries) is continuously varying in the course of charge or discharge. In order to analyse the functioning of such electrodes, the cycle is divided into stages, and the structure is assumed approximately constant within each of the stages.

The method of feeding the reactants. The natural convection of a liquid within the pores of cell electrodes is negligible. Two methods are available to supply the dissolved reactants to the internal reaction zones (or to remove the reaction products in the reverse direction): (a) diffusion through the stationary liquid, and (b) the flow of liquid through the porous electrode, driven by an external force. The electrodes of the first type are called diffusion-fed electrodes, while those of the second type are called flow-through electrodes.

The direction of component feed. The current-producing reaction on plane porous electrodes takes place either on one or on both sides. In the first case the working surface (the one facing the second electrode) is termed the front face, and the other the rear face. Two directions of reactant supply are possible: to the front face (Fig. 27(a)), or to the rear face (Fig. 27(b)). The rear-face feed is advisable if a direct contact of the dissolved reactant and the

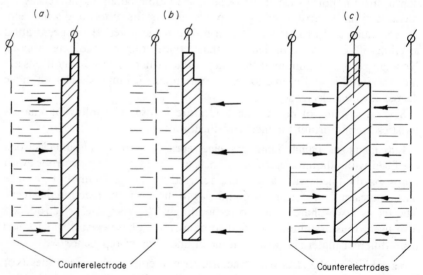

Fig. 27. Methods of reactant feed to porous electrodes: (a) front-face feed, (b) rear-face feed, (c) an electrode with two working surfaces.

second electrode is to be absolutely prohibited. The space between the electrodes is then occupied by a pure reactant-free electrolyte. Then either a pure reactant immediately dissolved in the electrolyte or a solution of the reactant in the electrolyte is fed to the rear face of the electrode. The reactant enters the reaction and is consumed within the electrode, not going beyond the front surface. Electrodes with rear-face feed of the reactant are often called diffusion electrodes; sometimes this term is used even when the reactant supply does not involve diffusion. If the reactant is a gas, the electrodes are referred to as gas-diffusion electrodes. The electrodes with two working surfaces (Fig. 27(c)) are usually symmetric with respect to the middle plane, and are conveniently treated as two juxtaposed electrodes of half the electrode thickness, each with the front feed of the components, or with the rear-face feed if the reactant is fed through a chamber in the middle of the electrode.

By convention, the term "macrokinetics" means simultaneously taking into account activation and concentration polarization, as well as ohmic voltage differences in a distributed-parameter system. It contrasts "microkinetics" which considers only the relationships governing the electrochemical reactions (such as that between the activation polarization and current density) and exclusively at a chosen point, without taking into account the influence of other regions of the electrode.

Considerable progress has been achieved during the last two decades in macrokinetics of processes in porous electrodes. The main goal was to find the current distribution inside the electrode pores and improve the electrode efficiency. The theory operates with a fairly cumbersome mathematics. In what follows, the discussion is therefore restricted to the simplest cases (first for the liquid and then gas–liquid electrodes); this approach enables us to elucidate the basic qualitative features.

(b) Macrokinetics of processes in liquid electrodes[4-6]

In chemical power sources, in most cases porous electrodes filled with liquid electrolytes are used. The lead storage cell is a typical example; its efficiency very much depends on the distribution of sulphuric acid concentration in the pores. The main reactant in methanol fuel cells with an alkaline electrolyte, namely methanol, is dissolved in the electrolyte and is continuously fed into a porous electrode.

Consider the macrokinetics of the process in a plane electrode of thickness d, assuming it to be uniform over the whole surface. Non-uniform current distribution then develops only normally to the surface along the x axis (a one-dimensional problem). The structure being too complicated, the calculations are based on the quasihomogeneous model, in which a liquid-filled porous electrode is treated as a homogeneous medium with effective transfer parameters: electrolyte resistivity ρ_{eff} and diffusion coefficient D_{eff} (see equation (5.6)). As a rule, resistance of the solid phase to the electron current is small compared to that of electrolytes, and is therefore neglected. The current generated per unit volume of the electrode at the depth x into it, $i_{v,x}$, is equal to $\sigma_v i_{\sigma,x}$ where $i_{\sigma,x}$ is the local (true) current density at the depth x, and σ_v is the true specific area per unit volume (σ_v is related to the roughness factor σ_s by the formula $\sigma_s = d\sigma_v$).

The potential of the porous electrode (and correspondingly its polarization) is a function of depth because of ohmic potential differences in the electrolyte. The acting potential of the electrode in a cell (in equations (1.8) and (1.9), for instance) is that at the surface of the front face.

Consider one particular case when the activation polarization and ohmic potential differences are not accompanied by concentration polarization (in other words, concentrations of components are assumed constant, i.e. independent of x). In this case the true current density is given as a function of depth by equation (4.24). The boundary conditions can be given in the form

$$\eta|_{x=0} = \eta_0, \qquad d\eta/dx|_{x=d} = 0 \qquad (5.10)$$

The first condition signifies that the electrode polarization is fixed and equals η_0. The second condition states that the potential gradient at the rear face of the electrode is zero since the total current toward the front face is very small here (in fact, tends to zero).

In order to find a solution of equation (4.24), one needs to know how the current density depends on polarization. Let us start first with a simple case of low current density, when the linear microkinetic equation (4.18) holds; a kinetic parameter in this equation is the polarization resistance ω.

The following expression is derived by a joint solution of equations (4.24) and (4.18) together with the boundary conditions (5.10). The local current density $i_{\sigma,x}$ as a function of depth into the electrode is

$$i_{\sigma,x} = i_0 \cosh\left[(d - x)/L_{\text{ohm}}\right]/\cosh\left(d/L_{\text{ohm}}\right) \qquad (5.11)$$

where

$$L_{\text{ohm}} \equiv (\omega/\sigma_v\rho_{\text{eff}})^{1/2}, \qquad i_0 = \eta_0/\omega \qquad (5.12)$$

Curves plotting $i_{\sigma,x}$ as a function of x for two electrode thicknesses are given in Fig. 28. The current density at the front face is equal to the value that a smooth electrode would have under the same conditions. As x increases, $i_{\sigma,x}$ is damped out.

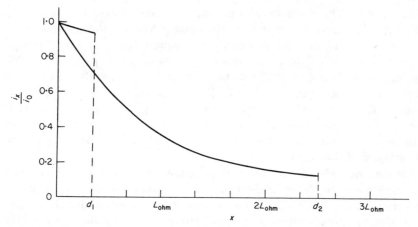

Fig. 28. Distribution of true current density over the thickness of a porous electrode (by equation (5.11)) for two electrodes, $d_1 = 0.33L_{ohm}$ and $d_2 = 2.5L_{ohm}$ thick.

The resultant apparent current density is found by integrating $i_{v,x}$ over the electrode thickness,

$$I = \sigma_v \int_0^d i_{\sigma,x} \, dx = i_0 L_{ohm} \sigma_v \tanh (d/L_{ohm}) \qquad (5.13)$$

The quantity L_{ohm} in these equations has the dimension of length and is referred to as the characteristic length of the ohmic process. It corresponds approximately to the depth x at which the local current density is diminished by a factor of e (e $\simeq 2.72$). Hence, L_{ohm} is a convenient characteristic of the damping out of current inside the electrode.

Substitution of equation (5.13) into (5.9) yields the efficiency of the porous electrode, h:

$$h = \tanh (d/L_{ohm})/(d/L_{ohm}) \qquad (5.14)$$

Apparently, h is a function of the ratio of the electrode thickness d to the characteristic length L_{ohm}. If $d < L_{ohm}$ ("thin electrode"), h tends to unity. Such electrodes operate uniformly over the whole cross-section and with maximum efficiency. The current itself is determined by the kinetics of the electrochemical stage proper, and not by ohmic factors; this mode of operation is often referred to as the inner-kinetics mode.

If, on the contrary, $d > L_{ohm}$ ("thick electrode"), h is a diminishing function of d. The damping out of the process means that the layers of the electrode located deeper than $2-3L_{ohm}$ contribute practically nothing to the total current. The characteristics are therefore not improved by increasing the electrode thickness above the cited value. According to equation (5.13), the limiting external current density I_{max} equals $i_0 L_{ohm} \sigma_v$. This mode of operation is referred to as the inner-ohmic mode. With L_{ohm} very small, the process is "squeezed-out" toward the frontal surface, so that the internal layers of the electrode are almost idle. In the limit, when L_{ohm} becomes smaller than the size of an individual pore or grain, a porous electrode operates as a smooth one with the total current of approximately i_0 (neither the quasihomogeneous model of the electrode nor the equations relevant to it are valid in this case).

In the case under discussion, that is for low current densities, L_{ohm} and h are determined by the specific properties of a system and do not depend on polarization η_0. As polarization rises, both the local current density $i_{\sigma,x}$ and the apparent total current I increase without changing the shape of the current's distribution over depth.

The situation is different in the range of high current densities when the kinetic equation (4.18) must be replaced by equation (4.19). In this case h and L_{ohm} can be shown to diminish as the total current increases (the intermediate manipulations are omitted). Figure 29 plots η_0 as a function of log I for a smooth (curve 1) and porous (curve 2) electrode; the situation is illustrated for a system operating at low currents in the inner-kinetic mode, that is at $h = 1$. The distance between the curves in this range of currents (i.e. within the segment AB of curve 2) is constant and equal to the logarithm of the roughness factor. As current (and with it ohmic factors) increases, h begins to diminish and the curves approach one another. The

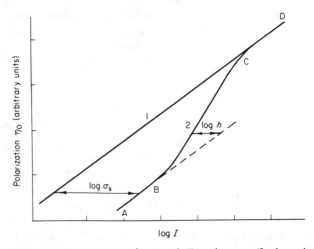

Fig. 29. Polarization curves of a smooth (1) and porous (2) electrode.

inner-ohmic behaviour is found between points B and C. The slope of the polarization curve in this range is typically twice that of smooth electrodes (formally, $2b$ must be substituted for b in an equation of type (4.19)). With the further increase in current, the process is squeezed out to the electrode surface, and the advantages of a porous electrode are lost (the outer-kinetic mode, segment CD). Sometimes, at still higher polarization, the outer-diffusion mode with a limiting current can be observed.

The relationships are similar if only the activation and concentration polarization components are taken into account, with ohmic potential gradients assumed negligible for the sake of simplification, owing to high conductivity of the electrolyte. We then have to resort to the differential equation (4.25). It is also necessary to know the true current density as a function of reactant concentration. In the simplest case, current is porportional to concentration (first-order reaction)

$$i_{\sigma,x} = k_r c_x \tag{5.15}$$

where the electrode reaction rate constant k_r is determined by the electrode potential or electrode polarization. For instance

$$\text{(a)} \quad k_r = \eta/\omega \qquad \text{or} \qquad \text{(b)} \quad k_r = k_r^0 \exp(\eta/b) \tag{5.16}$$

depending on whether kinetic equations (4.18) or (4.19) are valid. The boundary condition at the front surface is $i_0 = k_r c_0$, where c_b is the bulk concentration in the solution.

When solved with the appropriate boundary conditions (similar to (5.10)), equations (4.25) and (5.15) yield the same equations (5.11) and (5.13) for the current distribution within the electrode and for the total current, respectively. The process attenuation rate is a function of the characteristic diffusion length

$$L_{\text{diff}} = (nFD_{\text{eff}}/\sigma_v k_r)^{1/2} \tag{5.17}$$

As c_b rises, i_0 and I increase but the type of the distribution curve of the electrode remains unaltered. The mode is inner-kinetic and $h = 1$ if $d < L_{\text{diff}}$. This case differs from the previously analysed one in that, owing to dependence of k_r on polarization, L_{diff} is a diminishing function of polarization for high as well as low current densities. Increased current results in diminished h, and the electrode passes first from the inner-kinetic mode to the inner-diffusion mode ($d > L_{\text{diff}}$) and then to the outer-kinetic or outer-diffusion mode. As before, the slope of the polarization curve given by equation (4.19) in the inner-diffusion mode is twice that for a smooth electrode. Equations (5.11), (5.13) and (5.14) are valid in the case in question for the whole range of polarization and current density variation, provided the reaction remains of the first order (equation (5.15)).

If both the activation and concentration polarization and ohmic factors are taken into account, the corresponding equations are even more complicated, and in the general case can be solved only numerically, using computers. Therefore the efficiency of porous electrodes is often estimated by simply comparing the electrode thickness to the characteristic ohmic or diffusion length (equations (5.12) and (5.17)). The information provided by this evaluation is in most cases useful and quite sufficient.

Any factor raising L_{ohm} and L_{diff} increases the efficiency of porous electrodes. First of all, this efficiency depends on the nature of the electrochemical reaction. As a rule, efficiency is high for reactions with sufficiently low reaction rates (high ω, low k_r). Efficiency is considerably lower for high-rate reactions; this does not constitute an obstacle, though, because often no additional stimulation is required for such reactions. The efficiency of a reaction (or of a system) can be increased by increasing D_{eff} and diminishing ρ_{eff}, that is by such treatment of the porous electrode structure which lowers the resistance enhancement coefficient ε (see Section 5.1, c). The main factor conducive to this is the increased porosity of the electrode.

(c) Gas–liquid electrodes[6, 7]

The use of gaseous reactants, such as oxygen, hydrogen and so on, in electrochemical cells is now becoming more and more promising. Because of low solubility of oxygen and hydrogen in aqueous solutions, about 10^{-3} mole/litre, dissolved gases cannot provide sufficient inflow of gas to liquid porous electrodes. The process becomes sufficiently intensive only if porous gas–liquid electrodes are used.

On its front face a gas–liquid electrode is in contact with the electrolyte, and on the rear face it contacts the gas chamber. An important feature consists in one part of the pore volume being filled with the electrolyte and the other with the gas. In order for the electrode to operate effectively, the liquid- and gas-filled pores must be distributed uniformly throughout the electrode bulk. All the liquid-filled pores must be interconnected and contact the electrolyte on the outside of the electrode. These pores provide the paths for bringing the dissolved reactant particles to the inner reaction zones and for removing the reaction products. The gas-filled pores through which the reactant gas is fed must also be interconnected. A quite definite ratio of liquid to gas pores is required to optimize the electrode performance; a sharp reduction of either liquid or gas content cuts down the supply of the corresponding ingredient, with subsequent deterioration in the electrode characteristics. An electrode is said to have a high buffer capacity if the admissible liquid-to-gas ratio can be varied in a wide range without significantly impairing the electrode characteristics.

In porous electrodes, electrochemical reactions mostly proceed in the vicinity of the regions where gas-filled pores are in contact with liquid-filled ones, that is in the so-called three-phase interfaces. These are areas where the electrolyte forms menisci and thin films enveloping the pore walls. The gas dissolved in the electrolyte film diffuses rather rapidly through this liquid and is reacted at the solid surface. An electrode's efficiency is maximized if the total size of such reaction zones reaches maximum; this is possible for a specific ratio of gas- and liquid-filled pores in the electrode.

The interplay of the activation polarization and ohmic processes in gas–liquid electrodes is not principally different from that in purely liquid-filled porous electrodes. Not all of the pore volume, but only its liquid-filled part, must be taken into account in the corresponding calculations of these phenomena (one result is increased ρ_{eff}). Having this in mind, one can use equations derived in the preceding subsection.

The reactant supply process, however, is very different from the one described earlier. The gaseous reactant reaches the reaction zones not by diffusion from the rear face but by flowing through the gas channels. Usually this is a high-rate process, not limiting the net efficiency of the electrodes. Some concentration polarization occurs only if the gas reactant is fed at a reduced flowrate through a "gas cushion", that is in cases of a relatively low reactant concentration in the gas phase (such as the case of oxygen in the air) and of only a small fraction of gas-filled pores.

Two problems have to be solved when gas–liquid electrodes are designed: firstly, to produce and maintain a prescribed gas-to-liquid ratio in the pore volume, and secondly, to prevent electrolyte leakage into the gas chamber or the gas reactant bubbling through the electrolyte.

Most of the materials used as catalysts in gas–liquid electrodes (metals, oxides, etc.) are hydrophilic; hence, the pores of such an electrode are completely filled with the electrolyte when it contacts an aqueous solution. Two techniques may be employed to achieve a selective filling of pores with the liquid; elevated gas pressure, or electrodes with partially hydrophilic and partially hydrophobic properties introduced by special treatment.

Capillarity relationships dictate that with a gas pressure excess of Δp, all pores with radii equal to or above r_{crit} in a hydrophilic porous body are emptied, with r_{crit} given by

$$r_{crit} = \frac{2\pi\gamma}{\Delta p} \cos \theta \qquad (5.18)$$

where γ is the surface tension of the liquid, and θ is the wetting angle ($\theta < 90°$ in hydrophilic materials). By selecting Δp, and hence r_{crit}, it is possible to control the ratio of the gas- and liquid-filled pores. A typical excess pressure Δp does not exceed 0·1 MPa.

If the pressure difference across the electrode is high, the gas may bubble through it by chains of larger pores and appear at the face surface in the electrolyte. This effect is suppressed in two-layered (biporous) electrodes (Fig. 30(a)) in which the front face contacting the solution is covered by an additional fine-dispersion hydrophilic layer (a metal layer, for example). The pore radii in this layer are definitely below r_{crit}, so no gas leak is possible at the fixed Δp. The layer is termed the gas-tight layer. The total porosity of this layer must be sufficiently high to minimize the accompanying additional ohmic losses and diffusion restrictions.

In the second technique, an electrode is prepared by mixing a hydrophilic catalyst with a hydrophobic additive, such as Teflon. As a result, some of the pores are hydrophilic and thus fill with the solution while some are hydrophobic. The gas-to-liquid ratio is easily controlled by varying the blend composition. An important advantage of such partly hydrophobic electrodes consists in the possibility of eliminating elevated gas pressures. This factor is essential if aerial oxygen is used as oxidizer in electrochemical cells. In order to suppress the electrolyte leakage through a chain of hydrophilic pores into the gas chamber, the rear face is covered by a fine-dispersion hydrophobic layer (water-tight layer, Fig. 30(b)). This layer completely blocks the electrolyte flow while being easily penetrable for gases.

Unfortunately, the efficiency of many hydrophobic additives to the electrodes deteriorates with time, so that gradually the hydrophobic gas-filled pores are filled with the electrolyte. To compensate for this, combined electrodes are sometimes used, with partly hydrophobic pores but with the gas fed under elevated pressure. Electrodes of this type have two additional layers: a gas-tight layer at the front face and a water-tight one at the rear face.

A third method is also possible of producing a desirable liquid-to-gas ratio in the electrode. The method consists in using an electrolyte in the pores of an electrolyte-carrying matrix. The liquid is redistributed when the electrodes come into contact with the matrix. The liquid volume being limited, part of the pores remain gas-filled despite their hydrophilic nature.

(d) Large-scale macrokinetics[8]

The preceding subsections treated some phenomena caused by ohmic and transport inhibition in an individual porous electrode placed in contact with

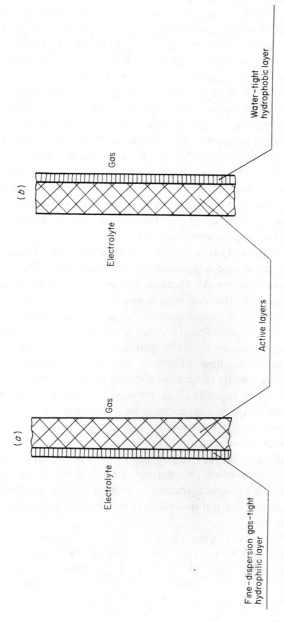

Fig. 30. Cross-sections of gas-diffusion electrodes with (*a*) gas-tight hydrophilic layers and (*b*) water-tight hydrophobic layers.

a liquid electrolyte and, in some cases, with a gas. Any interaction between this porous system and the ambient medium was, however, ignored.

It is rather frequent in electrochemical cells that several porous bodies are in contact. For instance, increasingly used are matrix electrolytes in which a liquid electrolyte is retained within a porous carrier matrix by capillary forces. As a rule, matrix electrolytes of cells are in contact with porous electrodes (this is the case in sealed storage cells, some fuel cells, etc.). The electrolyte distribution in the pores of each of these bodies is determined in this situation not only by the properties of the material itself (equation (5.18)) but by the surrounding objects as well. It is possible, for example, that with the total amount of liquid being limited, one of the electrodes may be "sucked dry" of its electrolyte by the second electrode and thus rendered useless because of large capillary forces. A pair of electrodes with high electrochemical characteristics may thus prove incompatible. The porous structure of the cell as a whole may often need optimization. Similarly, concentration changes in one porous element may affect concentration distribution in a neighbouring element.

Porous electrodes of other systems, for example in some fuel cells, actively interact with the ambient gas; water, the reaction product, is constantly removed by evaporation. An excessive removal of water diminishes the total amount of liquid in the cell with a matrix electrolyte, i.e. it reduces the number of liquid-filled pores; conversely, gas-filled pores may be flooded if insufficient water is removed. In order to avoid this, the transfer processes inside the cell must be carefully matched to the external mass transfer processes. However, these processes cannot be matched with absolute accuracy at all moments of time; as a result, electrodes in such systems must have a large buffer capacity. The processes in porous electrodes are also affected by the heat exchange conditions with the ambient, since all the parameters of electrode performance are strongly temperature-dependent.

The processes of concentration and temperature redistribution by convection in the liquid are very important in cells with liquid electrolytes. If convection is suppressed (by narrow interelectrode gaps in a compact electrode assembly), concentration and temperature gradients build up, affecting the concentration distribution in the porous electrodes and in the gaps between them.

Effects of this type were shown to be considerable in a number of electrochemical cell types; they are treated in the framework of so-called large-scale macrokinetics. The history of systematic studies is so far very brief.

REFERENCES

1. A. J. Salkind, The measurement of surface area and porosity, *in* "Techniques of Electrochemistry" (Yeager, E. and Salkind, A. J., eds.), Vol. 1, pp. 293–388. Wiley-Interscience, New York and London (1972).
2. P. B. Zhivotinsky, "Porous Separators and Membranes in Electrochemical Devices". Khimiya Publishing House, Leningrad (1978) (in Russian).
3. K.-J. Euler, The conductivity of compressed powders. A review, *J. Power Sources* **3**, No. 2, 117–136 (1978).
4. J. S. Newman and W. Tiedeman, Flow-through porous electrodes, *in* "Advances in Electrochemistry and Electrochemical Engineering" (Gerischer, H. and Tobias, Ch. W., eds.), Vol. 11, pp. 353–438, Wiley, New York (1978).
5. J. G. Gurevich, Yu. M. Volfkovich and V. S. Bagotzky, "Liquid Porous Electrodes". Nauka i Tekhnika Publishing House, Minsk (1974) (in Russian).
6. R. de Levie, Electrochemical response of porous and rough electrodes, *in* "Advances in Electrochemistry and Electrochemical Engineering" (Dehahay, P. and Tobias, Ch. W., eds.), Vol. 6, pp. 329–397. Interscience, New York (1967).
7. Yu. A. Chizmadzhev, V. S. Markin, M. R. Tarasevich and Yu. G. Chirkov, "Process Macrokinetics in Porous Media". Nauka Publishing House, Moscow (1971) (in Russian).
8. L. M. Pismen, Yu. M. Volfkovich and V. S. Bagotzky, Macrokinetics of processes in hydrogen–oxygen fuel cells and electrolysers, *J. Appl. Electrochem.* **6**, No. 6, 485–505 (1976).

Chapter Six

Design and Technology

6.1 Main features of design

(a) Design requirements

Electrical energy delivered by chemical power sources is produced at the expense of a chemical reaction between active reactants. The cell design must ensure the conditions required for the current-producing reaction and for the practical utilization of the electrical energy released. Among these conditions are:

(i) Separation of components (for example, by means of spacers).
(ii) Good and reliable contact between the electrodes and the electrolyte; a large contact area, and clamping in the case of contact between solid components.
(iii) Minimum ohmic losses in current collection from the reaction zone to the cell terminals.
(iv) Elimination of possible current leakage.
(v) Uniform loading of each of the electrodes.
(vi) Mechanical strength, spill-proof liquid electrolyte, sealing of the cell and so on.

Operation of fuel cells requires that precisely controlled reactant supply and reaction product removal be maintained in the course of discharge. A reserve cell must be equipped with an actuator.

Numerous elements and units are added to the cell design in order to satisfy all the above requirements: the container, the cell cover, terminals, separators, sealing and insulating elements, and so on, as well as mechanical devices for the reactant feed.

The total mass M of a cell is composed of the mass M_r of its reactants (fuel cells in which the reactants are fed from the outside are an exception), the mass of the electrolyte M_e and that of structural materials M_s: $M = M_r + M_e + M_s$. A rational cell design must minimize the relative mass M_s/M of structural materials. Usually, this fraction varies from 15 % to 75 % in different cell types.

(b) The balance of active ingredients

A problem which appears immediately when a specific cell is designed is that of the required ratio of the reactants for the positive and negative electrodes and for the electrolyte. An unnecessary excess in one of the components is undesirable. It is therefore typical for the reactants to be loaded in the stoichiometric ratio given by the current-producing reaction, with the actual utilization coefficient of each of the reactants taken into account. Difficulties may sometimes be encountered, since the utilization coefficients are functions of temperature, storage duration, and so on. Consequently, specific conditions of cell operation have to be taken into consideration when the reactant ratio is chosen. Sometimes a surplus of one of the reactants is provided to avoid undesirable side reactions. For instance, a surplus of mercury oxide is used in sealed mercury–zinc cells in order to consume the whole amount of zinc stored and to prevent hydrogen evolution at the cathode at the end of discharge. An excess of a reactant may be employed if this reactant also carries a structural function (like zinc cans in manganese–zinc cells). Obviously, the discharge capacity is then determined by the electrode with a shortage of the reactant.

Another factor, in addition to proportioning the masses of the reactants, is important for storage cell manufacturing, namely, the degree to which the electrodes are charged. If a storage battery is assembled, for instance, of charged positive plates and discharged negative ones, the subsequent operations with the battery are extremely awkward.

The electrolyte is an active participant of the overall current-producing reaction in some types of cells (sulphuric acid in lead storage cells, for example) and does not participate in others. In the latter case its composition is constant throughout the cell life and its volume can therefore be minimized. Since the electrolyte fills the interelectrode space, the volume reduction is only possible if interelectrode gaps are narrowed. The factors limiting this narrowing are cell reliability and the probability of shorting between the plates. A certain surplus of the electrolyte is sometimes required (normally above the upper edge of the electrodes) to compensate for the volume changes in the course of the chemical reaction and owing to temperature fluctuations, and to prevent the partial drying of the electrodes due to lowering of the electrolyte level.

If the cell operation changes the electrolyte composition, the concentration variation must be kept within the tolerance limits. The electrolyte composition may also be changed by secondary factors, such as carbonization of alkaline solutions owing to absorption of carbon dioxide from the ambient atmosphere. In this case it is necessary to take into account the intensity of these factors and the required service life of the cell.

(c) Electrode thickness

A correct choice of electrode thickness is important in cell design. The same amount of reactants can be spread over a small number of thick or a large number of thin electrodes. In the second case the structural parameter $s_V = S/V$, that is the total geometric surface area of the electrodes (of one polarity) per unit volume of the cell, is increased. The maximum admissible discharge current, that is the maximum admissible power, is enhanced for thinner electrodes and greater total surface. However, the Ah and Wh capacities are thereby diminished since the fraction of structural elements (current collectors, separators, etc.) in the total mass and volume are increased. Thinning of the electrodes also diminishes their mechanical strength. Therefore thicker electrodes are used in cells designed for moderate discharge currents but for maximum Wh capacity, and thin electrodes are employed in cells designed for high-current drain. The structural parameter s_V varies in a very wide range: from $0.2\ dm^2/dm^3$ in alkaline copper–zinc cells to $120\ dm^2/dm^3$ for cells with foil electrodes.

(d) Scale factors

Each cell type is normally manufactured as a series of models differing only in dimensions and mass (and naturally, in Ah and Wh capacity, etc.) but identical in structure and production processes. Not all parameters of cells are proportional to the cell volume, and so-called scale factors have to be introduced sometimes.

Miniature cells are characterized by an appreciably higher fraction of the mass of structural elements, M_s/M, since it is not always possible to reduce proportionally all the structural elements, such as valves, terminals, separators and so forth. Specific characteristics of miniature cells are therefore lower than those of larger cells.

As a cell's dimensions increase, its Wh capacity and power rise proportionally to its volume, that is to the third power of linear dimensions. At the same time, the rate of heat exchange, that of gas removal, as well as bar conductance are proportional to the cross-section or the outer surface area, that is to the square of linear dimensions. These processes are therefore slowed down in large-size cells. Thus, forced cooling may become necessary to eliminate overheating, and current collectors and buses may have to be of disproportionately high cross-section. As a result, high-capacity batteries (thousands of ampere-hours), and especially those for high current drain, may have lower specific performance characteristics than medium-capacity batteries.

Another problem involved in using large-size electrodes is that of uniform current distribution.[1] Normally, a current-collecting bus is fixed to the upper part of the electrode, and if the ohmic resistance of the current-collecting grid in the electrode is too high owing to the electrode's considerable height, the process will concentrate at the upper part of the plate. Moreover, the electrodes located near the container walls may be in different thermal conditions compared with those in the middle of the electrode group.

(e) Labelling

A large number of different cell packages are known: metal, plastic or glass containers, rectangular or cylindrical in shape, sealed or vented cells, and so on. Clear labelling (preferably indicating the useful life duration) and marking of terminal polarity constitute important prerequisites for subsequent successful performance. In some cell designs mistaken connection are avoided by installing terminals shaped as clamps or sockets of non-identical dimensions and geometries.

6.2. Ohmic losses

Chemical power sources are low-voltage sources so that even slight ohmic losses (0·1 V, for example) produce appreciable effects on electrical characteristics. Ohmic losses may be distributed over the whole current path inside the cell from one terminal to the other, both in the electrodes and metal connections and in the electrolyte. As a rule, the electrolyte resistance in the interelectrode gap is higher than the resistance of the electrodes, although the reverse situation may also be observed.

A cell electrolyte is characterized by its resistance referred to the unit cross-section area or, which is identical, to the unit electrode surface area, $r_s \equiv RS$; r_s is measured in ohm cm^2. If this characteristic is used, the ohmic potential difference in the solution regardless of the electrode size is given by the formula

$$\Delta\varphi_{\text{ohm, e}} = ir_s$$

The resistance r_s is a function of the electrolyte resistivity ρ (ohm cm) and the distance L between the electrodes of opposite polarity: $r_s = \rho L$. Resistivity of practically important electrolytes varies from 1 to 100 ohm cm, and L varies from 0·1 to 1 cm. In each specific case, the admissible value of r_s depends on the maximum current density. Such values of r_s as 10–100 ohm cm^2 are quite acceptable in cells with the maximum discharge current density of 0·001 A/cm^2, but r_s should not exceed 0·1–1 ohm cm^2 if the current density is above 0·2 A/cm^2.

In contrast to other components of ohmic resistance, the electrolyte resistance strongly depends on temperature. The use of separators and of porous electrodes soaked in the electrolyte (Sections 5.1 and 5.4) increases the resistance. Calculation of the electrolyte resistance is sometimes a very complicated task since the reaction products may be deposited on the surface and partially screen the electrode. The electrolyte resistance in the pores of this deposited layer is sometimes very high which results in deteriorated cell characteristics. A resistance increase is not so steep and is more easily taken into account if dissolving metal electrodes are used in cells: a gradual thinning of these electrodes increases L and correspondingly raises the electrolyte resistance. The electrolyte gap resistance is also increased if gas bubbles accumulate in it. Thus, the resistance increment is 60–80% if the bubble volume comes to 30% of the total volume.[2]

Ohmic losses in metal components of cells (buses, current collectors, terminals, connectors) can be found readily if the shape and resistivity of these components are known. The same is true in the case of welded junctions of conductors.

Conductors in cells are often connected by means of clamping, as for example in a clamp with a nut fixing the external cable. Dry clamped-on contacts are usually quite reliable and their contact resistance is low provided measures are taken against loosening of the clamp (by locking the nut). If, however, the contact is loosened or wetted by the electrolyte, the contact resistance will probably rise rapidly because of surface oxidation or salt accumulation. Furthermore, an increased contact resistance increases local heat generation, thereby accelerating the deterioration of the contact.

6.3. Separators[3, 4]

(a) Functions of separators and corresponding requirements

Separators are installed in the interelectrode gaps of practically all cells with liquid electrolyte. Their functions are several-fold: to be mechanical spacers between electrodes separated by a narrow gap, that is to prevent accidental contacts between electrodes of opposite signs (because of vibration, for example) and formation of electronic-conductivity bridges between them; to provide free access of liquid electrolyte to all regions of the electrodes; to support mechanically the active mass on the electrodes, preventing its crumbling; to suppress dendritic growth of metal deposit in the course of a storage battery charging; to inhibit mixing of electrolytes in the anode and cathode compartments, as well as the interpenetration of solutes, colloidal particles or suspensions from one electrode to another (separators of this type are also referred to as diaphragms).

A separator in a cell with a matrix electrolyte often functions as an electrolyte-carrier, that is retains the liquid electrolyte close to the electrode surface by capillary forces.

Separators are manufactured of dielectrics and form large-hole meshes (simple spacers), porous separators or ultraporous (swelling) membranes.

An ideal separator must introduce only a minimum resistance to ionic current and reactant transfer in electrolyte, including the solvent transfer. At the same time, a separator must create a maximum possible resistance to the transfer of other compounds which can participate in side reactions.

Separators meeting these requirements must have the following properties.

Conductance. The conductance attentuation coefficient ε must be small. It varies from 1·1 to 1·6 for simple spacers, and from 2 to 8 for porous and ultraporous separators (reaching 15 only in exceptional cases). Therefore, the net porosity of a porous separator cannot be below 40–50%. It is necessary for the pore space of the separator to be filled with the liquid electrolyte rapidly and fully. At the same time, a separator must prevent any electronic conductance which would produce internal shorting.

Filtration of the liquid. The rate of filtration is strongly dependent on pore radius (Section 5.1), being the higher the greater the number of large-diameter pores. Depending on the cell type, this rate must be either high (to ensure reactant supply by convection) or very low (to prevent mixing of solutions, as for example in the Daniell copper–zinc cell).

Selectivity. In some cases separators must function as semipermeable membranes letting through only one species of ion (realizing the required charge transfer) and blocking the passage to other particles.

Suppression of dendritic growth. When storage batteries are recharged (or simply stored), the metal deposited on one of the electrodes sometimes form thin needle-like crystals (dendrites) growing toward the second electrode and causing internal shorts. Separators are effective in controlling this process because they restrict the diffusion of ions to growing dendrites and thus slow down their growth. Moreover, separators mechanically limit the thickness of individual crystallites and are conducive to prevention of growth of dendrites in the network of narrow pores. The protective function of separators with respect to dendrites is a function of both their selectivity and the structure of their porous space.

Mechanical and chemical properties. Separators must withstand long-term chemical action of both the electrolyte (e.g. acid or alkali solutions) and the reactants possessing oxidizing or reducing properties. Their parameters must be stable in a wide temperature range. The separators must be sufficiently elastic so as not to break down in the course of assembling, and be sufficiently shockproof.

In addition to this variety of properties, separators must be cheap, simple to manufacture, with properties reproducible in large-scale production.

(b) Separators employed in cells

(1) *Simple spacers*

These are made of plastic or ebonite rods or of plastic cord (2 mm in diameter), thin sheets of perforated or corrugated vinyl plastic, synthetic fibre cloth (polycaprolactam, chlorinated PVC fibres), fibreglass felt and other materials. These separators have large pores or holes (from 0·1 to 2–4 mm) which hardly shield the electrodes electrically but at the same time have very poor filtering properties. Such spacer materials are chemically inert and sufficiently stable.

(2) *Porous separators*

The materials classified as porous separators have pores with radii from approximately 0·001–0·01 to 100 μm. By convention, these separators are sometimes classified into porous (with the mean pore radius above 3–5 μm) and microporous ones (having smaller pores). Separators of this type are most widely used in cells. Clay ceramics membranes or cups were used as separators in cells with liquid oxidizers popular in the first half of the 19th century. The materials employed nowadays have finer pores. A number of these materials are briefly characterized below; their parameters are listed in Table 1.

Veneer—thin layers of alder, cedar or poplar wood are subjected to a preliminary treatment by dilute solutions of alkalis or acids. Before World War II this material, with its fairly simple production techniques, was widely used in lead storage batteries. Veneer is nowadays almost completely substituted by synthetic materials, for reasons of its rather low chemical stability.

Microporous rubber (Mipore) is a material which is produced by means of a comparatively complex technology by curing a mixture of natural rubber with some additives. This is a high-quality material, with small pores and high total porosity. Its shortcomings are high cost and brittleness. Microporous rubber is used in some types of lead storage batteries.

Sintered PVC (Miplast) is prepared by coating a moving (metallic) strip with a thin layer of powdered PVC resin and then sintering it. This material has comparatively large pores, but the production process is simple and the final product rather cheap. The main shortcoming is brittleness. Sintered PVC is widely used in different cell types. More than 90 % of lead storage batteries manufactured in the USSR use separators made of this material.

Porvic, a separator material manufactured in Britain, is produced from PVC resin with starch added. The starch is hydrolysed in the process and leached out, which leads to high total porosity at low average pore size. Porvic's elasticity and mechanical strength are quite high. Potato and maize Porovinyl are two Porvic-like materials developed in the USSR. The corresponding production processes are somewhat different from that of Porvic.

Yumicron is a novel Japanese-made separator material.[5] The pore diameter of Yumicron is very small. The material has good mechanical properties. It is manufactured as a thin film and also as embossed waffle-shaped sheets.

Cardboard and paper are used to produce sufficiently cheap separators with low resistance. Chemical resistance is achieved by impregnating the material with resins enveloping the fibres. A large number of formulas of such materials were suggested. About 80 % of starter lead batteries produced in Japan are equipped with cardboard separators.

Asbestos cardboard is manufactured from fibres of chrysotile asbestos; the technology is that used in the paper industry. Difficulties are encountered in ensuring reproducibility of production of thin separators. Materials with similar structure in which asbestos fibres are replaced completely or partially by synthetic potassium titanate fibres show better reproducibility.

Table 1. Parameters of porous separators

Separator type	Thickness (without ribs) (mm)	Porosity (%)	Attenuation coefficient ε	Pore diameter (μm) Mean	Maximum
Veneer	1·5	75–80	2·5–4·0	10–15	—
Mipore	0·4–1·0	55–62	2·8–5·5	0·5–1·0	3–5
Miplast	0·5–1·0	40–55	2·7–5·0	10–15	25–50
Porvic	0·5–0·8	80–88	2·0–3·5	2–3	5
Maize Porovinyl	0·6–0·7	80–85	2·0–3·5	5	9
Yumicron	0·1–0·25	60–70	4	0·1–1·0	1–2
Cardboard	0·5–1·0	60–80	3–5	20–30	40–80
Asbestos cardboard	0·2–0·6	70–80	4–6	12–20	—
Non-woven polypropylene mat	0·2–0·7	40–60	4–6	10	30
Glass fibre mat	0·4–0·6	90–95	1·4–1·6	—	100

Non-woven polypropylene mats are obtained by spraying a thin jet of molten polypropylene. This gives fine fibres which are deposited by the air flow onto a moving belt. This material is characterized by high chemical resistance. Glass fibre mats are produced by a similar technique involving binders. Glass fibre mats have higher porosity and larger pores. They are mostly used as auxiliary separators combined with other types of separator material.

(3) Swelling separators (membranes)

Swelling separators are made of polymer materials without a developed system of pores but absorbing either aqueous or non-aqueous solutions. A solvent penetrating the polymer structure increases the distance between macromolecules (the membrane becomes swollen) which makes it possible for the ions in the solution to move across the membrane (to migrate). In contrast to porous separators, specific interaction forces between individual ions and macromolecules are highly pronounced. As a result, these membranes are typically selective, that is the coefficients of diffusion or migration attenuation are different for different ions. Cellophane (hydrated cellulose), polyethylene films with radiation-grafted acrylic acid (the acid interacts with aqueous solutions), and ion-exchange membranes are examples of swelling membranes.

(c) Application of separators

Separators either fill the whole interelectrode gap or take up only a part of it, depending on the type and function of the cell. In the second case the electrode surface is in contact with the free liquid electrolyte which sometimes is essential for the electrode reaction. Separators may have special ribs to ensure a gap between them and the electrodes. A separator must possess very good electrolyte retention properties if it fills the whole interelectrode gap.

Normally the separator size is equal to, or slightly greater than, the electrode size. Sometimes, however, this does not provide the necessary protection, and bypassing of the separator is possible by unwanted compounds or metal dendrites. In these cases one of the electrodes is completely wrapped up in a flexible separator film (Fig. 31). The open upper part of the separating wrap must reach above the liquid electrolyte level. This method of separation is used, for example, in alkaline cells with zinc electrodes where the tendency to dendritic growth is especially pronounced.

Multilayer separators made of different materials, with each one assigned

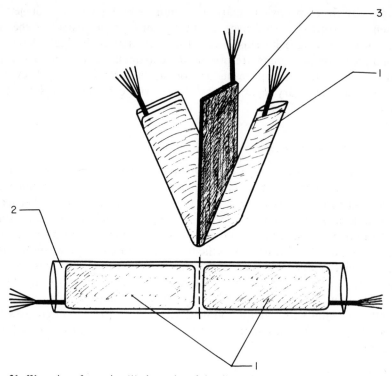

Fig. 31. Wrapping of negative (1) electrodes of the silver–zinc storage cell in a separator film (2); (3), positive electrode.

a specific function, are often used in cells. Combinations are known, for example, of chemically stable spacers with swelling membranes which are less stable but provide better separation. Several layers of a thinner material are more effective than a single thick layer even when only one separator material is used, because random defects in one layer are protected by the remaining layers. Moreover, "ingrowth" of dendrites into each subsequent separator layer is an additional obstacle to the dendritic growth.

6.4. Operation of batteries

As a rule, the voltage of an individual cell is between 1 and 2 V; in exceptional cases it may exceed 3 V. Most devices require higher supply voltage: 6–9 V for transistor circuits, 12 V for the automobile electric equipment, 24–28 V for on-board aircraft equipment, and even higher voltages for traction motors. Therefore cells are normally used connected in series into batteries. In order to increase the capacity and maximum admissible discharge current, individual cells are sometimes connected in parallel, and also in a combined series–parallel manner. Only identical cells can be connected into a battery in series (i.e. of the same electrochemical system, identical in construction, technological type and size). On the other hand, parallel circuits may include cells of different sizes.

A number of routine problems have to be solved when cells are connected to form a battery: (a) the choice of intercell connectors of minimum weight designed for the maximum current drain of a given battery; (b) the choice of the container design for the battery as a whole, with terminals, labelling, transportation handles (in the case of heavy batteries), and so on; and (c) electrical insulation of terminals, buses, and other conductors from other metal parts, designed for the maximum battery voltage.

Some specific and more complicated problems appear in batteries with cells connected in series.

(a) Overdischarging of individual cells

Owing to manufacturing tolerances, it is inevitable that the properties of individual cells are spread over a certain range. Therefore, although the current is the same in all cells connected in series, at least one of them will be discharged earlier than the others. These others being still operative, the battery continues discharging. Consequently, a discharge current passes through the "weak" cell even when its voltage drops to zero because of complete consumption of reactants or passivation. Any further passage of current is accompanied by new electrode reactions. Most often these are the reactions of electrolytic decomposition of the solvent or electrolyte, reactions similar to those taking place when storage cells are overcharged. For instance, hydrogen evolution (equation (1.5)) and oxygen evolution (equation (1.4)) reactions are known to begin in aqueous electrolytes. With this "overdischarging", however, the cathodic evolution of hydrogen takes place at the formerly positive electrode at which the oxidizer was cathodically reduced, and the anodic evolution of oxygen proceeds at the formerly negative electrode. As a result, overdischarging changes the polarity of the electrodes, that is the cell is reversed and the formerly positive electrode becomes negative, and vice versa (Fig. 32).

The reversal in a single cell lowers the battery voltage by 2–4 V (the cell discharge voltage prior to reversal plus the voltage of the electrolytic cell connected with the opposite polarity) which may not be readily apparent in

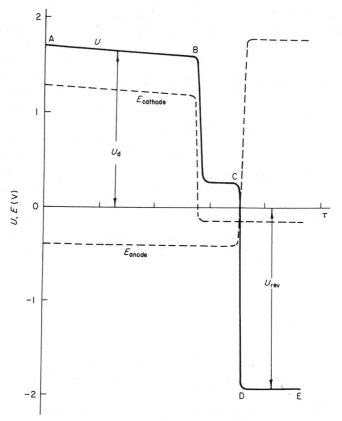

Fig. 32. Diagram of voltage and the cathode and anode potentials in the case of overdischarge. AB, normal discharge region; B, beginning of cathode overdischarge (hydrogen evolution); C, beginning of anode overdischarge (oxygen evolution); DE, the range of complete over-discharge (reversal) of the cell.

the case of a high-voltage battery. The consequences of the cell reversal may, however, be harmful: intensive gassing in the cell involved may rupture the cell and harm the neighbouring ones and even the load circuits. As a rule, cell reversal in storage batteries means that the battery is completely lost, becoming non-rechargeable. Cell reversal is especially harmful in cells with low-slope discharge curves in which the cell voltage drops sharply only at the end of discharge.

Consequently, a wise practice is to avoid complete discharge of high-voltage batteries, especially storage batteries, i.e. to use not more than 80–90% of their capacity. This restriction does not cover low-voltage batteries in which the cell reversal is immediately apparent owing to a considerable relative drop in voltage. In order to achieve high reliability of a system, each cell may be shunted by a diode; when the voltage of one of the cells drops to zero, the battery current bypasses the discharged cell through the diode connected in parallel.

(b) Charge and discharge of storage batteries

Storage batteries are often charged to a fixed end-of-charge (final) voltage $U_{c.fin}$. The total battery voltage is an indication only of the *mean* cell voltage. At the moment when the end-of-charge voltage is reached, some of the cells may be undercharged and others overcharged.

In principle, all the cells of a storage battery must always be charged to the same level, since the discharge or charge current passing through each one is the same. Unfortunately, unequal rates of self-discharge in individual cells may result in unequal states of discharge within a stored battery. The resulting "phase shift" may accumulate during prolonged cycling of the battery. The above-mentioned factors complicate the operations of storage batteries. The methods which serve to overcome these difficulties will be described in Section 7.2.

(c) Liquid junctions

The electrolyte potential is different in each cell connected in series. The difference between potentials in two neighbour cells equals the voltage of an individual cell. If a liquid junction forms between the electrolytes of adjacent cells (Fig. 33), the current flowing through it results in a gradual discharge of the battery. The intensity of this discharge depends on the resistance of the liquid junction. The rate of self-discharge is low if the liquid junction is a thin film with high resistance, otherwise the self-discharge may be quite fast. The process hardly depends on whether the battery is being discharged into an external circuit or not. Consequently, electrolytes in each cell of the battery must be carefully insulated from those of other cells.

Sometimes such contacts are inevitable, as in automatically activated batteries with bipolar electrodes in which the electrolyte flows into all the cells at the same time from a common container. After the filling, the liquid contacts are normally broken or remain in the form of very thin high-resistance films. Liquid junctions are permanent in batteries in which the forced circulation of the electrolyte through a common outer circuit is used to remove heat or reaction products.

Calculation of the leakage current through liquid junctions is difficult.[6] Even in the simplest circuit of two cells difficulties appear because the current distribution over the electrode is non-uniform owing to a non-symmetrical arrangement of the liquid junction. The extent of this non-uniformity depends, as in any distributed-parameter system, on the relative contributions of the polarization and ohmic effects.

As the number of cells N connected in series increases, the problem becomes more complicated since any combinations of cells through liquid junctions are possible. The general schematic of the processes is given by the simplified equivalent circuit shown in Fig. 34. In some cases the general leakage current for each cell is proportional to N^2, that is the leakage current in a battery of 40 cells is greater by a factor of 100 than that in a battery of four cells only. Leakage currents are not distributed uniformly along the chain of cells: self-discharge is maximum in the end cells, and diminishes towards the middle of the circuit.

6.5. Sealing

(a) Purpose of sealing

The types of cells most widespread in the last century were manufactured as open glass vessels containing electrodes and a liquid electrolyte. Obviously, such cells could operate only in stationary conditions. The development and wide use of portable equipment and of miniature cells necessitated the appearance of more convenient sealed designs.

Evolution of gases, mostly hydrogen, due to self-discharge of the metal of the negative electrode, is a factor making cell sealing a difficult problem. Gases (hydrogen and/or oxygen) are liberated at especially high rates at the end of charging and in overcharged storage batteries with aqueous electrolytes, as well as after cell reversals in the case of deep discharge of batteries (see Section 6.4). In order to release the gas, the electrode- and electrolyte-containing chamber must be open to the ambient medium.

In addition to such unwanted consequences of gas evolution as the necessity of periodic addition of water to balance out its loss by decomposition, and the possibility of producing explosive-hazard levels of hydrogen concentration in the surrounding air, some damaging processes are possible as a result of free contact of the electrolyte and electrodes with the ambient atmosphere. These include oxidation of the reducer by oxygen of the air; absorption of carbon dioxide from the air by the alkaline solution and the resulting formation of carbonic acid salts (carbonization of the

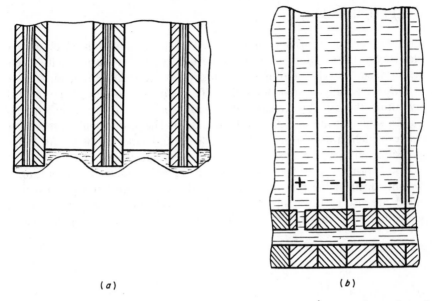

(a)   (b)

Fig. 33. Examples of liquid junctions: (a) along an electrolyte film across sealing gaskets of bipolar electrodes; (b) along the electrolyte channel of the filter-press battery.

electrolyte); penetration of moisture into non-aqueous electrolytes; drying of the electrolyte owing to vaporization of the solvent; formation of fog consisting of tiny drops of electrolyte transported by gas bubbles (this is harmful to personnel, and also results in a gradual loss of the electrolyte); release of such toxic components as sulphuric anhydride, hydrazine, vapours of organic solvents, and so forth, into the air.

Cells may be partially sealed, with the electrolyte rendered spill-proof and the damaging effects of each of the above processes reduced, or may be completely sealed.

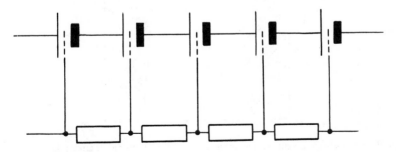

Fig. 34. Equivalent circuit of a battery with a liquid junction.

(b) Partial sealing

Primary cells are normally spill-proofed by using immobilized or matrix electrolytes. Utilization of immobilized electrolytes in storage cells meets with difficulties because of intensive gassing at the end of charging.

Liquid electrolytes are made spill-proof (meaning that they are not spilled by tilting the cell, by transportation vibration, etc.) by installing special valves which let through gas but block liquid. The most popular are rubber valves which in normal situations are closed and lock out the liquid. When pressure builds up in the storage battery, the valve opens up for a short period. Spring valves (Fig. 35) are more reliable. Valves incorporating a weight, open in the normal position of the battery and closed when it is tilted, are also used. Finally, hydrophobic porous filters easily permeable to gases but blocking the passage of aqueous electrolyte owing to water-

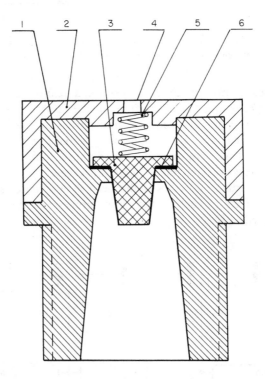

Fig. 35. Schematic of a vent with a spring valve. 1, vent housing; 2, cover; 3, valve; 4, vent hole; 5, spring; 6, rubber gasket.

repellent properties, may be utilized for the purpose. As a rule, the valves and filters in question are made in the form of plugs to be screwed into the filling hole.

If the electrolyte must never be spilt (from a cell tilted for a long time or operating in an upside-down position), a special "spill-proof trap" is used. The cell container has at the top a sufficiently large gas chamber communicating with the ambient atmosphere by an elongated vent, with a hole in the lower end (Fig. 36). The inlet hole in the vent is above the level of the liquid whatever the spatial orientation of the cell container, so that the electrolyte is not spilt. This design is usually avoided since it entails increased height and volume of the battery.

Prevention of spilling is not the only function of the devices described. A rubber valve also provides sufficiently effective protection of the electrodes and electrolyte against the ambient medium. A porous filter prevents the

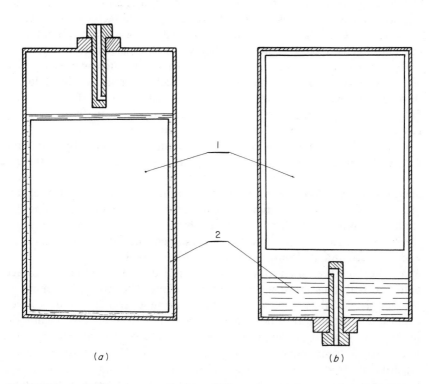

(a)   (b)

Fig. 36. Schematic of a storage cell with a spill-proof vent: (a) normal position; (b) inverted position. 1, electrode group; 2, liquid electrolyte.

escape of liquid drops, that is it suppresses venting of the electrolyte fog. Additional baffle screens are sometimes placed into the vents to increase efficiency of drop retrieval.

Gassing accompanying overcharging of a storage battery can now be reduced by using special plugs containing a catalyst (platinum black, for example) which facilitates the hydrogen–oxygen recombination reaction. The water formed thereby flows back into the electrolyte. The device does not achieve total suppression of gassing, especially if the ratio of reactants deviates from stoichiometry (due to one of the electrodes being charged earlier than the other).

(c) Complete sealing

Complete sealing is possible in cell types in which no gases are liberated. One example is the Weston cell in which evolution of gases is thermodynamically impossible and which can therefore be manufactured in a sealed glass vessel. Gassing can also be considered as practically absent in a number of cells (including storage cells) with non-aqueous electrolytes.

Hydrogen evolution rate in miniature primary cells, such as mercury–zinc disk cells, is very low, so that the gas is effectively removed by diffusion through packing layers, such as rubber or plastic gaskets. The hydrogen diffusion coefficient in these materials being greater than those of other gases, these cells can be considered as completely sealed.

If gas evolution in a cell is considerable, complete sealing is only possible if pressure build-up of the accumulating gas is avoided by making the gas enter some chemical reaction. If, for example, hydrogen and oxygen were liberated in stoichiometric proportion at any moment at the end of charging, it would be possible to have them form water. For complete sealing this method cannot be realized, however, because gases do not evolve in stoichiometric ratios when a battery is recharged; the gas liberated in excess would be accumulated until the rupture of the container.

Another method, first developed for sealed nickel–cadmium storage batteries, is the one most widely used nowadays. The store of reactants in the cell is calculated to be such that during charging anodic oxygen evolution at the positive (nickel oxide) electrode starts long before cathodic hydrogen evolution at the negative one. To achieve this, an excess of the non-charged component, namely cadmium oxide, CdO, is added to the negative electrode (Fig. 37). The cell design is such that oxygen evolved at the anode easily reaches the surface of the cathode to react with metallic cadmium formed during charging. The result is two-fold: gaseous oxygen enters a reaction and thus is not accumulated inside the battery, while metallic cadmium in the negative electrode is partially oxidized. Consequently, the degree of charging of the negative electrode increases at a rate not exceeding that of the positive one. On completion of charging of the positive plate, the degree of charging of the negative one stops increasing since the rates of formation and oxidation of metallic cadmium become equal. Hence, no hydrogen can evolve at the cadmium electrode. If charging is continued, no additional chemical changes appear in the cell, since oxygen participates in a closed cycle, evolving at the anode and reacting at the cathode. The whole energy of the charge current is then released as heat. This is why intensive heat generation and temperature build-up are observed at the end of charging of sealed storage batteries.

Oxygen pressure in a sealed storage cell is somewhat elevated and stable, owing to the balance of oxygen formation and consumption rates. The access of the gas to the cadmium electrode is facilitated since the volume of the electrolyte is rather small; moreover, the electrolyte is in the pores of a cloth separator with part of the pores unfilled and serving to transport the gas.

With the charging current below a certain limit, sealed nickel–cadmium storage batteries permit prolonged (unlimited) overcharging, which is especially important for battery operation in non-serviced devices with auto-

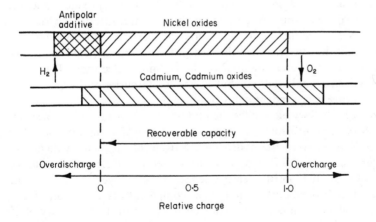

Fig. 37. Processes involved in overcharge and overdischarge of sealed nickel–cadmium storage cells.

matic recharge units. The critical current level is determined by the rate of oxygen migration to the cadmium electrode surface. As a rule, $j_{crit} \approx 0{\cdot}02$. The admissible duration of overcharging has to be restricted if high charge currents are used.

Instead of the oxygen-based cycle, a similar hydrogen cycle could be used in sealed storage cells: an excess of the reactant at the positive electrode, hydrogen evolution at the negative one, and hydrogen reaction with the oxidizer. The oxygen cycle proved more convenient in the case of nickel–cadmium cells.

Gas evolution is dangerous not only at the end of charging but also in the cases of overdischarging of individual cells within batteries (Section 6.4). When a nickel–cadmium cell is overdischarged, either oxygen is first evolved at the cadmium electrode or hydrogen at the nickel oxide electrode, depending on which of the electrodes limits the capacity. These processes can be suppressed in the same manner as gas evolution in overcharged cells. The ratio of active masses and the degrees of charging are selected so that the cadmium anode is the capacity-limiting one. This means that gassing at this electrode starts earlier than at the nickel oxide cathode (Fig. 37). Oxygen evolved at the anode is transported to the cathode to be reduced there. Unfortunately, the potential of the nickel oxide electrode is too positive for the cathodic reduction of oxygen. The problem is solved by adding to the active mass of this electrode some cadmium oxide (the so-called antipolar additive). When the nickel oxide in the electrode is completely discharged, overdischarging results in reduction of cadmium oxide to metallic cadmium and hydrogen is not liberated. The potential of this process is much more negative than that of nickel oxide reduction. The amounts of reactants are chosen so that the transition from reduction of nickel oxide to reduction of cadmium oxide takes place before oxygen starts evolving at the anode. When this last process starts oxygen is easily reduced at the cathode at the potential of the cadmium electrode. This leaves the cathode mass unaltered, so that at moderate currents the storage battery may overdischarge indefinitely.

Special auxiliary electrodes facilitating reactions involving gases evolved are introduced into some cell types. For instance, a small porous hydrogen electrode (similar to the hydrogen electrode of fuel cells with liquid electrolytes, see Section 17.5), partially immersed in the electrolyte solution and connected to the positive electrode can be employed in sealed lead batteries.[7] If hydrogen appears in the gas compartments as a result of lead self-dissolution, it is immediately oxidized to water on the auxiliary electrode. This reduces an equivalent amount of lead dioxide, so that capacity losses are identical on both electrodes.

(d) Structural and technological problems of sealing

A cell is sealed reliably if the joints between individual parts of the cell (container, cover, terminals, valves, etc.) are tight with respect to liquids and gases. Depending on the materials used, the joints are sealed by gluing, soldering, welding, filling with sealing compounds and so on. Moreover, the joints must be chemically and thermally resistant (for instance, must not crack because of non-uniform thermal expansion). These problems are usually solved without too much effort. But in the case of high-temperature cells (and especially with molten electrolytes which normally are chemically very agressive) they are often very serious and hinder the development of new cell types.

Sealing around cell terminals is especially difficult. In cells with a metal container at least one terminal must be electrically insulated from it. It is therefore always necessary to assure gas-tight sealing between terminals and non-metal insulator materials (Fig. 38). Terminals with rubber or plastic sealing gaskets are often employed in low-temperature cells. These materials provide sufficiently effective sealing of joints. Difficulties appear in cells with alkaline electrolytes. This electrolyte has an ability to "creep" through the tiniest cracks and the rubber-metal joints.[8] Creeping along the container surface, alkali forms large white patches of carbonic salts (as a result of absorption of carbon dioxide from the air). The process of alkali creepage depends strongly on the electrode potential, and usually it is more intensive at the negative terminal of the cell.

Elastic gaskets cannot be used for sealing and insulation in high-temperature cells; for this purpose these cells employ insulating ceramic materials.

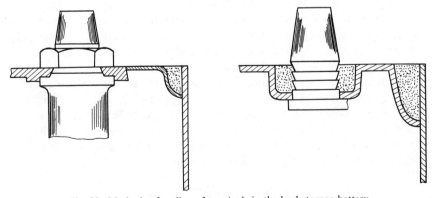

Fig. 38. Methods of sealing of terminals in the lead storage battery.

6.6. Thermal processes in cells

Thermal energy is inevitably released together with electrical energy when a cell is discharged. Heat is also produced when a storage cell is being charged. The power of heat generation, that is the amount of heat released per unit time, can be found from equation (1.16) (Section 1.5):

$$\text{(discharge)} \quad P_{\text{therm}} = I_{\text{d}}(\bar{\mathscr{E}} - U_{\text{d}})$$
$$\text{(charge)} \quad P_{\text{therm}} = I_{\text{c}}(U_{\text{c}} - \bar{\mathscr{E}}) \quad (6.1)$$

In principle, it is possible for some electrochemical systems to discharge at $U_{\text{d}} > \bar{\mathscr{E}}$ or to be charged at $U_{\text{c}} < \bar{\mathscr{E}}$, thus absorbing rather than releasing heat. In practice, however, such cases are extremely rare.

Obviously, the ratio of thermal to electrical power P_{therm}/P of a current-delivering cell is equal to $(\bar{\mathscr{E}} - U_{\text{d}})/U_{\text{d}}$. In most electrochemical systems and in normal discharge conditions this ratio varies from 0·2 to 0·4. Heat release is especially intensive in cells with negative electrodes made of magnesium and aluminium, since polarization losses in these electrodes are high and the above ratio exceeds unity.

If the current-producing reaction changes in the process of discharging or charging, the value of $\bar{\mathscr{E}}$ changes correspondingly. When sealed nickel–cadmium storage cells are overcharged (see Section 6.5), no overall chemical reaction occurs since oxygen and metallic cadmium immediately react and form cadmium oxide. Consequently, $\bar{\mathscr{E}}$ equals zero, and the electrical energy is completely transformed into heat. The energy (enthalpy) of side processes (such as corrosion) which are not current-producing and hence have U_{d} formally equal to zero, is also released only as heat. Obviously, heat generation may be very high in the case of shorting, especially in high-capacity cells with active reactants (for instance, in silver–zinc cells).

The heat generation described above gradually raises the temperature of the current-delivering cell. A steady-state thermal situation is ultimately reached, when heat generation is balanced out by heat transfer to the surrounding medium (Fig. 39, curve 1).

The steady-state temperature difference relative to the ambient, $\Delta T_{\text{s}} = T_{\text{s}} - T_{\text{amb}}$, depends on the intensity of heat generation and on cell dimensions. It is convenient to introduce the specific bulk current $i_{\text{V}} = I/V$, where V is the cell volume, in order to obtain a convenient comparison of similar cells of unequal volume V operated in identical modes, that is with identical j. The heat generation intensity is then

$$i_{\text{V}}V\Delta U \quad (\Delta U \equiv \bar{\mathscr{E}} - U_{\text{d}})$$

The total intensity of heat removal is apparently given by $\alpha S_{\text{rem}} \Delta T_{\text{s}}$, where α

is the heat transfer coefficient, and S_{rem} is the heat removal surface area. Equalization of these quantities yields

$$\Delta T_s = \alpha^{-1} i_v L \, \Delta U \qquad (6.2)$$

where $L \equiv V/S_{rem}$ is a characteristic linear dimension of the cell. In similar operating conditions, therefore, the temperature increase is proportional to linear dimensions: low in miniature cells and high in large-size batteries.

Equation (6.2) is a crude approximation. It disregards the non-uniformity of heat production over the cell volume and hence the temperature distribution non-uniformity within it. The latent heat $\bar{q}_{entr}$ and heat polarization losses are released within the reaction zone, that is at the electrode surface, while the Joule heat due to ohmic losses is mostly generated in the electrolyte. Local overheating at the electrode surface may be observed even

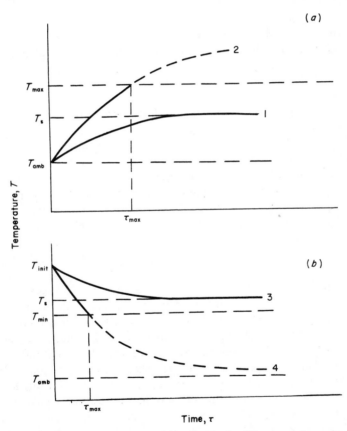

Fig. 39. Temperature of the cell as function of time in different modes and at ambient temperature (a) above and (b) below the minimum working temperature.

in steady-state conditions. On the one hand, this is a factor improving cell characteristics; on the other hand, it may intensify the electrode corrosion. Temperature is somewhat levelled off by the liquid electrolyte with its high specific heat, circulating freely in the interelectrode gap.

Each cell is characterized by a certain maximum admissible temperature T_{max} above which all the destructive processes (corrosion and ageing) are accelerated. There exists, therefore, a critical steady-state current I_{crit} above which the cell overheats (the maximum admissible current I_{adm} may be lower than I_{crit} if it is limited by other factors). The higher the ambient temperature, that is the smaller the admissible temperature increase ΔT_c, the lower I_{crit}. A number of methods may be used to increase I_{crit}:

(a) Gaps for free access and circulation of air are left between the cells of a battery (an increase of S_{rem}).
(b) Air cooling of the cell by a fan (increased α).
(c) Cell design with external heat-exchange fins (increased S_{rem}).
(d) Forced circulation of the liquid electrolyte in an external circuit with a cooling unit.
(e) Built-in coolers with liquid coolants.

Normally such measures are used only for comparatively large batteries, with powers of 1 kW and more.

Cells are also operated in non-stationary thermal modes. In particular, this concerns short-duration discharges by currents exceeding I_{crit}. This is the case, for example, with starter batteries for internal combustion engines. The battery current drain at the moment of car engine starting corresponds to a j from 3 to 5, with the discharge duration not exceeding 10 to 20 s. The admissible current is higher the shorter the discharge pulse (Fig. 39, curve 2). It is necessary to take into account, however, the fact that limitations on the duration of high-current pulses may be imposed by factors other than temperature build-up.

In certain infrequent cases the thermal instability may increase with time. Charging of sealed storage batteries at constant charge voltage is one such case. Normally the charge current in these batteries is at first high but later drops off to a safe level (Section 7.2). At the end of charging, however, heat release in sealed batteries rises steeply (Section 6.5). Increased temperature diminishes polarization, that is accelerates electrode reactions and increases the charge current in the battery. As a result, thermal runaway sets in, when a gain in current results in further increase in temperature, which enhances the current, and so on, until the explosion of the battery.[9] The same picture may be encountered though even less frequently, when a cell is discharged through a constant low-resistance load. Chain thermal processes may also develop in batteries: intensive heating-up of one cell caused by shorting may propagate to neighbour cells, resulting in their breakdown and intensification of heat release.

Specific problems are encountered when the ambient temperature T_{amb} is below the minimum operating temperature T_{min} of the cell. This is often the case when batteries operate in winter conditions, and also when high-temperature cells are used (those having working temperature above $50°C$). In these cases the cell, prior to loading, is heated up to the working temperature T_{init}. Various methods are known for preheating a battery: exposure to the room air of a heated building, filling of the battery with warmed liquid electrolyte, heating with combustible compounds (see Section 16.5 on thermal batteries), and so on. In some cases it is permissible to use heaters plugged into the electric mains network. The heating intensity must be moderate since many cell materials have low heat resistance (plastics are a good example). As a result, a certain time interval is necessary for a battery to reach the desired temperature, usually from 15 minutes to 2–3 hours. Induction heaters have been developed recently to heat primarily the metal electrodes rather than the structural plastic elements; the time required to reach the prescribed temperature is consequently shortened. Storage batteries may also be heated up by passing through them a small-amplitude AC current (causing alternating charging and discharging of the plates). One must, however, be very careful with this method, because of possible deterimental effects on the service life.

If no heating means are available, so-called self-heating is employed, when a cell is shorted by a low-resistance load so that despite the low temperature the discharge current is not zero. The heat released thereby slowly increases the cell temperature, which increases current, and accelerates the process. Self-heating is not interrupted until the working temperature is reached. The method must be applied with caution, and even then not in every cell type.

With the heating-up phase completed and the working conditions established, the problem of temperature stabilization has to be solved. The cell service life is quite long when heat generation balances out heat loss in the cell, and this is realized for currents above a certain minimal value I_{min} (Fig. 39, curve 3). If current is below I_{min}, the cell temperature ultimately drops below a permissible level (curve 4). Heat losses and I_{min} may be lowered by various methods of cell insulation, for instance, with cloth, foam plastic or other materials.

On the average, the cell specific heat is of the order of 0.8–1.0 J/g K. This characteristic does not affect steady-state processes, that is such quantities as I_{crit}, I_{min} and ΔT_s, but influences the kinetics of transient thermal processes.

6.7. Reserve batteries

The term "reserve battery" is usually applied to single-time non-rechargable cells or batteries with comparatively active reactants. The concept does not cover those dry-charged storage cells which are not filled when manufactured and must be filled with the electrolyte before operation. On being filled, these batteries allow long-term cycling and in fact are identical to ready-to-use battery types.

Several versions of reserve batteries are distinguished, depending on the method of activation.

Manually activated reserve batteries. These batteries are manufactured without electrolyte. Liquid electrolyte is introduced before operations through the filling hole in the battery cover. Filling is carried out either manually or by an auxiliary feeder unit.

Sometimes a battery is filled with water, instead of the electrolyte (the so-called water-activated batteries). In some reserve cells, an anhydrous solid alkali or salt is placed into the electrodes, or in special bags into the interelectrode gaps, to be quickly dissolved and form an electrolyte of the required concentration when water is added. In other cell designs, the cell reaction produces dissolved salts which form the electrolyte; the cell characteristics are somewhat lowered at the start of discharge owing to the high resistivity of water (sea water is typically used). In particular, this principle is realized in water-activated cells with magnesium anodes (Section 13.3).

Automatically activated reserve batteries. A battery of this type (ampoule battery) differs from manually activated ones in that the electrolyte is kept in a special container (ampoule) inside the battery. Special devices feed the solution into the electrode compartments at the moment of activation.

Batteries of this type are most often activated pneumatically (Fig. 40). Gas from a high-pressure bottle is released by rupturing a special membrane with an explosive cartridge. The gas pressure compresses an elastic ampoule (bladder) containing the electrolyte; the liquid breaks through a membrane at the ampoule neck and fills the electrode chamber. Designs of ampoule batteries are known in which the total time of activation does not exceed 0·1–0·2 s.

Another version is used in artillery proximity fuses. A glass ampoule containing an electrolyte and placed in the central part of a cylindrical cell is broken at the moment of the artillery shot. The centrifugal force due to the projectile's revolution around its longitudinal axis immediately throws the electrolyte to the electrode group placed concentrically around the ampoule.

Automatically activated batteries are very expensive. They are applied in

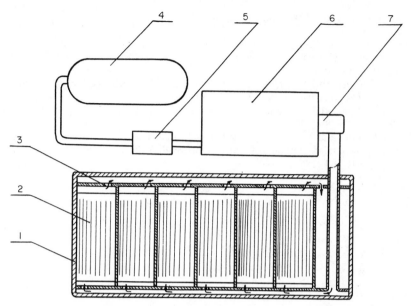

Fig. 40. Schematic diagram of an automatically activated battery. 1, monoblock; 2, cell; 3, air escape valve; 4, compressed gas bottle; 5, membrane with an explosive cartridge; 6, ampoule with electrolyte; 7, reverse valve.

a number of military devices which require a combination of long shelf-life with constant readiness to speedy activation (activation time below one second).

Thermal batteries. The principles of thermal battery design are discussed in Section 16.5.

REFERENCES

1. N. E. Bagshaw, K. P. Bromelow and J. Eaton, The effect of grid conductivity on the performance of tall lead-acid cells, *in* "Power Sources 6" (Collins, D. H., ed.), p. 1–14. Academic Press, London and New York (1977).
2. R. E. Meredith and Ch. W. Tobias, Conduction in heterogeneous systems, *in* "Advances in Electrochemistry and Electrochemical Engineering", (Tobias, Ch. W., ed.), Vol. 2, pp. 15–47. Interscience, New York, London (1962).
3. G. R. Robinson and R. L. Walker, Separators and their effect on lead-acid battery performance, *in* "Batteries" (Collins, D. H., ed.), pp. 15–41. Pergamon Press, Oxford (1963).
4. P. B. Zhivotinsky, "Porous Separators and Membranes in Electrochemical Devices". Khimiya Publishing House, Leningrad (1978) (in Russian).
5. S. Itoh, Characteristics of "Yumicron" batteries for automobiles, *in* "Rechargable

Batteries in Japan" (Miyake, Y. and Kozawa, A., eds.), pp. 175–196. JEC Press, Cleveland (1977).

6. B. P. Nesterov, T. A. Razevig and N. V. Korovin, Distribution of stray currents in batteries with common collectors—a discrete problem, *Sov. Electrochem.* **10**, No. 7, 1033–1035 (1974).

7. P. Rüetschi and J. B. Ockerman, Sealed cells with auxiliary electrodes, *Electrochem. Technol.* **4**, No. 7/8, 383–393 (1966).

8. M. N. Hull and H. J. James, Why alkaline cells leak, *J. Electrochem. Soc.* **124**, No. 3, 332–339 (1977).

9. A. J. Salkind and J. C. Duddy, The thermal runaway condition in Ni/Cd cells and performance characteristics of sealed light weight cells, *J. Electrochem. Soc.* **109**, No. 5, 360–364 (1962).

Chapter Seven

Operational Problems

7.1. Discharge and maintenance of primary cells

One of the important advantages of primary batteries consists in very simple maintenance procedures. Most such cells require no servicing. Before a battery is switched on, its appearance and the remaining service life are checked; sometimes actual parameters are measured (o.c.v. and the initial discharge voltage). Correct polarity and reliable contacts must be ensured; a violation of polarity correspondence may result in serious disorders and even in the breakdown of load circuitry, especially circuits involving transistors and electrolytic capacitors.

The electrical conditions of cell operation are determined by the load schedule. Typically, primary cells are used with complicated, and often arbitrary, loading schedules. The current drain of a transistor radio battery, for example, is a function of the volume setting; the times of turning the set on and off are arbitrary. Batteries in electronic watches and pacemakers are loaded continuously but the discharge current is pulsed. The cases of continuous discharge to a constant load are rather infrequent.

As a rule, primary cells are used until completely discharged. Sometimes, however, this should be avoided; especially harmful may be leaving the discharged battery in the unit it has powered. Leakage after complete discharge is observed in some cells, and in particular in the most widely used manganese–zinc cells. The lost electrolyte may cause corrosion of the consuming unit. Explosions of sealed cells, especially of the mercury–zinc cells, may sometimes (although very rarely) be observed if spent cells are stored. Batteries with a great number of cells connected in series must be watched with special care, for signs of cell reversal (see Section 6.4). In order to avoid reversal, a voltage check on each cell is desirable. This being impossible in most cases, the battery must be disconnected in advance.

Most of the primary cells have high internal resistance, so that brief external shortcircuiting is unlikely to produce extensive damage.

Certain types of primary batteries require more complicated handling. Thus, reserve batteries need to be activated before discharging may begin. In low-temperature conditions a battery often has to be warmed up. Detailed manuals are supplied with cells and batteries for all these situations.

7.2. Maintenance of storage batteries

(a) Maintenance modes

Storage cells and batteries are serviced in a much more complex way than primary cells, owing mostly to the charging procedure. While the discharge mode is determined, in the first place, by the specifics of the consumer circuits, the charging mode depends primarily on the specifics of the storage cells and strongly affects cell lifetimes. Because of gassing at the charging phase, most storage cells and batteries are not sealed, which entails additional complications in maintenance in comparison with primary cells.

Storage batteries operate in three distinct modes: alternating charge–discharge mode, buffer mode and standby mode. In the first of them battery charge regularly alternates with discharge. This operation is typical for traction batteries and light portable batteries. In the buffer mode, a battery is connected in parallel to another current source; when the load rises, the battery is partially discharged, to be recharged again when the load diminishes. The buffer mode is characteristic for automobile starter batteries, batteries in aerospace systems, and some others. A storage battery in the standby mode is always maintained ready to take over but is connected to the load circuit only in emergency situations, when the main power source breaks down.

In addition to conventional charge–discharge cycles, some special cycles may be used: forming cycles, employed with freshly manufactured plates or after filling the battery with electrolyte, to achieve the optimum operating state of the plates (sometimes only a forming charge is used), and training cycles, employed when the battery is used on an irregular basis or is never discharged fully. Training cycles consist of a complete discharge followed by a complete charging at a rated current.

(b) Charging of storage batteries

As a rule, charging devices have a decreasing current–voltage characteristic, that is the charge voltage is the lower the higher the supplied current. This means that unless special control circuits are used, the charge current falls off as the battery voltage increases. Charging devices are in most cases equipped with systems that control and stabilize one of the electrical parameters, either voltage or current. Correspondingly, two basic methods exist: constant-current charge, and constant-voltage charge.

Constant-current charging (Fig. 41, curve (a)) requires comparatively simple equipment, since current stabilization is normally less complicated than that of voltage. Another advantage is that the total electrical charge is easily found as the product of current and charging time. The method unfortunately has its shortcomings. Charging time is high if current is small. With high charge currents, chargeability suffers at the end of charging, because of non-uniform distribution of current over depth into the porous electrode: intensive gassing occurs in the outer, fully charged layers while the inner layers are still slowly accumulating charge.

In the case of constant-voltage charging, current is high at the beginning of the process and then slowly diminishes (Fig. 41, curve (b)). The total charging time is quite high since the end-of-charge current is very small. The choice of the charge voltage becomes critical: lower voltage means suppressed side processes and gassing but longer charging time. The optimum charge duration is a function of temperature and the state of the battery so that this technique is rather difficult to use. Another shortcoming consists in battery overheating by high initial currents.

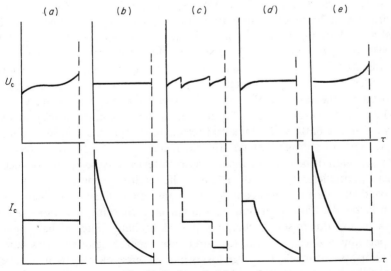

Fig. 41. Various charging modes.

Various combined charging procedures are widely used to obviate the defects inherent to the simple constant-current and constant-voltage techniques. In all these procedures, charging time is shortened and gassing diminished by using large currents at the begining of charging and small currents at the end of the process.

Stepwise-current charging is a comparatively simple procedure: a battery is first charged by rated current until the prescribed cutoff voltage is reached, after which the charge current is decreased by a factor of 2 to 3 and the process goes on until the prescribed voltage is reached again. Three- and even four-step constant-current charging is possible (Fig. 41, curve (c)).

Overheating of the battery in the case of constant-voltage charging is often prevented by restricting and stabilizing the initial charge current. After a certain interval, charging is switched over to the constant-voltage mode (Fig. 41, curve (d)). In other cases, charging starts with constant voltage and is continued in the constant-current mode after current drops to a prescribed level. This shortens the total charge duration (Fig. 41, curve (e)). Various other combinations of these modes are possible. Sometimes the transition from one method (or step) to another is done by hand, and sometimes it is realized automatically, when the preset current or voltage is reached. The dropping current–voltage curve of a charging unit can be used to diminish current in the process of charging, without resorting to control circuits. This requires, however, that the characteristics of the charging device and those of the battery be matched; this is a restriction on the universality of the method.

In some cases storage batteries (especially alkaline batteries) can be charged with asymmetric alternating current, that is by superposition of DC and AC charging modes.[1] This procedure modifies the structure of the active mass formed in the process, and therefore affects the operational characteristics of storage batteries (the discharge capacity may be slightly increased or discharge voltage stabilized). The observed effects greatly depend on the ratio of AC and DC current and on the frequency of the former. Asymmetric-current charging intensifies heating and may increase gassing in the battery. This method results in sharply reduced service life of some types of storage batteries (in particular, some lead-battery systems), and thus cannot be used for these. These batteries may be adversely affected even by insufficiently filtered current at the rectifier output.

We distinguish between complete charging of a battery which was almost completely discharged previously, and partial recharging to compensate for capacity loss due to self-discharge or partial discharge. Such recharging may be periodic or continuous. In the latter case, recharging at very low current to compensate for self-discharge losses in the course of long-term storage of a standby battery is called compensation recharging.

In some types of storage batteries, so-called equalizing (levelling) charging is sometimes performed using small charge current ($j = 0.03–0.05$). Equalizing charges serve to regenerate active materials on all electrodes of all cells of the battery, in order to level off the differences in the degree of charging.

In charging a battery, one faces the problem of determining the moment when the battery as a whole and its individual cells are charged to completion. Long service life of batteries is only realizable if this moment is determined correctly. For instance, undercharge is harmful to lead batteries, and therefore they are charged until stable gas evolution. On the contrary, overcharging is unacceptable in the case of silver–zinc storage batteries.

Several techniques are used to determine the moment of charge completion. Often a battery is simply charged with a predetermined amount of electricity calculated on the basis of the known degree of discharge. It is sufficient to monitor the duration of charging if constant-current charging is employed. Coulometers (calibrated in Ah) are used to measure the cell capacity as a function of time in constant-voltage charging, with current diminishing continuously. In other cases, the completion of charging is derived from the voltage at the predetermined charge current. For example, charging of silver–zinc batteries is terminated immediately after the cutoff voltage is reached. In order to avoid overcharging, not only the net voltage of the battery but also that of each cell has to be monitored. Lead batteries require slight overcharging; consequently, an additional charge around 10 to 20% of the rated capacity is delivered to the battery after the voltage hike to 2·6–2·7 V. Some commercially available charging devices automatically cut off the charge current immediately after the end-of-charge state is diagnosed by one of the above-mentioned methods.

If charging is conducted at low temperatures, one has to take into account that charge voltage is a function of temperature. Thus, the end-of-charge voltage of nickel–cadmium batteries increases, for rated-current charging, from 1·85 V at an electrolyte temperature of 20°C to 2·35 V at −40°C. The use of thermal insulation is recommended in winter charging of batteries, because of the decreasing chargeability at low temperatures. Conversely, summer charging has sometimes to be interrupted to avoid excessive heating of the battery.

Systematic use of booster charges works to reduce the capacity and cycle life of the batteries. One must therefore be very careful in applying such a charging technique. It is advisable, for instance, to alternate booster charges with training cycles (see above).

Reduction of charge time and of battery sensitivity to booster charging constitutes one of the important problems in storage cell development.

(c) Storage batteries in the alternating charge–discharge mode of operation

This mode of operation (Fig. 42(a)) is characterized by deep discharges. Monitoring of the batteries at the end of discharge must therefore be more careful than in the case of primary batteries. Reversals of storage cells are intolerable since they result either in substantial shortening of service life or in complete failure of the unit (lead and silver–zinc storage batteries are especially sensitive to cell reversal). Excessively deep discharge, even if it does not cause cell reversal, may also be harmful to some battery types. For instance, regular deep discharges produce an extremely undesirable sulphatation of lead battery plates. Consequently, the methods and devices for determination of the degree of charging of a battery (calculation of the capacity delivered, the use of coulometers, determination of the electrolyte concentration in lead batteries, etc.) are very important for the operation and maintenance of storage batteries.

Alternating charge–discharge operation creates the most favourable conditions for monitored charging. Normally, batteries are charged in special rooms with normal temperature conditions, by optimal charging methods and with all the required monitoring.

Spare batteries have to be used if the battery-powered equipment cannot be switched off for sufficiently long intervals. Discharge of one battery is accompanied by charging of another battery (or batteries).

(d) Storage batteries in the buffer mode of operation

In this mode (also referred to as "floating") a battery is connected to the loading circuit parallel to another (main) source of electrical energy (Fig. 42(b)). The buffer mode is used in two cases: (a) when the main energy source is working intermittently, such as a wind generator, or the car generator operating only when the engine is running, and (b) when the generator's power is not sufficient to meet periodically occurring extreme power demands (this is encountered in aerospace systems powered by solar panels). With small loads and the generator running, the generator voltage is somewhat higher than the battery o.c.v. so that the battery is being charged; at high loads or with the generator off, the battery is discharging. Discharges in the buffer mode are typically not deep. A system can operate in the buffer mode for long periods only if the generator power corresponds to the mean integral consumed power (taking into account energy losses in cycling of the storage batteries).

Typically, fluctuations of voltage are harmful for load circuits. In the case

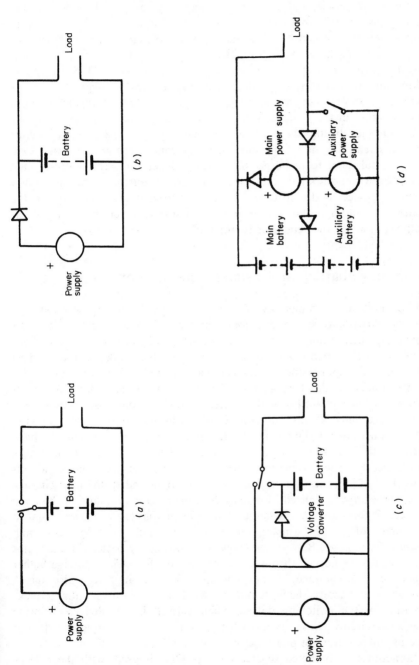

Fig. 42. (a) Alternating charge–discharge mode; (b) buffer mode; (c) standby mode; (d) combined mode.

of the buffer mode, voltage varies from the value characteristic for the battery discharge to that characteristic for its charging. Lead batteries, with their inherently low-slope charge and discharge characteristics, and with the charge and discharge voltages differing only by 10–15%, have definite advantages in this respect. The difference reaches up to 20–30% in nickel–cadmium batteries (this also reflects their lower rated voltage).

When a battery is operating in the buffer mode, it must be protected from voltage hikes in the generator, that is from the possibility of overcharging (for example, during idle runs of the generator, with all external circuits disconnected). On the other hand, the generator voltage at an average load must be sufficient to recharge the battery. Generators are therefore often equipped with additional voltage control devices. Furthermore, the battery must be protected (for instance, by diodes) from useless discharge to the generator winding when the generated voltage is lower than that of the battery, or when the generator is stopped.

(e) Storage battery in the standby mode of operation

In this mode all consumer circuits are powered by the main power source—electric network or an autonomous generator. The storage battery starts delivering current only in an emergency, that is when the main source fails. Several circuit versions are possible in the standby mode. In the simplest case, the main source and the standby battery are connected in parallel (as in the buffer mode, Fig. 42(b)). The main source provides continuous compensation recharging of the battery whose discharge starts when the main source fails. At this moment the voltage in the system drops jumpwise, which constitutes a significant shortcoming of this arrangement. A sharp drop of 10–20% is not crucial in some cases (such as emergency lighting), but is clearly unacceptable for more critical equipment. In the second version (Fig. 42(c)), the standby battery is not connected to the load circuits. The connection is realized by a relay only in the case of failure of the main source. In this case the discharge voltage of the battery is chosen equal to the rated voltage of the main source, so that switching causes only small changes in voltage. Between emergencies the standby battery is either not recharged or recharged through a converter which provides voltage higher than the rated voltage of the generator. This arrangement also has a shortcoming, namely, the finite time of switching. This time is equal to pick-up time of the switching device, when current is not delivered to the consumer circuits. Some types of equipment, such as computers, do not tolerate interruption of power supply even for 1 ms.

Sometimes a combined scheme of Fig. 42(d) is used, with the battery consisting of two subsections connected in series.[2] One subsection is constantly connected to the main electric source while the second subsection (its voltage is 10–20% of that of the first subsection) is connected to the first in series in the case of emergency to compensate for the drop in voltage.

7.3. General aspects of battery maintenance

Chemical power sources need certain servicing which includes preparation procedures, monitoring, periodic checks and other operations. Following correct maintenance procedures is very important for storage batteries designed for long service life. Maintenance is aimed at keeping batteries in their "best shape" and at preventing premature breakdown.

The required maintenance operations may be different with different battery types, and sometimes the differences in regulation operations may be very substantial. In any case, these operations must never deviate from those specified for each battery type. In this section we discuss only some general aspects of maintenance.

The first requirement is that a battery must be kept clean. The outer surface must never have liquid electrolyte on it. Alkaline batteries must be visually checked with special care because of the tendency of alkaline solutions to "creepage" (see Section 6.5). Electrolyte on a battery cover leads to shorting and wasteful discharge, as well as to corrosion. The parts that have to be kept specially clear are terminals, vents, valves and filling orifices. Electrolyte and other contaminants getting into terminal contacts cause corrosion which results (especially if tightening is insufficient) in formation of transition resistances and in sparking. It is advisable to corrosion-proof contacts with special acid-free greases.

Measuring of the o.c.v. and on-load voltage is an important method of battery checking. In some cases even very crude measurement confirms that the battery is still operational. In other cases, however, such as in quantitative determination of the degree of charging of a lead battery, the o.c.v. must be measured by means of high-resistance voltmeters (precision grade not worse than 1·5%). In order to minimize ohmic-loss error when measuring the on-load voltage, the voltmeter must be connected to the battery terminals or very close to them. A very convenient device for checking a battery is the so-called load prong. This device comprises contact legs, a load resistor and a voltmeter. The pointed contact legs provide reliable electrical connection, while the load resistance is such that the current is typical for the battery to be checked.

With the exception of sealed storage cells, the main element of maintenance is the control of the amount and concentration of the electrolyte. When the electrolyte is poured into a battery, its purity must be checked. All the accessories required for maintenance operations, such as vessels, thermometers, funnels, areometers, and so on have to be very clean because both acids and alkalis of battery grade are chemically very pure. Contacts of alkaline electrolytes with air must be minimized to avoid absorption of carbon dioxide (carbonization of the electrolyte).

In order to compensate for partial decomposition of water due to overcharge and corrosion, water must be regularly added. Normally addition of acid or alkali solutions is inadmissible since the resulting increase in electrolyte concentration impairs battery characteristics. Only deionized water (if unavoidable, pure rainwater or melted snow may be acceptable) is to be added to the electrolyte. This water must not be kept in metal containers since even very small amounts of iron ions (or ions of other metals) introduced into the electrolyte are very harmful to batteries. Chlorine ions present in normal tap water are also very dangerous. It should be kept in mind that boiling of water does not remove either chlorine or iron ions. When water is added, the electrolyte must come to the level indicated in the manual for each specific battery (usually it is 3–5 mm above the upper edges of the plates). A certain time is required for the concentration to level off in the bulk of the electrolyte.

The electrolyte concentration should be periodically checked. Sometimes it has to be corrected by adding an acid or alkali, for instance, after spilling part of the liquid or after prolonged charges (in the last case gases carry away tiny droplets of the electrolyte fog). Sometimes the electrolyte concentration is changed in the transition from summer to winter conditions of operation (or vice versa).

Usually electrolyte concentration is determined by measuring its density with special areometers. Temperature dependence of density of solutions must not be overlooked. Battery servicing manuals always indicate the temperature recommended for density measurements.

The monitoring and maintenance of the level and concentration of electrolyte in non-sealed storage cells constitute fairly labour-consuming operations. New versions of storage cells, the so-called low-maintenance and maintenance-free cells and batteries have been developed in recent years; the rate of gas evolution and water decomposition in such cells is very much reduced so that they can operate for a long period without regular additions of water (see Section 9.6).

The conditions in which batteries are stored must also satisfy certain requirements. Again batteries must be clean, and especially their venting holes. Unfilled batteries must be stored with plugs tightly screwed. It is advisable to store unfilled lead batteries and all alkaline batteries in discharged state. Electrolyte-filled lead batteries should be stored after charging them to completion.

Certain safety regulations have to be followed when using storage batteries. Facilities with a large number of storage batteries, such as electric power stations, communication centres, transport depots, and so on, must have special space for servicing batteries. This space must have an appropriate layout and equipment (ventilation, water pipes, sewer system, electric power mains, fire-prevention equipment, etc.). Since hydrogen is produced in battery charging, ventilation must be sufficiently intensive to prevent accumulation of detonating concentrations of gas. The electrolyte fog also enters the atmosphere. Therefore, the personnel of large battery maintenance stations must be supplied with individual protection kits. Conventional safety measures must be taken when electrolytes are prepared or anything else is done with acids or alkalis (rubber gloves, aprons and protective goggles are compulsory; each working site is to be equipped with neutralizing agents and a first-aid kit; acids and alkalis are not to be poured into common sewers). It must be remembered that corrosive activity of these solutions is very high.

Alkaline and acid storage batteries must be serviced separately (this covers storage, charging and repair). Alkali batteries of all types are damaged irrevocably by addition of sulphuric acid and lead batteries—by addition of alkaline solutions.

All rules covering operations with electrical units are valid in the case of storage batteries as they often have voltages above 36 V. Batteries must be reliably protected from shorts which could be caused not only by defects of the external circuit but also by tools, pieces of bare wire, or other metallic objects left around by careless personnel. Wiring of all electric circuits must be reliable, and contacts must be kept clean. Sparking in contacts may result in explosion of the liberated hydrogen.

7.4. Charging devices[3]

Motor generators and semiconductor AC–DC converters (rectifiers) are used as sources of controlled direct current for charging storage cells and batteries.

Motor generators. Two types of motor generators are employed: diesel engine/generator sets and single-armature converters. A diesel engine/generator set consists of a mechanically coupled AC motor (usually an induction motor with cage rotor) and a DC generator (with separate or shunt excitation). Converters with shunt-excitation generators have rather narrow ranges of voltage variation at the rated current, i.e. the maximum-to-minimum voltage ratio is not more than 1·5. Such sets can be used to charge only a narrow spectrum of battery types. The output voltage of a separate-excitation generator can be varied, for constant load current, from zero to some maximum value. Converters with these generators can also be used for discharging storage batteries. This means using the reversibility of electrical machines which can function as generators or motors, depending on voltage. In the case of charging, the generator voltage exceeds that of the battery; the converter consumes the electrical energy from the AC network, converts it into the DC energy, and sends it to the battery. In the case of discharge, the generator voltage is less than that of the battery so that it becomes a motor; the AC motor then becomes a generator, so that the battery sends the energy to the converter which send it into the network after converting DC to AC. Diesel engine/generator sets are either manufactured as single-casing converter units with the rotor of the AC motor and the generator armature sitting on a common shaft, or the sets are assembled from standard brands of induction motors and DC generators coupled by a clutch or a belt drive.

A single-armature converter is an electrical machine with armature coils having taps at both ends: collector at one end of the armature (as in DC machines) and slip rings at the other (as in a synchronous AC machine). On the side of slip rings, the converter is connected to the three-phase or single-phase AC network, and on the side of the collector to the storage battery. A single-armature converter can be used both to charge and to discharge batteries. In the first case the network-supplied energy is used to rotate the armature. This generates DC voltage across the collector, so that the converter operates as an AC motor and a DC generator at the same time. In the second case the converter operates as a DC motor and an AC generator, that is it transmits energy into the network. For equal power, single-armature converters are smaller than diesel engine/generator sets but their output voltage control circuits are much more complicated.

Another advantage of single-armature converters is their high efficiency. Power losses in electrical machines are caused by ohmic losses in winding and contacts, mechanical losses on friction, and losses in magnets due to eddy currents. In the case of single-armature converters, losses are rather low, and the efficiency is from 80% to 85%. Efficiency of single-casing diesel engine/generator sets is about 70%. It should be taken into account that underloading considerably lowers efficiency of motor generators.

Semiconductor diodes. The types used in charging devices are selenium and silicon diodes. The relevant schematics are given in Fig. 43. A cross-section of a selenium diode is shown in Fig. 43(a). A layer of cadmium and

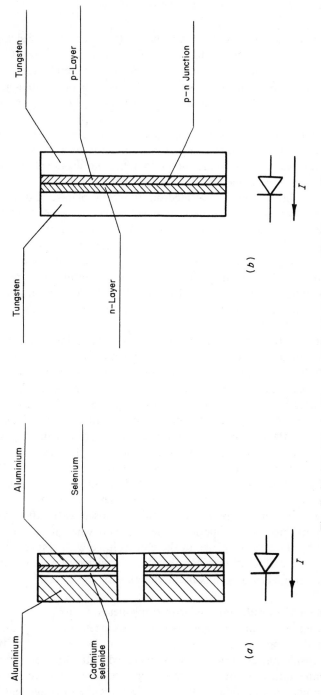

Fig. 43. Semiconductor rectifier schematic: (a) selenium and (b) silicon rectifier.

then a layer of selenium are deposited on an aluminium plate. Selenium is a semiconductor with hole conductivity (p-type semiconductor). Cadmium selenide, which has electron conductivity (n-type semiconductor) is formed at the cadmium–selenium interface. This forms a rectifying p–n junction: current can flow from selenium to cadmium but not in the opposite direction.

Silicon rectifiers (Fig. 43 (b)) are made of silicon wafers 0·4–0·5 mm thick, with one side turned into a p-type semiconductor by diffusion of boron dopant, while the other is made n-type by diffusion of phosphorus dopant. The p–n junction is formed in the middle of the wafer. The wafer is then protected on both sides by welded-on tungsten plates, and the whole sandwich is placed in a sealed housing which serves as one of the terminals.

Both selenium and silicon diodes can function at temperatures from $-50°C$ to $+50°C$. Selenium diodes operate at rated current densities of the order of $0·05 A/cm^2$. Heat released in a diode is usually dissipated without special cooling. Silicon rectifiers are much more powerful: they operate with current densities of the order of $1 A/cm^2$. It is typical to enclose silicon diodes into finned radiators or equip them with water cooling systems, in order to intensify heat transfer.

Components that are increasingly popular nowadays are controlled diodes, the so-called thyristors. A thyristor comprises four semiconductor layers with p- and n-conductivity arranged in the order p–n–p–n. Outer layers are usual diode plates, and the middle p-conducting layer is the control electrode (gate). The control voltage is applied between this electrode and the cathode (that is the outer n-conducting layer). When control voltage is zero, the thyristor is "blocked", that is current cannot pass either in direct or reverse direction (unless the voltage applied exceeds a certain threshold). As the gate voltage increases, the threshold diminishes. It means that the thyristor can be kept in "on" or "off" position by varying the gate voltage. Usually the gate signal is given as pulses of the same frequency as that of the alternating current to be rectified. Consequently, the thyristor is "on" only during one half-period. The duration when it is "on" can be controlled by the relative phase of the gate voltage with respect to the AC signal; the output DC voltage can therefore be varied in a very wide range (practically from zero to a maximum value).

Charging devices with semiconductor components. A diode in an AC circuit is conducting in one direction only, so that the current sine wave has its part below the abscissa axis cut off. A plot of current as a function of time is a sequence of sine pulses of one polarity, repeated at the frequency of the AC signal. Semiconductor gates of charging devices are used as components of different circuits minimizing fluctuations of the rectified voltage and current.

The principles on which these circuits are based are illustrated in Figs. 44 and 45.

Half-wave rectification circuits are used only in domestic systems, to charge small-capacity cells. In full-wave circuits, diodes are in the secondary winding of a transformer having a tap from the neutral point. The battery to be charged is connected to this tap, and is charged for one half-wave through one of the gates, and through another for the second half-wave. The battery is therefore supplied with rectified current pulsing at twice the frequency of the input voltage. A bridge circuit with two diodes for each half-wave functions in a similar manner.

Three-phase current rectification circuits make it possible to obtain current with higher frequency but lower amplitude of pulsation than single-phase rectifiers. In a circuit with zero tap, each phase wire of the secondary winding of the transformer contains a diode which opens only during one half-wave of the relevant phase. The voltage at the battery input is therefore oscillating at tripled frequency. The three-phase bridge circuit uses six diodes, with one diode of each phase open during one half-wave, and the other during the following half-wave. As a result, the frequency at the battery input is further doubled and the pulsation amplitude further diminished.

Further smoothing of pulsations in the rectified current is achieved by connecting an inductance coil or a more complicated filter in series with the battery. Amplitude of voltage pulsation in good charging devices available on the market is around 0.1% of the output DC voltage.

Devices for charging with asymmetric current. The simplest system for generating asymmetric current is a resistor-shunted diode (Fig. 46(a)). This circuit is connected in series with the battery to the AC source. The AC and DC components in the charge circuit are independently controlled by two ammeters. This device is used only with small batteries; it effeciency is low because of energy lost irretrievably in the shunt. Practical chargers for charging with asymmetric current are more complicated and have higher efficiency. Fig. 46(b) shows a schematic of one such device transforming AC current to asymmetric current. The circuit uses a transformer with two secondary windings. One of the windings contains a diode D_1 comprising a half-wave rectification circuit. When this circuit is conducting, transistor T is blocked. The diode blocks the current during the second half-wave while the current in the second winding and the transistor is in the opposite direction and restricted by resistor R_2 (resistor R_1 controls the transistor base current).

Figure 47 shows a schematic of a device for asymmetric-current charging using a DC voltage source. The circuit is an inverter comprising a trans-

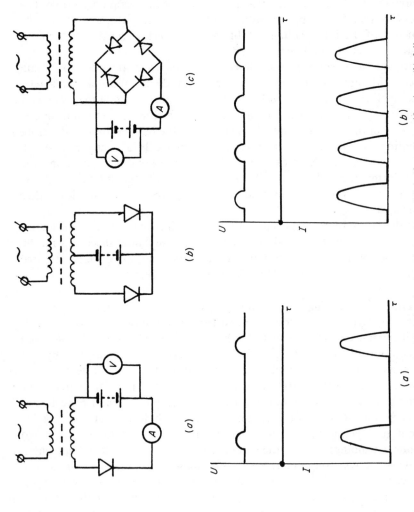

Fig. 44. Single-phase rectification circuit and plots of voltage and currrent: (a) half-wave rectification; (b) full wave rectification; (c) bridge rectification.

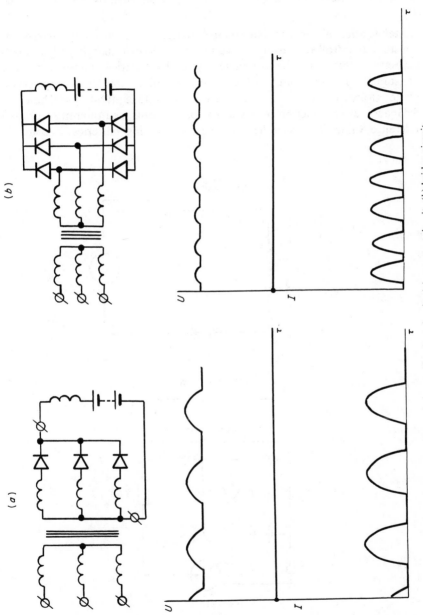

Fig. 45. Three-phase rectification circuit: (a) zero tap circuit; (b) bridge circuit.

former with two transistors T_1 and T_2 and controlling two transistor gates T_3 and T_4. The direct and reverse currents are adjusted by resistors R_1 and R_2.

Stabilization of current and voltage. Batteries are normally charged at constant (controlled) current or voltage. Occasional charging of a small-capacity battery may be conducted with the electrical parameters kept constant by manual control. Regular charges (like those at battery service stations) are run with automatic current or voltage stabilizers. These are feedback devices reacting to deviations of the controlled quantity from the reference value. The reference voltage for charging batteries at constant

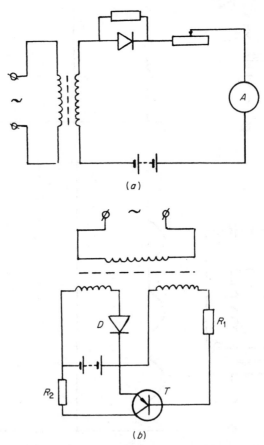

Fig. 46. AC-to-asymmetric current conversion circuits: (*a*) elementary circuit; (*b*) optimized circuit.

voltage can be given, for example, by a stabilizer with adjustable output voltage. A comparator gives the difference between the reference voltage and the signal coming from the battery voltage sensor. The mismatch signal is first amplified and then input into the power converter. The output of the converter is the stabilized voltage. Voltage dividers or magnetic amplifiers are normally used as sensors in such circuits.

7.5. Transient processes

When a chemical power source is connected to a constant load, its voltage in the first moments is somewhat higher than the steady-state value that is established later. Similarly, voltage immediately after disconnecting the load is somewhat lower than the steady-state o.c.v. Behaviour is similar if load is changed. The characteristic time required for the voltage to change from the initial value to the new steady-state level is termed the transient, or relaxation time (typically up to 1 s). If a cell's internal resistance is measured with DC and then AC current, the results are different, with the AC value

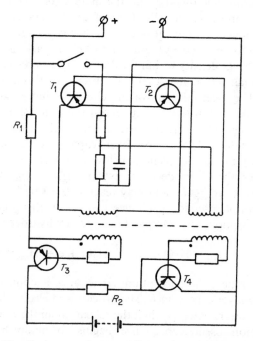

Fig. 47. Asymmetric current charger for storage batteries.

being a function of frequency. This signifies that electrochemical cells cannot be represented by an equivalent circuit consisting of a constant e.m.f. source and an ohmic resistor in series with it. The circuit must include reactive elements as well. The type of the equivalent circuit is an important factor; knowledge of it is necessary for evaluation of the significance of voltage changes under pulsed load, as well as for correct utilization of electrochemical cells in electronic circuits in which high-frequency components of current or voltage can occur.

In the process of charge or discharge of some types of cells voltage dips or hikes may be observed which are characterized by longer relaxation times than ordinary transient processes (times may reach into tens and even hundreds of seconds). These slow effects are related to specific features of electrochemical systems of each cell type. They characterize each individual system and therefore are discussed in detail in Part Two of the book.

(a) The nature of cell reactance[4]

The reactive component of cell impedance is in most cases capacitive in nature; the reason for this lies in the nature of electrochemical and physicochemical processes in cells. One of the factors giving rise to the capacitive impedance component is the familiar capacitance of the double layer at the electrode–electrolyte interface. This capacitance is caused by changes in the charge density within the double layer in response to changed electrode potential, and comes to 10–$40\ \mu F/cm^2$ (with respect to the true surface area). The electrode equivalent circuit includes the double-layer capacitance in parallel with the active resistance R_r of the reaction. R_r reflects the passage of Faraday current across the electrode–electrolyte interface (Fig. 48(a)). The double-layer capacitance is independent of frequency (at any rate for frequencies below 10 MHz).

In some cases the electrode–electrolyte interface capacitance is much greater than the above figure, and may reach 400–$1000\ \mu F/cm^2$. This is the so-called adsorption pseudocapacitance caused by the processes of electrochemical adsorption, for example, of oxygen or hydrogen:

$$2OH^- \rightarrow O_{ads} + H_2O + 2e \qquad H_2O + e \rightarrow H_{ads} + OH^-$$

In equivalent circuits, the adsorption capacitance is connected in parallel with the double-layer capacitance. The processes of electrochemical adsorption are much slower, however, than the relaxation in the double layer. As a result, adsorption pseudocapacitance begins to diminish at quite low frequencies (hundreds or tens of hertz). This retardation of adsorption

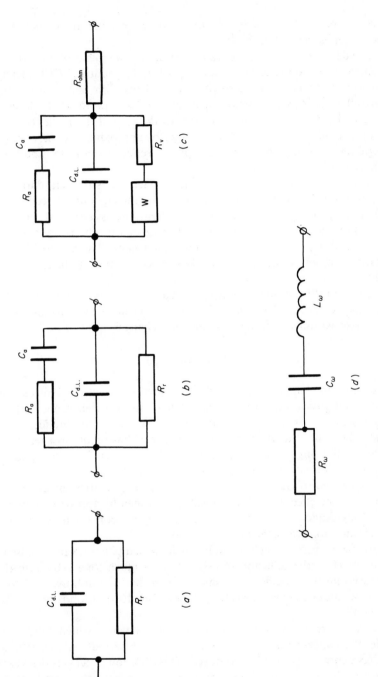

Fig. 48. Equivalent circuit for the electrode processes. $C_{d.l.}$, double layer capacity; R_r, reaction resistance; C_a, adsorption pseudocapacity; R_{ohm}, ohmic resistance; R_a, adsorption resistance; C_a, adsorption pseudocapacity; R_{ohm}, ohmic resistance; W, Warburg impedance.

processes is reflected in equivalent circuits by connecting an ohmic resistance R_a in series with C_a (Fig. 48(b)).

To an order of magnitude, the transient time τ in a system of a resistor R and a capacitor C connected in series is equal to the product RC. Typically, R_a is about 20 ohm cm^2 for adsorption of hydrogen on platinum catalyst (in fuel cells). Adsorption pseudocapacitance is typically 5×10^{-4} F/cm^2, which gives 10^{-2} s for the transient time. In the case of the nickel catalyst and hydrogen adsorption, resistance is larger by approximately three orders of magnitude at nearly the same adsorption pseudocapacitance. The transient time in this case is about 10 s.

The third factor responsible for capacitive reactance is the retardation of diffusion. Electrolyte concentration is the same throughout the cell when current is not being delivered. Once DC current is switched on, the concentration in the vicinity of the electrode starts changing and the concentration distribution becomes non-equilibrium. The process of redistribution continues until a new steady-state concentration distribution is established.

The concentration equilibration time, that is the transient time of diffusion, τ_d, is determined by the diffusion coefficient D and the thickness δ of the layer within which the concentration changes:

$$\tau_d \approx \delta^2/D$$

Concentration changes in cells with free liquid electrolyte take place over a fairly thin diffusion layer of the electrolyte close to the electrode surface. The thickness of this layer is a function of flow rate of the liquid, and is normally about 10^{-3} to 10^{-2} cm. Coefficients of diffusion of most reactants in aqueous solutions are approximately equal to 10^{-5} cm^2/s. This gives for the diffusion transient time a figure of the order of 1 s.

If a matrix electrolyte is employed, δ is equal to the matrix thickness (see Section 4.2). The effective diffusion coefficient is then by approximately one order of magnitude smaller than in free electrolytes. For a matrix 10^{-2} cm thick, the diffusion transient time is around 100 s.

If an electrode process includes diffusion in a solid phase (as in the case of manganese cells), the diffusion transient time is much longer because the solid-phase diffusion coefficient is smaller than that in solutions by 3 to 6 orders of magnitude. This means that steady-state distribution is practically never reached.

When AC current passes through an electrode system in which diffusion is possible, the concentration profile also assumes a periodic shape: the so-called "concentration wave" is produced. This diffusion process is described mathematically by the Warburg impedance which consists of frequency-dependent resistance R_W and capacitance C_W, connected in series:

$$R_W = W/\sqrt{\omega}, \qquad C_W = 1/W\sqrt{\omega}$$

where $W \equiv RT/n^2F^2c\sqrt{2D}$ is the Warburg constant, and R is the universal gas constant. In equivalent circuits the Warburg impedance is connected in series with the active resistance R_r (see Fig. 48(c)).

The total ohmic resistance of the electrolyte, current collectors and so on, R_{ohm}, is connected in equivalent circuits in series with the net complex resistance which reflects the processes directly on, and in the vicinity of, the electrode surface (Fig. 48(c)).

The inductive component of the cell impedance is caused by the current redistribution in massive metal parts. This component is mostly appreciable in large-size cells with high discharge capacity.

Summarizing, the final equivalent circuit of a cell is a complex combination of active resistances, capacitances and inductances. This is explained by the diversity of processes taking place in cells, as well as by the distributed character of these processes. Owing to these difficulties, equivalent circuits of cells cannot be found theoretically and have to be determined by direct measurements. Sometimes an equivalent circuit of a cell is given as a combination, in series, of one ohmic resistance R_ω, one capacitance C_ω, and one inductance L_ω (Fig. 48(d)). In this case all details in the behaviour of a real electrode are given by dependence of R_ω, C_ω and L_ω on frequency ω.

The dependence of the total impedance on frequency, naturally, varies with relative contributions of the capacitive and inductive components.[5,6] If inductance is small, the cell equivalent circuit practically contains only active and capacitive elements and the total impedance is a decreasing function of frequency. This is the case of manganese–zinc and mercury–zinc cells carrying no massive conductors. If the capacitive component is very small, and inductance appreciable, the total impedance increases with increasing frequency. This behaviour is typical for high-capacity lead batteries. Finally, if the capacitive and inductive components are comparable, the total impedance as a function of frequency has a minimum at resonance conditions, that is for equal capacitive and inductive components. This situation is observed in lead and alkaline batteries of low and medium capacity.

An analysis of the frequency dependence of cell impedance makes it possible not only to determine the general type of the impedance but sometimes to extract information on the kinetics and mechanism of the processes in the cell.

The resistance and reactance of cells are best measured by AC bridge instruments. A phase-sensitive voltmeter, giving the magnitude of the impedance $|Z|$ and the phase shift φ between the cell voltage and current in the circuit, is a simpler but somewhat less accurate method.

Useful information about relaxation processes in cells can be obtained by applying to a cell small rectangular current pulses and observing voltage as a function of time on an oscilloscope screen. This method directly yields the transient time.

(b) Inherent noise of chemical power sources[7]

In principle, electrochemical cells generate direct current. The voltage contains, however, an oscillating (fluctuating) component called noise. Normally the noise amplitude is quite low—from a fraction of one millivolt to several millivolts; in some cases, however, even these levels are excessively high, as in some battery-powered electronic circuits. Resulting from statistical fluctuations, noise in cells and in other complicated electrochemical systems is characterized by a very broad frequency spectrum; this distinguishes noise from steady-state self-excited oscillations, and also constitutes an obstacle to its filtration. Electrochemical processes which cause noise in cells are metal crystallization, formation of films of different phases in anodic processes, and other factors.

7.6. Reliability of cells and batteries

Reliability of cells and batteries is defined as their capability for maintaining characteristics as specified by the manufacturer for a guaranteed time and in specified conditions.

Manufacture conditions, and among them variations of characteristics of the initial raw materials, result in rather widely spread cell parameters. Normally technical documentation takes this into account, by guaranteeing parameters which are lower than actual mean values. For this reason, reliability margins of cells are often quite wide.

High reliability means low probability of failures in a batch of cells after the initial screening in the process of manufacturing. Two types of failures are distinguished: sudden failures, and drift failures. Sudden failures result from visible or hidden manufacturing defects overlooked by quality inspection, such as microcracks in separators, poor sealing in the joints, and so on.

Such defects are often revealed at the initial stages of operation, and cause rapid premature failure of the cell.

By definition, drift failure occurs when a parameter in a still operable cell drifts out of the acceptable range. Usually drift failures are caused by low material quality or by minor deviations from the manufacturing standards, and as a rule are detected by the end of the guaranteed service life. Drift failures are typical for storage batteries and shorten the cycle life.

Reliability is defined as the product of probabilities of the absence of failures, for all possible failure types. It is therefore a statistical quantity, so that more or less "reliable" data on reliability of equipment can be obtained only by way of analysing the results of a sufficiently large number of tests or statistical data on actual operation. Detailed reliability data are available only for mass-produced cells.

A number of factors make it difficult to organize production of high-reliability cells. There is strong influence from minute amounts of impurities on the active mass properties, large numbers of stages in manufacturing of electrodes and cells, utilization of very different raw materials (metals, plastic and rubber parts, sealants, and so on). An increase in reliability can be achieved, first of all, by very thorough inspection of all raw materials, by keeping all production equipment in sound condition, and by adhering to technological standards during production and product inspection.

Cell reliability is greatly reduced when a large number of cells are assembled into a battery. In the case of in-series connection, disconnection of one of the cells means that the whole circuit is broken, and with parallel connection, shorting of one of the cells shunts and discharges all the cells in the battery. An additional factor not associated with reliability of individual cells appears in high-voltage batteries: the possibility of leakage and self-discharge through insufficient intercell insulation, via liquid junctions and so on. This factor may constitute a very serious problem in small-size batteries. It is often preferable, therefore, to use low-voltage batteries coupled to solid-state or other types of voltage converters.

The reliability of reserve batteries delivered in non-activated state deserves special consideration. A number of essential parameters of these batteries (including the o.c.v. and on-load voltage) cannot be checked by the manufacturer. Quality can therefore be ensured only by monitoring successive production processes. The requirements for reliability are quite high because reserve batteries are mostly used in highly critical applications.

The reliability of complex power sources, such as electrochemical generators (fuel cells), reserve batteries, cells with electrolyte circulation or with external heating, also depends on the reliability of auxiliary and control mechanisms.

REFERENCES

1. V. V. Romanov and Yu. M. Hashev, "Chemical Power Sources." Soviet Radio Publishing House, Moscow (1978) (in Russian).
2. R. Kinzelbach, "Stahlakkumulatoren," S. 91–107. VDI, Düsseldorf (1968).
3. R. A. Harvey, "Battery Chargers and Charging." Iliffe, London (1953).
4. P. Delahay, The study of fast electrode processes by relaxation methods, *in* "Advances in Electrochemistry and Electrochemical Engineering," Vol. 1, pp. 233–318. Interscience, New York and London (1961).
5. R. J. Brodd and H. J. de Wane, "Impedance of Commercial Leclanché Dry Cells and Batteries." Government Printing Office, Washington (1963).
6. R. J. Brodd and H. J. de Wane, Impedance of sealed nickel-cadmium dry cells, *Electrochem. Technol.* 3, No. 1–2, 12 (1965).
7. V. A. Tyagay, Electrochemical Noise, *in* "Itogi Nauki i Tekhniki, Elektrokhimiya," Vol. 11, pp. 109–175. VINITI Publishing House, Moscow (1976) (in Russian).

Chapter Eight

Applications of Cells

8.1. Present-day applications

(a) Automotive equipment (starter and auxiliary batteries)

Storage batteries are employed in all vehicles with internal combustion engines (cars, motorcycles, airplanes, diesel locomotives, etc.) to start engines and for auxiliary needs. Three functions are assigned to the battery in a car with a carburettor engine: energizing the starter, energizing the ignition system while the generator is not yet operative, and supplying energy to auxiliary circuits (lighting, radio, etc.) when the engine is off. A convenient abbreviation is SLI-battery: starting, lighting, ignition. In addition, aircraft storage batteries stand by for emergencies, that is they supply current to all essential electric systems on board aircraft when the generator breaks down. Such emergency operation may last till the emergency landing, that is they may exceed half an hour.

Certain specific features are typical for starter batteries. First of all, starter currents are quite high: j_d reaches 3 to 5 in car batteries and may be even higher in aircraft batteries. A battery must guarantee starting of the engine in winter, i.e. at low temperatures. Starter discharge is usually of short duration, not more than 20 s in automobiles and 60 s in airplanes. Consequently, the starter battery discharge is normally not deep, with not more than 3–10% of capacity drained by a single starting procedure, immediately followed by recharging. At the same time, a battery must have sufficient Wh capacity (for lighting, emergency operations, etc.). A battery designed for repeated starting must be sufficiently quickly recharged. This is carried out by a generator whose output voltage is kept constant by special control facilities, with an accuracy of $\pm 3 \cdot 5$–8%. Starter batteries are required to have long service life and low self-discharge. Starter batteries may be idle for long periods of time, but must be in constant readiness for

165

operation. Finally, vehicle-born batteries have to withstand substantial mechanical loads (vibration and shock loads).

Starter batteries belong to the category of cells which require servicing. The personnel involved has, as a rule, no special training (such as car drivers). For this reason there are special demands for simple maintenance procedures.

Battery types used nowadays for starter batteries are mostly lead batteries, sometimes nickel–cadmium, and silver–zinc ones (the last of these only in aircraft applications). The rated voltage of automobile starter batteries is 12 V, with the Ah capacity from 40 to 200 Ah. The rated voltage of aircraft batteries is 24 V. Engines of heavy aircraft not equipped with starter batteries are started by high-capacity landbased batteries mounted on trolleys.

A considerable number of vehicle-borne batteries are not meant for engine starting but power auxiliary systems. Such types are railroad (carriage-born), ship-borne, and some other batteries. These are used for lighting, and for supplying current to air conditioners, transceivers, signalling equipment, and so on, during stops when the generator is off. Batteries used for the above applications are lead and nickel–iron types. They operate in extreme conditions: wide temperature range (practically from $-50°C$ to $+50°C$), high humidity, elevated contamination levels (railway carriage batteries), substantial mechanical loads. In contrast to starter batteries, ship and carriage-borne batteries undergo rather deep discharges (during long stops) and long overcharging (during long runs between stops). The typical Ah capacity of carriage-borne batteries is from 70 to 500 Ah.

(b) Traction batteries

Performance characteristics of internal combustion engines are higher than those of a combination of a battery and an electric motor. Consequently, autonomously powered vehicles are mostly powered by internal combustion engines. There are fields, however, where application of these engines is inconvenient or unacceptable. Such are submarines and submersible craft in which internal combustion engines cannot be used (in the submerged position) owing to the lack of oxygen and difficulties involved in the disposal of exhaust gases; vehicles in mines where internal combustion engines would constitute an explosion hazard and pollute the atmosphere; transportation systems in factories and in other enclosed spaces where a vehicle must have high manoeuverability, no toxic exhaust, and so on. In all these cases traction electric motors and special traction batteries are employed.

The operation mode of traction batteries is very different from that of starter batteries. As a rule, traction batteries are discharged with a wide range of loads (j_d from 0·05 to 1·0) and to a considerable depth. Moreover, they must provide high power in the first several seconds of motion. Traction batteries are usually recharged in special shelters, and charge conditions are optimized. In contrast to starter batteries, traction units are supervised by skilled personnel. As with all vehicle-borne batteries, traction systems operate in conditions of high mechanical load. As with starter batteries, service and cycle life must be long, and repair servicing is needed.

Lead and nickel–iron storage batteries are used nowadays as traction batteries. Their capacity varies from 40 Ah to 1200 Ah, and reaches several thousand ampere-hours in the case of submarines.

(c) Stationary batteries

Conditions of operation of stationary batteries are much less severe than those found in transportation systems. They are invariably in a stable position so that careful sealing becomes unnecessary. The requirements on battery and electrode design are greatly simplified by the absence of vibrations and other mechanical loads. Stationary units are often located in special buildings and operate at optimal temperatures. In some applications, however, such as automatic weather stations, stationary batteries are under severe climatic conditions and operate at temperatures from $-50°C$ to $+50°C$. The installed capacity often reaches into thousands of ampere-hours. Usually neither dimensions nor mass are critical parameters: in other words, high specific characteristics are not required. Many stationary units are designed for long-term operation, and thus must have long service life, high reliability, and low level of self-discharge. As a rule, servicing of such batteries is carried out by skilled personnel.

Stationary batteries installed for long-term discharge by small currents often consist of primary cells, such as alkaline copper–zinc cells. Batteries of this type are employed in railway systems to energize signalling equipment, switch networks, and so on. Sometimes they are placed in wells which results in better thermal conditions in winter. Long-term operation with no servicing is typical for automatic weather stations, relay stations, and the like. Stationary storage batteries are still in use as high-stability DC sources in telephone exchanges, since high-quality rectification of AC voltage remains a problem without a satisfactory solution. High-capacity stationary units (usually based on buffer lead batteries) are employed in power stations to supply power to auxiliary equipment. And finally, emergency (standby) storage batteries, providing power for lighting and uninterrupted functioning of important systems when the mains supply fails, are now used more and more widely. Such batteries are kept in constant readiness; they are designed to operate for several hours until the emergency is over or a diesel generator is assigned to the job; in the last case the stationary battery is used to start the diesel unit. Emergency batteries are installed in hospitals (to ensure uninterrupted functioning of surgery rooms), in entertainment centres and other public places, in shelters of various types, in important industrial plants, in computer centres, and so on. In normal conditions emergency batteries are hardly ever discharged; self-discharge losses are continuously or intermittently compensated for by recharging.

(d) Portable and domestic systems

The widest field of cell use, both in the number of cells, and in their diversity, is that of portable units and domestic appliances. Typically, this involves cells and batteries of low and medium capacity, that is from 0·01 to 100 Ah.

Cell-powered electrical units include various radio and TV devices (transistorized radio sets, tape recorders, TV sets, radio transmitters) for civil and military use, portable flashlights for underground and other use, miniaturized devices (electronic watches, hearing aids, pocket calculators), toys, some household appliances (electric shavers), measuring instruments (potentiometers, testers, radiation monitoring instruments, etc.) and various medical devices (cardiac pacemakers, etc.).

The diversity in cell-powered devices entails a great variety of operational conditions. Thus, electric torch batteries operate in a wide temperature range, while those of cardiac pacemakers are used in strictly isothermal conditions. Electronic watch batteries discharge uninterruptedly by low current pulses, while those in transistor radios operate with randomly variable load. Batteries for military and medical applications must be highly reliable, which would be redundant in electrified toys. Cost is often an important characteristic of a cell; performance characteristics and shelf-life may have to be lowered somewhat for the sake of cost reduction. In other cases one of the cell characteristics, such as stability of on-load voltage of standard cells, specific energy of miniaturized cells, and so on, may be the most important of all.

With very few exceptions, cells for portable and domestic appliances must be spill-proof and must be transportable any way up.

Cell types now most widespread in this field are "dry" manganese–zinc primary cells and sealed nickel–cadmium storage batteries, as well as mercury–zinc, manganese–magnesium and other types of cells.

(e) Special applications

Specialized application of chemical power sources covers military and space technology and some scientific instruments. Examples of such cell-powered devices are satellites, space probes, rockets, bathyscaphes and other submersible craft, electrically driven torpedoes, meteorological balloons, and so forth. Specialized cells may be subsumed under the two subdivisions: power sources for heavy short-term loads (typically for a single discharge), and those for long-term low-drain discharge.

Power systems used in space ships typically combine semiconductor solar panels and buffer storage batteries.[1,2] Storage batteries installed on earth satellites energize the systems while the satellite is in the Earth's shadow, and also when the loads exceed the power of solar panels. For the rest of the time of flight solar panels recharge the batteries. Typically, the operation schedule of such batteries may be 20–40 min of discharge, and 50–70 min of recharging, which means about sixteen charge–discharge cycles (partial, as a rule) in 24 hours.

As with every item of equipment for critical use, cells must satisfy exacting requirements concerning specific parameters and reliability. They must be designed for prolonged warehouse storage. Throughout their shelf-life, these cells must be permanently ready for discharge; the reserve battery represents, therefore, a logical solution to a number of problems. As a rule, these power sources are expected to operate in a wide range of climatic conditions, from arctic to tropical, and at high mechanical loads. Cost considerations normally play only a secondary role when such power sources are selected.

8.2. Possible future fields of application

(a) Large-scale power production (electrochemical power plants)

Ever since electric generators based on the principle of electromagnetic induction took over the field of power production, chemical power sources have been used only as autonomous units. They were discarded as potential sources of energy in power plants owing to the high cost of energy in comparison to that of energy produced by heat engines. At the present moment, however, a potential role for electrochemical cells in large-scale power production is again being widely discussed in the literature; this stems both from the improvement in the cell parameters and from the gradual decrease in natural fuel resources.

Three main approaches are being discussed. The first is the large-scale direct conversion of the chemical energy of natural fuel (or products of fuel processing) into electrical energy by means of fuel cells. The idea of producing electrical energy by fuel cells with efficiency much higher than that achievable in thermal power plants (see Section 1.5) is very attractive. Considerable attention was paid to this problem in a number of countries for quite a few years. Work has been started in recent years to design large-scale pilot installations with powers of several megawatts; they can be considered as first prototypes of fuel-cell power plants of the future (for details see Section 17.11).

The second approach to utilization of cells in large-scale power production consists in levelling the load of power plants.[3] Load fluctuations on ordinary power plants are determined by daily, weekly and seasonal fluctuations of energy consumption. Generation of energy by units of novel types (solar, wind or tide generators, and other similar setups) is governed by natural phenomena and thus cannot be uniform. An opportunity to level off these fluctuations by temporary accumulation of some of the excess energy and subsequent utilization of it would constitute an extremely important economic factor. This can be realized, for example, by means of giant storage batteries. Neither scarce nor expensive materials will be acceptable for these batteries because of their enormous size. This makes attractive such novel electrochemical systems as sulphur–sodium and chlorine–zinc ones, to name only two. Construction of storage batteries with a capacity of several megawatt-hours is planned in the USA for 1981–82.

Projects are also discussed for constructing gigantic systems accumulating oxygen and hydrogen, and consisting of electrolysers, hydrogen tanks and fuel cells.[4] It is possible to use for this purpose low-temperature oxygen (air)–hydrogen fuel cells with an alkaline electrolyte which by now have reached a high degree of perfection.

And finally, the third approach to cell utilization in large-scale power production involves realization of thermoelectrochemical cycles. A thermoelectrochemical cycle enables us to convert thermal energy into electrical energy without resorting to heat engines. A cycle based on copper chlorides is an illustration of a very simple thermoelectrochemical cycle. At high temperatures (above 500°C) copper (II) chloride dissociates into chlorine and copper (I) chloride, absorbing heat:

$$CuCl_2 \underset{\text{in FC}}{\overset{500°C}{\rightleftarrows}} CuCl + \tfrac{1}{2}Cl_2$$

The reaction is spontaneously reversed at lower temperatures (at room temperature, for instance), and thus can be used to produce electrical energy in a fuel cell. Thermoelectrochemical cycles of a different type use heat to produce not electrical energy but energy-rich fuel, namely hydrogen. The underlying idea is to utilize primarily the heat generated by nuclear reactors.

At the moment all these approaches to cell utilization for large-scale power production do not pass the levels of research, preliminary design and technical and economic evaluation. Economic factors are decisive in large-scale power generation, the most important being the cost of power. This quantity is determined by the cost of the primary energy (the cost of fuel, primary electrical energy and thermal energy, respectively, in the three approaches mentioned), conversion efficiency, capital investment in construction of the convertors and the projected life of the convertors affecting capital depreciation costs. The economic parameters of the projects involved are not yet clear, especially in comparison with such anticipated competing schemes for transformation or accumulation of energy as the MHD-generator, hydroaccumulating stations, and so on. Nonetheless, it seems quite probable that some of the projects introducing chemical power sources into large-scale power production will be realized as early as the eighties of this century.

The soaring price of oil and the menace of the approaching depletion of resources of organic fuels have led to intensive discussion in recent years of the prospects of the so-called hydrogen-based economy.[5-7] This concept involves the use of hydrogen as a universal fuel and energy carrier. The growth of nuclear power plants which are sources of enormous amounts of hard-to-utilize heat, constitute prerequisites for the large-scale production of cheap hydrogen. The methods likely to be employed in the hydrogen generators are the high-temperature electrolysis of steam, various thermoelectrochemical cycles and others. It may prove competitive, in the future, to produce hydrogen by utilizing solar energy (for instance, photochemically, using biocatalysts). Hydrogen will be used both as an energy source (in fuel cells, heat engines, flameless catalytic burners, etc.) and as a reducer in the chemical industry and metallurgy. Moreover, hydrogen is very convenient for transportation of energy over large distances, by pumping through pipelines. A growth of the hydrogen economy will undoubtedly be conducive to a large-scale use of the oxygen(air)–hydrogen fuel cells, forming, when necessary, large-scale stationary power plants.

(b) The battery-driven car[8-13]

A car driven by an electric battery, or electric car, was the main type of vehicle at the beginning of this century, and it ran on lead, nickel–cadmium or nickel–iron storage batteries. The total output in 1912 was six thousand battery-powered cars and four thousand trucks. Vehicles with carburettor engines quickly replaced electric cars immediately after the electric starter was invented (1912), since petrol provides much more power per unit weight than do battery reactants, and so provides a much longer run after a refuelling. Furthermore, petrol engines have much higher specific power, and make possible higher speeds. This did not mean, however, the end of the electric car. Various types of vehicles powered by lead batteries are built and used in some countries even now, although on a very limited scale. In Britain, for example, there are approximately thirty thousand battery-driven lorries employed in cities for delivery of milk and school lunches, and similar tasks.

The problem of the electric car became increasingly interesting over the last decade, first because of the energy crisis which demands that oil use be economized and its efficiency increased (the average efficiency of a car engine in a city is only about 10%), and secondly owing to intensive pollution of the air within the cities by engine exhaust gases.

At present the efficiency of electric vehicles is limited by the insufficient specific energy of batteries. Electric-car-oriented research and development is being conducted in a number of countries, with a view to improving the available traction batteries and to design new cell types acceptable to the automotive industry.

The existing electric vehicles are mostly equipped with lead storage batteries having specific energy of 20–28 Wh/kg for $j \sim 0.2$ and 16–20 Wh/kg for $j = 0.7$ (i.e. for specific powers of 4–6 and 14–20 W/kg, respectively). These batteries provide a range for a car of 60–85 km, and the maximum speed of 40–80 km/h. This is acceptable for certain types of vehicles, although definitely insufficient for a versatile electric car. Calculations demonstrate that in order to have a multipurpose car for city use, the specific energy of storage batteries must be from 70 to 140 Wh/kg (depending on the type and function of the vehicle), and the specific power from 40 to 70 W/kg. Even higher specific energy is required for the out-of-town use of a car, so that the foremost problem now is that of turning out an electric vehicle for the cities, where atmospheric pollution is significant.

It can be anticipated that the specific energy of lead traction batteries will ultimately reach 35–40 Wh/kg which means a 50% increase of range of an electric car. There is a point, though, that should not be ignored. Namely, the mass of lead in a car's traction battery is greater by a factor of 20 to 30 than that in a starter battery, so that a transition to lead-battery traction in a substantial fraction of the world's cars would automatically require that the global output of lead be drastically increased. Improved models of nickel–iron batteries which are being developed now in a number of countries and which are expected to provide 50–60 Wh/kg at $j_d = 0.3$–0.5, may prove better than lead batteries in some types of electric cars.

Among the novel types of storage batteries which seem closest to the stage of technical realization are the nickel–zinc and nickel–hydrogen ones, and, to a smaller extent, the air–iron batteries. These systems are expected to have specific energy of about 60 Wh/kg, which is greater than that of lead batteries by a factor of 1·5–2·0. The systems anticipated in the more remote future are the high-temperature sulphur–sodium and iron sulphide–lithium batteries with specific energy of 100–150 Wh/kg, which is quite close to the level required for a versatile electric car. Much hope was generated by the advent of the air–zinc batteries with specific energy of about 100 Wh/kg, but substantial difficulties, not yet overcome, were encountered in the way to developing acceptable storage batteries of this type.

The prospects for practical use of various types of batteries in electric vehicles are determined not only by their specific electrical parameters but also by their economic characteristics (which depend, among other factors, on the battery cycle life) and by maintenance procedures. For example, high-temperature batteries requiring preheating can only be used in electric vehicles which run on a fixed schedule (delivery vans, buses, office cars, etc.).

Storage batteries intended for use in electric vehicles must allow booster charging. Therefore, high-power charging units have to be designed for rapid recharging of traction batteries (tens of kilowatts per car). In principle, rapid recharge may be substituted by the replacement of discharged batteries by recharged ones (at special exchange stations). This is not always convenient, however, especially if a traction battery has a mass of 0·5 to 1·5 ton.

Another important aspect of battery operation in an electric car is the possibility of constant monitoring of the degree of charging of the cells.

The prospects of installing fuel cells or the so-called hybrid systems on electric vehicles are also a subject of discussion.[14] A hybrid system includes, in addition to a battery of fuel cells, a storage battery operating in the buffer mode and delivering current when the engine is started or when the load is excessive. Judged on the basis of specific parameters only, alkaline air–hydrogen fuel cells with specific power of 80–100 W/kg could be used to power a car. The basic difficulty lies, however, in the necessity to store hydrogen in the car. The possibility of fast refuelling with hydrogen (compared to the protracted procedure of battery recharging) is an essential advantage of fuel cells. Nevertheless, acceptable methods of storing hydrogen on an electric vehicle are still to be worked out.

A great deal of research and development work has to be carried out, therefore, before acceptable power sources can be installed on mass-produced electric vehicles.

(c) The use of cells in medicine (implanted devices)

Electric pacemakers, that is devices feeding current pulses to the cardiac muscle, are now often implanted into the patient's body.[15] These pacemakers are powered by mercury–zinc or other types of cells with service life of about five years and discharging at a current of several microamperes.

The problem of designing an artificial heart received considerable attention in the nineteen seventies, and large sums have already been spent on its solution. At the present moment the problem is essentially that of designing a power source which would energize the blood pump for several years and which could be implanted into the body. An artificial heart requires a much higher power than a pacemaker, namely several watts. A fuel cell consuming the arterial blood oxygen and a fuel contained in body liquids, such as glucose, appears to be the only suitable cell for the purpose. Only very modest progress was reported in this field. Basic difficulties are singled out: development of suitable semipermeable membranes, development of selective electrodes not poisoned by the compounds in the body liquids, the problem of energy balance of the human body as a whole, and so on.

It is to be expected that considerable new effort will be put into the development of cells for medical purposes.

8.3. Economic problems[16, 17]

It is not easy to evaluate today's total output of different types of cells. Much of the relevant data are not published, being kept secret by firms or military agencies. When published, the data do not always allow comparisons. Sometimes the output is given as the number of cells or batteries having different capacities, and sometimes as the total Ah capacity which is equally awkward because the number of cells in a battery is omitted. A better description of the global output of chemical power sources could be obtained in terms of the energy capacity of the cells given in Wh. A very rough estimate shows that the total Wh capacity of the output of the world's cell industry is approximately 150 million kWh per annum. Two thirds of this capacity is accounted for by lead batteries. The total number of individual units is about ten billion a year, with 85–90% of them being of the manganese–zinc type.

Economic problems in the production and operation of cells involve the cost of electrical energy they deliver, the availability of raw materials, links with other industries, and so forth.

The cost of electrical energy produced by large power plants is 1–3 cents per kWh. The costs of energy delivered by primary cells (those to be thrown out after a single discharge) are much higher. For instance, the cost of electrical energy delivered by mass-produced Leclanché cells is 30–80 dollars per kWh.

In the case of storage cells, the cost of electrical energy is determined by the cost of the cell itself and the cost of the electrical energy the cell reconverts. The first component of the cost is drastically reduced as the number of charge–discharge cycles increases, that is as the total amount of the delivered energy rises. The second component is calculated as the ratio of the cost of the electrical energy supplied during charging, to the energy efficiency of the storage cell, μ_w. In the case of lead storage cells with a cycle life of 800 cycles, the overall cost of energy is 15–20 cents per kWh.

Reduction of the cost of the energy below the level achieved at thermal power plants may be expected in the future when sufficiently effective, long-lived, and inexpensive fuel cells are developed (owing to a higher conversion coefficient).

This shows that at the present stage electrochemical cells are used not to generate cheaper energy but to obtain it in the conditions in which the cheap electrical energy of the mains network cannot be utilized.

Primary cells generate relatively expensive electrical energy and are mostly utilized in devices with low energy consumption (such as transistor radios), so that the energy cost is of secondary importance. Storage cells with their less costly energy are used both in low- and higher-power setups (with total consumption up to hundreds of kilowatt-hours).

The cost of the chosen battery type and its ratio to the total cost of the unit powered by this battery are additional significant factors. As a rule, this ratio is quite small in the case of transistor sets (not more than 2 %). On the other hand, the cost of the power unit in future electric vehicles will be a significant fraction of the total cost.

The structure of the cost is different for different cell types. For example, Leclanché manganese–zinc cells are mass-produced by high-productivity machinery and relatively simple production processes; as a result the cost of such cells is mostly determined by that of raw materials. Likewise, the cost of silver–zinc storage cells is made up mostly of the cost of silver. In the case of nickel–cadmium disk cells, however, the cost is mostly determined by manufacturing expenses.

The cost of raw materials depends, to a great extent, on their processing costs. Manganese–zinc cells used to be manufactured from manganese ore subjected only to simple concentration procedures. Now that the deposits of high-quality manganese ores are being depleted all over the globe, manganese dioxide is often prepared from low-grade ores, with the ensuing complex power-consuming processing greatly increasing the cost of the final product.

The problem of supplying the cell industry with raw materials has recently become rather pressing. A substantial fraction of these materials are non-ferrous metals (lead, manganese, zinc, nickel, cadmium). Estimates indicate that natural deposits of these metals may be depleted within several decades. Two problems therefore become important: recovery of non-ferrous metals from used cells, and development of cell types not utilizing scarce materials (limited by the available natural resources).

Silver from scrapped silver–zinc batteries is nowadays almost completely recovered. Collection and total recovery of lead from used lead batteries is also organized in many countries. The recovery of non-ferrous metals proves economically justified in the case of large-size batteries (those in industrial plants, transportation depots, etc.). Metal recovery from small-size cells is a more complicated and sometimes economically unjustified operation.

The sulphur–sodium storage cell is an example of a promising cell system not using scarce active materials. The global resources of sulphur and alkali metals greatly exceed those of lead, nickel and cadmium. It must be taken into account, however, that costly corrosion-resistant materials (niobium, tungsten, etc.) are still employed in the high-temperature cells, but it is hoped that they will later be replaced by cheaper ones.

Fuel cells provide another example of systems with a bright future. The metal requirements in fuel cells are considerably less pressing than in other cell types, owing to their specific design features and to the use of liquid and gaseous reactants. No significant difficulties with raw materials for fuel cells are in view once active and stable catalysts not requiring precious metals are developed.

REFERENCES

1. P. Bauer, "Batteries for Space Power Systems", NASA, Washington (1968).
2. M. G. Gandel, Energy storage requirements for spacecraft, *J. Power Sources* **3**, No. 3, 277–289 (1978).
3. N. P. Yao and J. R. Birk, Battery energy storage for utility load levelling and electric vehicles. A review of advanced secondary batteries, *in* "Proceedings of the 10th Intersociety Energy Conversion Engineering Conference, Newark, 1975", pp. 1107–1119.
4. J. M. Burger, P. A. Lewis, R. J. Isler, F. J. Salzano and J. M. King, Jr., Energy storage for utilities via hydrogen systems, *in* "Proceedings of 9th Intersociety Energy Conversion Engineering Conference, San Francisco, 1974", pp. 428–434.
5. J. O'M. Bokris, "Energy. The Solar-Hydrogen Alternative". Australia–New Zealand Book Company, Sydney (1975).

6. T. N. Veziroglu (ed.), "Hydrogen Energy", Parts A and B. Plenum, New York (1975).
7. T. N. Veziroglu and W. Seifritz (eds.), "Hydrogen Energy Systems. Proc. 2nd World Hydrogen Energy Conf." Vol. 1–4. Pergamon Press, Oxford (1978) (Advances in Hydrogen Energy I).
8. J. D. Busi and L. R. Turner, Current developments in electric ground propulsion systems, R. and D. worldwide, *J. Electrochem. Soc.* **121**, No. 6, 183C–190C (1974).
9. K. V. Kordesh, The electric automobile, *in* "Batteries, Vol. 2, Lead Acid Batteries and Electric Vehicles" (Kordesh, K. V., ed.), Ch. 2. Dekker, New York (1977).
10. W. Vielstich, Zur Entwicklung elektrochemischer Stromquellen für die Electrotraktion, *Elektrotechn. Z. (Ausg. A)* **Bd. 98**, No. 1, S. 79–82 (1977).
11. K. Salamon and G. Krämer, Batterien für Elektrostrassenfahrzeuge—heute und morgen, *Elektrotechn. Z. (Ausg. A)* **Bd. 98**, No. 1, S. 69–74 (1977).
12. K. V. Kordesh, Power sources for electric vehicles, *in* "Modern Aspects of Electrochemistry" (Bokris, J. O'M. and Conway, B. E., eds.), No. 10, pp. 339–444. Plenum, New York and London (1975).
13. D. A. J. Raud, Battery systems for electric vehicles—a state-of-the-art review, *J. Power Sources* **4**, No. 2, 101–143 (1979).
14. K. V. Kordesh, Hydrogen-air/lead battery hybrid system for vehicle propulsion, *J. Electrochem. Soc.* **118**, No. 5, 812–817 (1971).
15. K. A. Gaspar and K. E. Fester, Cardiac pacemaker power sources, *in* "Proceedings of 10th Intersociety Energy Conversion Engineering Conference, Newark, 1975", pp. 1205–1213.
16. A. I. Harrison and G. R. Lomax, Applications and costs of electrical energy sources, *in* "Power Sources" (Collins, D. H., ed.), pp. 577–591. Pergamon Press, Oxford (1967).
17. K. V. Kordesh, 25 years of fuel cell development, *J. Electrochem. Soc.* **125**, No. 3, 77C–91C (1978).

Part Two

Various Cell Systems

Chapter Nine

Manganese–Zinc Cells with Salt Solution Electrolyte

9.1. General

For over a hundred years primary manganese–zinc cells with salt solution as electrolye (Leclanché cells), and batteries of them, have been used as the main type of primary cells. Every year 7–9 billions of such cells are manufactured in the world. The wide use of manganese–zinc cells is due to their favourable combination of properties—relative cheapness, satisfactory electrical parameters, suitable storage life and convenience of use. Their disadvantage is a sharp drop in voltage with discharge, depending on the load the final voltage is 50–70% of the initial value.

Manganese–zinc cells are manufactured as "dry" cells with leakproof electrolyte. Their capacity ranges from 0·01 Ah to 600 Ah and their mass (for a single cell) from 0·5 g to 7 kg, mostly small-size cells with a capacity up to 5 Ah are produced.

The first manganese–zinc cell, made in 1865 by a French engineer G.-L. Leclanché, was a glass jar containing an aqueous solution of ammonium chloride into which there were immersed a zinc rod (the negative electrode) and a porous earthenware jar packed with a mixture of manganese dioxide and powdered coke and containing a carbon-rod current collector at the centre (the positive electrode). Though this first cell had poorer parameters than the Daniell and Bunsen cells which were popular at that time, the Leclanché cells soon moved into first place. The simplicity and safety of manufacture and use, the wide range of working temperatures and other advantages resulted in a rapid rise in production of these cells. As early as 1868 more than twenty thousand Leclanché cells were being manufactured.

Further development of the cell led to the replacement of the zinc rod with a zinc can serving simultaneously as the anode and the cell container. The earthenware jar containing the active mass of the positive electrode was

181

replaced with a paper or gauze wrap. In the eighties of the nineteenth century thickening the electrolyte to a paste was suggested, and dry Leclanché cells were produced. In the first half of the twentieth century the parameters of the manganese–zinc cells were markedly improved by adding acetylene black to the active mass of the positive electrode. In about 1935 production began of a new type of manganese–zinc chemical power source—the flat cell batteries.

Close analogues of the Leclanché cells are manganese–zinc cells with alkaline electrolyte and manganese–magnesium cells with salt solution electrolyte which will be discussed in Section 12.4 and 13.2, respectively.

9.2. Electrochemical and physicochemical processes

(a) Current-producing reactions

In the manganese–zinc cell the active materials are manganese dioxide and zinc. The electrolyte is an aqueous solution of the chlorides of ammonia and zinc and, sometimes, calcium. Owing to partial hydrolysis of these salts the solution is weakly acidic with pH about 5. Since the buffer capacity of the solution is low the electrode reactions result in changes of the pH value of the solution in the vicinity of the electrodes, increasing it to 8–10 near the cathode (manganese dioxide) and decreasing it to 3·5–4 near the anode.

The complicated mechanism of the electrochemical reduction of MnO_2 is still a subject of numerous studies. It is most probable that the reaction occurs via a solid-state mechanism by transport (diffusion) of electrons and protons from the surface into the MnO_2 grain producing partial reduction of the Mn^{4+} ions of the crystalline lattice to Mn^{3+} ions:

$$MnO_2 + H^+ + e \rightarrow MnOOH \qquad (9.1)$$

At the initial stage of this penetration of hydrogen the crystalline lattice of MnO_2 is somewhat stretched but its structure is not changed and a homogeneous phase of variable composition $yMnOOH \cdot (1-y)MnO_2$ is formed. During discharge y continuously increases and the electrode potential varies accordingly. It is important that owing to slow proton transport the composition of the surface layer of the MnO_2 grain differs from the composition of the deeper layers—near the surface the degree of discharge (that is, y) is higher.

These properties of the positive electrode explain in part the characteristic discharge features of manganese cells. Owing to continuous variation of the potential of the positive electrode the voltage of the cell drops considerably during discharge, particularly with high currents (Fig. 49). Simultaneously, the o.c.v. drops. When the current has been switched off the voltage at first jumps to an intermediate value (elimination of the ohmic drops); after that the o.c.v. increases only slowly to the value corresponding to the given degree of charge. During this rest period there occurs levelling-off of the concentration (the value of y) in the solid phase.

Other causes of voltage drop with discharge are the gradual increase of the internal resistance (both ohmic- and polarization-induced) and development of a pH gradient in the cell as alkalinization of the solution near the cathode shifts its potential to the negative.

When a certain critical value of y has been reached formation of the proper crystal structure of MnOOH (manganite) starts and two solids appear in the system (sometimes the reaction at this stage is referred to as heterogeneous). With further discharge the relative amounts of both phases change but their composition does not; owing to this, voltage drop becomes slower.

At the end of discharge at sufficiently negative electrode potential (that is, low voltage of the cell) manganite may be further reduced:

$$MnOOH + H^+ + e \rightarrow Mn(OH)_2 \qquad (9.2)$$

This process does not lead to formation of a phase with variable composition, it too is heterogeneous and the change of the electrode potential at this stage is small. This part of the discharge curve is not used in practice.

Manganese dioxide exists in various crystal modifications (see below) and each of them has different electrochemical characteristics as regards the initial potential, behaviour of the potential during discharge, etc.

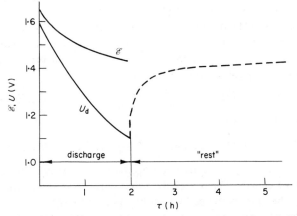

Fig. 49. Discharge curve and recuperation curve for the manganese–zinc cell. Variation of the o.c.v. with discharge is shown.

The character of the discharge curves is also influenced by the pH of the solution near the electrode. In more acidic solutions, with pH not exceeding 4·5, MnO_2 is reduced to the Mn^{2+} ion:

$$MnO_2 + 4H^+ + 2e \rightarrow Mn^{2+} + 2H_2O \qquad (9.3)$$

Since the product of the reaction is soluble the composition of the solid phase is not changed and the electrode potential remains constant during the entire course of the discharge. Unfortunately, corrosion of the zinc electrode in the acidic solutions is considerable.

Anodic oxidation of zinc in salt solutions primarily gives rise to the Zn^{2+} ion. However, in practice, discharge in cells is accompaned by various secondary chemical reactions which result in formation of barely soluble complex compounds containing zinc in the electrolyte, the separator and even in the pores of the positive electrode. As the concentration of the zinc ions near the anode increases, their hydrolysis is enhanced:

$$Zn^{2+} + H_2O \rightarrow Zn(OH)^+ + H^+ \qquad (9.4)$$

so that the pH of the solution decreases. The zinc ions diffuse to the zones with high pH and precipitate there as various oxychlorides $ZnCl_2 \cdot xZn(OH)_2$ (typically, $x = 4$) or hydroxide $Zn(OH)_2$. Owing to alkalinization of the solution, near the positive electrode the ammonium ions partially decompose giving rise to free ammonia (the odour of ammonia sometimes appears with discharge of the cells). This contributes to precipitation of crystals of $Zn(NH_3)_2Cl_2$ which partially shield the active materials of both electrodes, increase the internal resistance and the pH gradient and produce deterioration of the cell parameters. The zinc ions can also react with the product of discharge of the positive electrode giving rise to a new solid phase—the hetaerolite $ZnO \cdot Mn_2O_3$.

Thus, the electrode processes in manganese–zinc cells are complicated and their thermodynamic analysis is difficult. In a rough approximation not taking into account all secondary processes, the current-producing reaction can be described by the following equation:

$$Zn + 2MnO_2 + 2H_2O \rightarrow Zn(OH)_2 + 2MnOOH \qquad (9.5)$$

The often used equation

$$Zn + 2MnO_2 + 2NH_4Cl \rightarrow Zn(NH_3)_2Cl_2 + 2MnOOH \qquad (9.6)$$

also does not give an exhaustive description of the process since the actual capacity of the cells can be higher than that corresponding to the amount of ammonium chloride in the cell (given by equation (9.6)).

The o.c.v. of undischarged freshly manufactured manganese–zinc cells varies between 1·55 V and 1·85 V depending on the type of manganese dioxide being used and the composition of the active materials. The o.c.v. decreases in the process of discharge and formation of the variable-composition mass. In principle, every phase composition is characterized by its own thermodynamic value of the e.m.f. However, such a correspondence is difficult to establish since even in one particle the compositions of the surface and deeper layers differ. In the process of heterogeneous transformations, when the phase composition does not vary, the o.c.v. remains practically constant. With prolonged storage (over a month, say) the o.c.v. of the undischarged cells gradually decreases.

(b) Self-discharge

Both electrodes of the manganese–zinc cells are thermodynamically unstable and can react with aqueous solutions liberating hydrogen and oxygen respectively.

Though an excess of zinc is used in the cells and their discharge capacity is limited by the positive electrode, corrosion of zinc results in deterioration of the cell parameters. Corrosion gives rise to the same primary and secondary products as discharge. Formation of the large-grain precipitate of $Zn(NH_3)_2Cl_2$ in the diaphragm with slow self-discharge increases the internal resistance of the cell and decreases its capacity. Corrosion of zinc noticeably increases on intermittent discharge when two factors are combined—on the one hand, during discharge the electrolyte near the anode is acidified and, on the other, owing to the interruptions, the total time of use is increased. Corrosion of zinc is sharply decreased by its amalgamation. If a cell has not been sealed carefully enough, zinc can corrode owing to reaction with atmospheric oxygen.

The rate of spontaneous decomposition of manganese dioxide giving rise to oxygen and a certain amount of MnOOH is usually very low. The complete decomposition of MnO_2 to MnOOH is quite impossible thermodynamically; on the contrary, MnOOH readily reacts with oxygen giving rise to a mixed phase containing a large proportion of MnO_2; this fact is utilized in manganese–air–zinc cells.

At the same time, some reactions can occur between MnO_2 and the gelling agents used to immobilize the electrolyte (starch, flour) so that they are partially oxidized to CO_2 and the capacity of the positive electrode is decreased.

The loss of capacity in stored manganese–zinc cells can also be caused by drying and flaking of the electrolyte paste, development of intercell short-circuits, and other effects. "Drying" of the electrolyte can be caused not only by loss of water (for instance, due to evaporation) but also by its binding in crystalline hydrates.

(c) Electrolyte leakage

At the last phase of discharge or after discharge of manganese–zinc cells the electrolyte often leaks from the cells, producing salt deposits on the cell jacket. Especially heavy electrolyte leaks occur after high-current discharges or short-circuiting of the cell.

One of the causes of electrolyte leakage is the increase in the volume of the active materials of the positive electrode with discharge, the decrease of its porosity and the ejection of electrolyte from the pores of the active materials. Moreover, electro-osmotic effects in the separator can cause flow of the electrolyte from the cathode towards the zinc anode. These effects develop mostly with high-current discharges when a concentration gradient is established in the separator. It has been shown recently that electrolyte leakage can be decreased by employing an electrolyte containing only zinc chloride without ammonium chloride. In this case secondary processes give rise mostly to precipitates of zinc oxychlorides which bind large amounts of water in crystalline hydrates, for instance, $ZnCl_2 \cdot 4ZnO \cdot 5H_2O$.

(d) Multiple use of cells

Manganese–zinc cells tolerate a certain number of charge–discharge cycles subject to the condition that not more than 20–25% of the capacity is used in one discharge (that is, the cell is discharged to a final voltage not below 1·1–1·2 V) and that charge is started immediately after discharge. The secondary processes of formation of various precipitates after discharge or during deeper discharge greatly hinder recharging. Charging is also more difficult after prolonged storage of the cell prior to discharge. It should be borne in mind that cycling enhances electrolyte leakage and sharply decreases the service life and that charging can cause rupturing of the cell. In view of all these obstacles manganese–zinc cells are rarely recharged.

9.3. Types of manganese–zinc cells

(a) Cell and battery constructions

There are two basic types of the manganese–zinc cells, round cells (can cells) and flat cells with bipolar electrodes.

The round cells with small or medium capacity have cylindrical shape, that is, their cross-section is circular; can cells with high capacity typically have rectangular cross-section. The cans are manufactured mainly from zinc and serve simultaneously as the cell container and the negative electrode. The cylindrical zinc cans are produced by impact extrusion from slugs preheated to 180–200°C on heavy-duty toggle presses; the rectangular cans are made from zinc sheets by soldering or welding.

At the central part of the can (1) (Fig. 50) there is the positive electrode (2) pressed from the cathode mix with a centrally disposed carbon rod (3) (the current collector). The cathode is round or rectangular according to the shape of the can. The cathode is physically separated from the bottom of the can by an insulating disk or cup (10). The free space under the cardboard washer (5) is known as the expansion chamber (4) and serves for accommodation of the gaseous products of discharge and self-discharge—hydrogen and ammonia. A sealing mixture is poured into the top of the cell to form the seal (6). The metal contact cap is pressed onto the carbon rod top (7).

In the older cell constructions (Fig. 50(a)) the cathode was wrapped in thin fabric (calico) which was secured by string; this cathode is known in the trade as a "dolly". The space (8) between the dolly and the zinc can (1–3 mm) is filled with electrolyte; after short-time heating the liquid solution gels under the effect of the gelling agent. At present this construction is used in the large-size cells and sometimes in the cells designed for high-current discharge. The cathode is wrapped in paper or gauze which is often secured by glueing rather than tying with a thread.

In the paper-lined cells (Fig. 50(b)) the zinc can is lined with a paper separator (14) carrying a layer of the electrolyte paste on the outside. The cathode is inserted into the can and compressed from above tightly pressing the separator to the zinc can (the so-called "packing" technique). In this construction the electrolyte gap is sharply decreased (down to 0·15–0·2 mm) and the amount of manganese dioxide in a cell of given size is increased, resulting in a marked increase of the cell capacity.

Cells not intended for use in batteries are enclosed in a cardboard jacket (9) bearing the manufacturer's label (Fig. 50(a)). Round cells are now often enclosed in a jacket (15) of thin steel (Fig. 50(b)) which is insulated from the zinc can by a plastic tube (16). The lid (11) and the steel bottom disc (17) are retained in position by swaging the steel jacket. The ring (12) insulates the can from the lid and seals the cell. The insert (13) is used to make the expansion chamber. The main advantage of metal-clad cells is efficient sealing, improved shelf-life and lack of electrolyte leakage. For these reasons they are widely used despite complicated construction and higher cost.

The sizes of the round manganese–zinc cells are standardized. Table A.4 gives the cell dimensions and markings used in various countries.

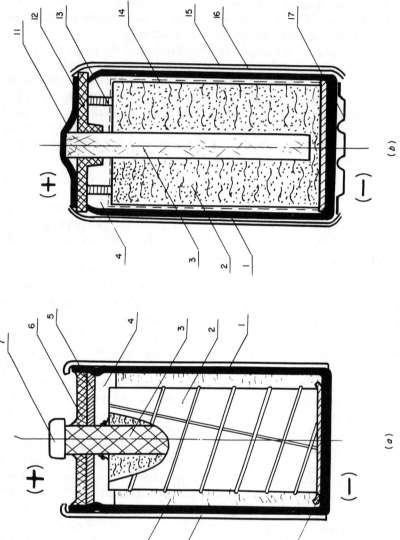

Fig. 50. Construction of cylindrical manganese–zinc Leclanché cells: (a) cell with "dolly"; (b) paper-lined cell.

Figure 51 shows the flat cell construction. The negative electrode is a zinc plate (1) coated on one side with a conductive film (2) consisting of carbon and polymeric binding agents and impermeable to the electrolyte. The conductive film acts, in fact, as the separator for two neighbouring cells. The separator (3) pasted with electrolyte (similar to the separator of the paper-lined cells) is pressed to the other side of the zinc electrode. The flat cathode mix (4) with a projection to be pressed to the conductive layer of the neighbouring cell in the battery, is pressed to the separator. The cathode mix is wrapped in tissue paper (5) to prevent mix particles from breaking away and to rule out short-circuiting between the cells. The components of the flat cell are enclosed in a PVC tube (6) which provides for internal contact between cell components and prevents electrolyte leakage.

The flat cells are typically assembled into batteries. The individual cells are stacked and retained in position by tying. Batteries of flat cells make a considerably better use of volume than batteries of round cells so that their specific energy is higher. Moreover, the consumption of zinc in the flat cells can be lower by almost a factor of three per unit capacity than in the round cells since in the flat cells zinc does not serve as a structural material and can be dissolved "right through". The flat cell batteries do not require intercell connections, thus eliminating the associated consumption of brass and solder. Most manganese–zinc batteries are therefore now manufactured from flat cells. Only low-voltage high-capacity batteries or batteries designed for high discharge currents (for instance, the standard "flat" flashlight battery) are made from the round cells.

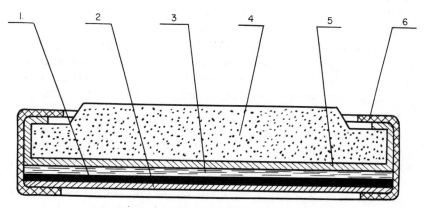

Fig. 51. Construction of the flat manganese–zinc cell.

(b) Composition of the electrodes and electrolyte

Modifications and types of manganese dioxide

Manganese dioxide has a large number of crystal modifications denoted by the Greek letters α, β, γ, δ, ε and η. The naturally occurring modifications are cryptomelane α-MnO_2, pyrolusite β-MnO_2 and ramsdellite γ-MnO_2. Some modifications contain the ions K^+, Ba^{2+}, etc. (α- and δ-MnO_2) or 4–6% of structural water (α-, γ-, δ- and η-MnO_2). The stoichiometric composition is described by the formula MnO_n where n varies between 1·9 and 2.

Four types of manganese dioxide are used in manufacture of the manganese cells:

Natural ore. The pyrolusite deposits are the most important ones. Concentrated pyrolusite ore contains 85–90% of β-MnO_2; it is the cheapest electrode material but with a comparatively low activity. It has a little tendency to spontaneous decomposition and provides for good storageability of the cells.

Activated manganese dioxide. This is obtained by heating the ground natural ore giving rise to Mn_3O_4 by partial reduction of MnO_2 on the grain surfaces. Subsequent treatment with sulphuric acid results in dissolution of the lower manganese oxides and admixtures and highly porous γ-MnO_2 is obtained. The activated manganese dioxide has a more positive initial potential (approximately by 0·15–0·2 V) and a higher utilization coefficient than the starting ore.

Electrolytic manganese dioxide (EMD). Obtained by anodic deposition from $MnSO_4$ solutions on graphite anodes, EMD consists of γ-MnO_2 and has high purity and high activity. EMD is increasingly used in cell manufacture owing to these advantages and to the possibility of using low-content manganese ores as the raw material.

Synthetic manganese dioxide (SMD). This is obtained by chemical processes which yield products with varying properties depending on the process. An important product is the highly hydrated SMD produced by thermal decomposition of permanganates. It has the η-MnO_2 structure and a fairly stable discharge potential.

The conductivity of MnO_2 powders measured under a pressure of 100 MPa varies from 10^{-3} to 5×10^{-2} ohm^{-1} cm^{-1}. To increase the conductivity the naturally occurring flaky graphite ("battery-grade graphite") and/or acetylene black are added to the active materials. Another important function of carbon black is to increase the moisture content of the active mass and to retain the electrolyte near all the electrode particles.

Other types of carbon black or artificial graphite give poorer results. The content of the carbon additives varies from 8 % to 20 %. The cells designed for high-current discharge contain up to 20 % graphite. Minimum amounts of additives are used in the cells designed for low currents and prolonged storage.

Negative electrode

Manganese–zinc cells contain zinc of purity of no less than 99·94 % with a relatively high corrosion resistance. Zinc can contain impurities on which the rate of hydrogen evolution is low, for instance, cadmium or lead. Sometimes, a special lead admixture is added to zinc to improve its structure and to facilitate extrusion of cans.

Electrolyte

The main electrolyte components are ammonium chloride (sal ammoniac) and zinc chloride along with the gelling agents, flour or starch. Both chlorides take part in the secondary reactions and thus greatly influence the character of the cell discharge process. The increase of NH_4Cl content in the electrolyte increases its conductivity but at the same time decreases the pH of the solution thus enhancing zinc corrosion. Therefore, cells with elevated NH_4Cl content have poorer storageability. Zinc chloride has a strong effect on the thixotropic properties of electrolytes gelled with flour or starch: in the presence of $ZnCl_2$ the electrolyte gels much faster. Moreover, the $ZnCl_2$ solutions have antiputrefactive and partial buffering properties. In the presence of $ZnCl_2$ the NH_4Cl solutions have a lower tendency to leak and to form thick crusts of crystalline ammonium chloride above the liquid surface, as is typical of the pure solutions. If a cell is designated for operation at low temperatures calcium chloride, which lowers the freezing point of the solution, is often added to the electrolyte. In some cases lithium chloride is used for this purpose.

Since the individual components of the electrolyte have different effects on the positive and negative electrodes, various mixtures are typically used for impregnation of cathodes and separators. In particular, the elctrolytes for the pasted separators contacting with the zinc electrode contain from 5 to 15 g/litre of mercury chloride ($HgCl_2$) to decrease self-discharge. Mercury is deposited on the zinc surface forming an amalgam. Sometimes small amounts of potassium bichromate, which serves as inhibitor of zinc corrosion, are added to the electrolyte. Tanning agents—chrome alum or chromic sulphate—are added to some electrolytes to prevent liquefaction with increasing temperature.

(c) Manganese–zinc cells with stable voltage

In the sixties modifications of synthetic manganese dioxide were produced which made it possible to develop a positive electrode with a stable discharge voltage. These modifications (η-MnO_2) contain a fairly large amount of structural water, that is, a proportion of the oxygen atoms in the lattice are replaced by hydroxyl groups. These modifications of manganese dioxide have good ion-exchange properties and some of the protons in the crystal lattice can be replaced by zinc ions. Discharge of the positive electrode using hydrated manganese dioxide probably proceeds at first according to equation (9.3). At the very beginning of discharge the pH increases slightly and the voltage is somewhat decreased. Soon there starts the secondary reaction in which the zinc ions are incorporated into manganese dioxide and a new phase (hetaerolite) is formed:

$$MnO_2 + Mn^{2+} + Zn^{2+} + 2H_2O \rightarrow ZnO \cdot Mn_2O_3 + 4H^+ \qquad (9.7)$$

Owing to this reaction alkalinization of the solution stops. The hetaerolite formed in the reaction is not isomorphic to manganese dioxide and does not form a phase of variable composition with it. Owing to this the potential of the positive electrode is independent of the degree of discharge. The total reaction in the cell including the two stages (9.3) and (9.7) is described by the following simple equation:

$$Zn + 2MnO_2 \rightarrow ZnO \cdot Mn_2O_3 \qquad (9.8)$$

During this reaction the composition of the electrolyte is not changed.

Figure 52 shows the discharge curve of the cell with hydrated manganese dioxide in comparison with the curves for conventional cells. After initial decrease of the voltage owing to alkalinization of the electrolyte, voltage again increases as a result of the increase in the number of crystallization centres of the hetaerolite. In the subsequent course of discharge the voltage decreases very slowly. It can be seen that the hydrated manganese dioxide not only improves the discharge curve but also increases the cell capacity. The first stage of reduction of manganese dioxide (to the trivalent form) is utilized almost fully. However, since the stage (9.7) is very slow these conditions prevail only on discharge with very low current densities; when the discharge current is increased there starts a parallel process giving rise to a variable-composition phase and resulting in a shift of the potential. The cells with hydrated manganese dioxide are used, in particular, as a power supply for electronic wrist watches.

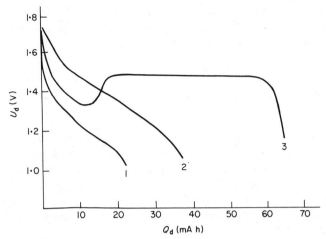

Fig. 52. Discharge curves of the managanese–zinc cells for electronic watches with various modifications of manganese dioxide. 1, non-hydrated pyrolusite (MnO_2); 2, weakly hydrated activated manganese dioxide (MnO_2); 3, hydrated manganese dioxide (MnO_2); normalized discharge current $j_d = 0.005$.

(d) Manganese–air–zinc cells

The hydroxide MnOOH of trivalent manganese produced on discharge from manganese dioxide can, in principle, be again partially oxidized by the oxygen of the air giving rise to a mixed phase rich in MnO_2. Therefore, free access of the air to the active materials of the positive electrode increases the cell capacity. Moreover, the carbon materials in the cathode mix—carbon black and graphite—can adsorb oxygen and, to a certain extent, act as oxygen electrodes. Hence, a relatively widely used type of cell is the manganese–air–zinc cell in which the cathodic process consists in concurrent reduction of manganese dioxide and the oxygen of the air. In such cells the cathode mix contains a larger proportion of the carbon additives and the carbon black is often replaced by activated charcoal which has a highly extended surface and readily adsorbs oxygen. Such a cathode mix can have, for instance, the following composition: 35–40% manganese dioxide, 45% graphite and 15–20% activated charcoal.

The manganese–air–zinc cells have special channels for improving the supply of air to the cathode mix. Prior to discharge these channels are covered with glued-on paper which should be ruptured when the cells are discharged. Under the conditions of low-current drain these cells act, primarily, as air cells; with medium and high-current drain the main process is reduction of manganese dioxide.

Manganese–air–zinc cells are employed as radio cells and batteries. For some discharge regimes they have a specific capacity which is approximately twice that of the conventional manganese–zinc cell.

9.4. Performance

The discharge characteristics of manganese–zinc cells are determined by the compositions of the active mass of the positive electrode and of the electrolyte. The typical discharge curves of a round D-size cell on continuous discharge are given in Fig. 53. For low-current drain the initial voltage is 1·6–1·65 V and for high-current drain it drops to 1·2–1·3 V. Depending on the load features and the current drain the cell is discharged to the cut-off voltage of 0·7–1·0 V.

Manganese–zinc cells are distinguished by a significant dependence of the discharge capacity on the current. Starting from currents as low as $j_d = 0·002$ the capacity markedly decreases with the current. Therefore, the concepts of "rated capacity" or "rated discharge current" are rarely applied to these cells. The characteristics for each cell type are specified and tested typically for a given, somewhat arbitrary, regime related to a certain field of use. Sometimes the external resistance R_{ex} is specified rather than the discharge current.

Figure 54 shows that the capacity of the manganese–zinc cells increases in the case of intermittent discharge at medium- and high-current drains in

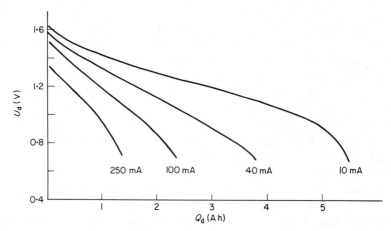

Fig. 53. Discharge curves of a D-size manganese–zinc cell at room temperature and various drain rates.

comparison with the capacity for continuous discharge at the same current. If the interruptions of discharge are long enough the capacity is considerably increased. Therefore, these cells are typically used in intermittent-operation units, for instance, flashlights, transistor radios, toys, etc. However, intermittent prolonged discharge with low-current drains ($j_d < 0.002$) leads to a marked decrease of the capacity owing to self-discharge caused by corrosion of zinc (the normalized current j here is referred to the maximum capacity at low-current drain).

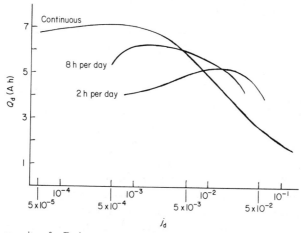

Fig. 54. The capacity of a D-size manganese–zinc cell as a function of load under continuous and intermittent discharge conditions at room temperature.

For continuous discharge with low-current drains ($j_d \approx 0.002$) or intermittent discharge with medium-current drains the specific energy of manganese–zinc cells amounts to 60 Wh/kg or 100–140 Wh/dm³. For continuous discharge with current drains corresponding to $j_d = 0.05$–0.1 the specific energy drops to 10 Wh/kg. Manganese–air–zinc cells have a lower initial voltage, 1·30–1·35 V. For discharge currents corresponding to $j_d = 0.001$–0.002 their specific energy is about 85–100 Wh/kg.

Figure 55 shows the discharge curves for the D-size cells at low temperatures. Even for low-current drains ($j_d = 0.002$) the capacity can be seen to decrease noticeably even at 0°C while at -40°C it amounts only to 20% of the capacity at room temperature. The temperature of -20°C is regarded as the lower limit for cell operation for high-current drains. The normal electrolytes freeze below -20°C and various additives are used to decrease the freezing temperature. Since these additives affect performance at higher temperatures different electrolyte compositions are sometimes employed in

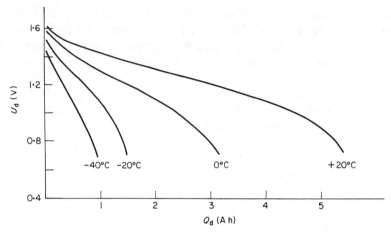

Fig. 55. Discharge curves for a D-size manganese–zinc cell for a current of 10 mA and various temperatures.

cells designed for summer conditions (from $-20°C$ to $60°C$) and cold-resistant cells (from $-40°C$ to $40°C$).

The concept of the effective internal resistance of the cells is not quite definite owing to the increase of the slope of the voltage–capacity curves with increasing current drain (Fig. 53). It can be estimated that the normalized internal resistance (also corresponding to the maximum cell capacity at low-current drains) is 5–10 ohm Ah at the beginning of discharge at room temperature and increases by a factor of 2–2·5 at $0°C$. These resistance values are high compared with the values for other types of chemical power sources.

The rated parameters of the manganese–zinc cells correspond to the freshly manufactured cells, that is, stored for not more than a month from the time of manufacture.

Storageability of manganese–zinc cells and batteries varies according to the size, construction and compositions of the active materials and electrolyte, being between 3 months and 2–3 years. During storage the processes of ageing and self-discharge are occurring in the cells resulting in decrease in the capacity and discharge voltage and increase in the internal resistance. By the end of the specified shelf-life the capacity decreases by 30–40%. Storageability is greatly affected by the quality of sealing as it inhibits evaporation of water and penetration of the air oxygen to the zinc electrode. The round cells with steel cans are especially good in this respect.

Temperature conditions strongly affect the processes of self-discharge and ageing. Two to three months of storage under tropical conditions (say, at

45°C) are regarded as being equivalent to one-year's storage under normal temperature conditions (20–25°C). At low temperatures (say, −20°C) the cells and batteries can be stored for long periods without any significant deterioration of their parameters.

BIBLIOGRAPHY

1. J. Brenet, Electrochemical behaviour of the dioxides of manganese, in "Batteries" (Collins, D. H., ed.), p. 357. Pergamon Press, Oxford (1963).
2. R. Huber, Leclanché batteries, in "Batteries" (Kordesch, K. V., ed.), Vol. 1, Manganese Dioxide, pp. 1–239. Dekker, New York (1974).
3. N. C. Cahoon, Leclanché and zinc chloride cells, in "The Primary Battery" (Cahoon, N. C., Heise, G. W., eds.), Vol. 2, pp. 1–148. Wiley, New York (1976).
4. V. N. Damye and N. F. Rysukhin, "Production of Galvanic Cells and Batteries". Vysshaia Shkola, Moscow (1970) (in Russian).

Chapter Ten

Lead (Acid) Storage Cells

10.1. General

The lead cell is the most widely used type of storage cell at present. The world output of starter batteries alone is more than 100 millions a year. The yearly consumption of lead for manufacture of lead cells is about two million tons, more than a half of the total world output. The widespread use of lead cells is explained by their cheapness, reliability and good performance. They produce a high and stable voltage which varies only slightly with temperature and load currents. The cycle life of lead cells is a few hundred charge–discharge cycles and for some cell types it is more than a thousand cycles.

Lead cells are used to manufacture starter batteries (with capacities from 5 Ah to 200 Ah), traction batteries (40–1200 Ah) and stationary batteries (40–5000 Ah). Lead storage cells are also employed in various electronic and communication devices.

The first working lead cell manufactured in 1859 by a French scientist, Gaston Planté, consisted of two lead plates separated by a linen strip, coiled and inserted into a jar with sulphuric acid. With first charge, electrochemical reaction produced a surface layer of lead dioxide on the positive electrode. To increase the capacity of the cell, charge and discharge were repeated many times producing an increase in the surface area of the electrodes. Planté called this procedure the forming of the electrodes. In 1880 C. Faure proposed making the electrodes by coating lead plates with a paste of lead oxides and sulphuric acid. This technique considerably increased the capacity of the lead cells. In 1881 E. Volckmar suggested applying the paste to a lead grid rather than plate; later G. Sellon patented a grid of a harder alloy of lead and antimony as a replacement for the lead grid.

As early as the eighties of the nineteenth century lead cells were manufactured in batches in the developed countries. This was assisted by the spreading use of electric generators for charging the storage cells (earlier the cells were charged by batteries of primary cells). In the first half of the twentieth century the technology of lead cells made considerable advances resulting in improvement of their parameters. An important advance was the use of expanders in the negative electrodes, starting from 1920, as it contributed to a sharp increase of the cycle life of the cells.

In recent years new types of maintenance-free and hermetically sealed cells have been developed.

10.2. Electrochemical and physicochemical processes

Current-producing reactions. In the charged lead cell the negative electrode contains sponge lead and the positive electrode contains lead dioxide, PbO_2; the electrolyte is a sulphuric acid solution. The current-producing reactions are described by the following reactions:

$$(+) \quad PbO_2 + 3H^+ + HSO_4^- + 2e \underset{\text{charge}}{\overset{\text{discharge}}{\rightleftharpoons}} PbSO_4 + 2H_2O$$

$$(-) \quad Pb + HSO_4^- \underset{\text{charge}}{\overset{\text{discharge}}{\rightleftharpoons}} PbSO_4 + H^+ + 2e$$

$$(\text{cell}) \quad PbO_2 + Pb + 2H_2SO_4 \rightleftharpoons 2PbSO_4 + 2H_2O$$

(At the concentrations used in the cells sulphuric acid dissociates practically only into H^+ and HSO_4^-.) Thus, charging of the cell consumes sulphuric acid and produces barely soluble lead sulphate on both electrodes. This reaction mechanism was suggested as early as 1883 by Gladstone and Tribe in their theory of "double sulphatation".

The o.c.v. of the lead cell practically coincides with the thermodynamically predicted e.m.f. and is equal (in V, at 25°C) to

$$\mathscr{E} = 2 \cdot 047 + \frac{RT}{F} \ln \left(a_{H_2SO_4} / a_{H_2O} \right) \quad (\text{accuracy} \pm 0 \cdot 002 \text{ V})$$

where $a_{H_2SO_4}$ is the activity of sulphuric acid and a_{H_2O} is the activity of water. Table 2 presents the activities of sulphuric acid and water and the e.m.f. as a function of the concentration of sulphuric acid. The concentration of sulphuric acid can be expressed in various units the relationships between which are given in the table and in Fig. 19(b). In this chapter we shall measure the concentration as percentages by mass (g) equal to the

Table 2. The activities of sulphuric acid and water and the thermodynamic values of the e.m.f. of the lead cell for solutions of sulphuric acid of various concentrations (at 25°C)

Solution density (g/cm^3)	Concentrations		Activities		E.m.f. (V)
	% by mass, g (grams of H_2SO_4 per 100 g of solution)	Molar, μ (moles/ litre)	$a_{H_2SO_4}$	a_{H_2O}	
1·050	8	0·86	0·0069	0·96	1·922
1·078	12	1·32	0·021	0·94	1·951
1·106	16	1·81	0·060	0·91	1·979
1·136	20	2·32	0·159	0·88	2·005
1·167	24	2·86	0·424	0·84	2·031
1·200	28	3·43	1·14	0·78	2·059
1·231	32	4·03	3·28	0·72	2·088
1·264	36	4·66	10·8	0·65	2·121
1·300	40	5·31	34·6	0·57	2·154
1·334	44	6·00	118	0·48	2·190

Notes: (1) The concentration in gram/litre is equal to 98 μ (the molar mass of sulphuric acid is 98); the concentration in molalities m (moles of sulphuric acid per 1000 g of water) is equal to $100g/98(100-g)$. (2) In the literature one typically meets the mean ionic activity coefficients $\gamma_\pm$ (molal) which are calculated under an assumption that sulphuric acid completely dissociates into the ions SO_4^{2-} and H^+ and are related to $a_{H_2SO_4}$ by the following equation: $a_{H_2SO_4} = 4(\gamma_\pm m)^3$.

number of grams of the acid in 100 g of the solution. The concentration of a sulphuric acid solution is often determined by measuring the density of the solution.

Lead dioxide has two crystalline modifications, the orthorhombic α-PbO_2 and the tetragonal β-PbO_2. The equilibrium potential of α-PbO_2 is more positive by 0·01 V than that of β-PbO_2. Typically, a charged electrode contains both crystalline modifications. Neither α-PbO_2 nor β-PbO_2 are fully stoichiometric; their composition may be given by PbO_x where x varies between 1·85 and 2·05.

Discharge and charge. Lead sulphate is formed on both electrodes during discharge. Its conductivity (in contrast to that of lead and lead dioxide) is very low—less than 10^{-8} ohm^{-1} cm^{-1}. Lead sulphate, as lead dioxide, is somewhat soluble in sulphuric acid; dissolution gives rise to the ions Pb^{2+} and $PbO(OH)^+$, in concentrations of 10^{-6}–10^{-5} mol/litre (when the concentration of sulphuric acid is increased the concentration of the former ions decreases and that of the latter increases). Therefore, the current-producing reactions mostly proceed via solution with intermediate formation of soluble products.

Of a great significance for operation of the electrodes is their porous structure providing for penetration of sulphuric acid into the bulk of the electrodes. The porosity of charged electrodes can be as high as 50%; the average pore diameter for the positive electrode is 1–2 μm and 10 μm for the negative electrode. During discharge the porosity decreases considerably since the specific volume of lead sulphate is larger than the specific volumes of lead and lead dioxide.

A typical effect is high dilution of the electrolyte during discharge owing to consumption of sulphuric acid and formation of water. In charged storage cells the concentration of sulphuric acid is 30–40% (depending on the cell type). The lower the volume of electrolyte in comparison with the amount of the electrode active materials the larger the decrease of concentration with discharge, at the end of discharge the concentration varies between 12% and 24%. Accordingly, the o.c.v. of the charged storage cell is 2·06–2·15 V and that of the almost discharged cell varies between 1·95 and 2·03 V. For a given cell the relative decrease of acid concentration is directly related to the amount of electricity passed. Therefore, a convenient and accurate method for determining the degree of charge of a cell is to measure the concentration or density of the electrolyte. This is an advantage of the lead cell compared to other storage cells. During discharge the volume of electrolyte decreases by approximately 1 ml for each ampere-hour.

Figure 56 presents the discharge and charge curves for one type of lead cell for $j = 0·1$; the dashed line shows variation of the o.c.v. with discharge and charge caused by variation of the acid concentration. Immediately at the beginning of discharge the voltage can be somewhat higher owing to formation of a small amount of PbO_x where $x > 2$. After the beginning of discharge there is a small voltage drop which is caused by the difficulties with initiation of crystallization of sulphate on lead and lead dioxide and the resulting temporary supersaturation of the solution with Pb^{2+} ions. After stabilization of these rapid processes the voltage slowly decreases in the course of discharge. One of the causes is the decrease of the o.c.v. Furthermore, as the reactions spread into the bulk of the active materials and their porosity decreases the concentration polarization and the ohmic losses in the porous electrodes increase.

Under normal conditions the discharge of storage cells is limited by the positive electrode. In some cases, for instance at low temperatures, the negative electrode limits the process. If the volume of electrolyte is small, discharge can be limited by decrease of the concentration of sulphuric acid. If g_1 is the initial concentration and g_2 is the lowest permissible concentration (in per cent) then 1 kg of the starting solution according to the reaction equation provides a discharge capacity of

$$Q = 10^3(g_1 - g_2)/(366 - 3g_2)\,\text{Ah} \qquad (10.2)$$

During charge the voltage gradually increases owing to the increase of the o.c.v. and the spread of the process into the bulk of the electrode. After deep discharge a distinct voltage peak appears at the very beginning of charge; this peak is caused by the ohmic resistance of the dense layer of lead sulphate which rapidly cracks after the beginning of reduction. Charging of the negative electrode is not accompanied by any appreciable side processes. After transformation of the bulk of sulphate the electrode potential (and the cell voltage) jumps sharply and evolution of hydrogen begins. A small amount of oxygen is liberated at the positive electrode even during charge. Therefore, the electrode output is 85–90% of the capacity. To obtain the full discharge capacity after the hike in voltage during charge an additional 10–20% of the capacity is delivered to the cell. This overcharge is accompanied by massive evolution of hydrogen and oxygen (the "boiling" of electrolyte).

Passivation. In discharge the active materials are not fully utilized, at small drains the degree of utilization is 40–60% and at large drains it drops to 5–10%.

The decrease in the degree of utilization at large drains is caused,

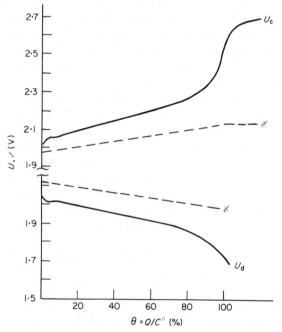

Fig. 56. Typical charge–discharge curves for the lead storage cell.

primarily, by concentration polarization, i.e. a sharp decrease in the concentration of sulphuric acid in the pores of the positive electrode (tending to zero). At small drains the premature decrease of the discharge voltage is caused by passivation of the electrodes. Passivation consists in shielding of the active materials, both lead and lead dioxide, by the dense fine-grained layer of lead sulphate formed with discharge.

To reduce passivation of the negative electrode special additives are introduced into its active materials, e.g. barium sulphate, potassium lignosulphonate, humic acid, various tanning agents and so on. The organic additives are adsorbed at the surface of lead or lead sulphate where they hinder formation of new nuclei of lead sulphate and promote growth of larger crystal grains and formation of a looser layer. Crystallization of barium sulphate is isomorphic to that of lead sulphate so that addition of $BaSO_4$, on the contrary, increases the number of nuclei of lead sulphate crystals. However, if the small grains of barium sulphate are distributed sufficiently uniformly in the bulk of the active materials they also contribute to formation of a loose layer.

Depassivating additives simultaneously play another role, that of expanders of the active materials. Storage and cycling of cells result in gradual sintering of the sponge lead, that is, increase in the size of crystal grains and decrease in the true surface area and porosity. This results in increased true current density and enhanced passivation. Introduction of 1 % of additives into the active materials prevents sintering and setting of the sponge lead (hence the term "expander"). At high concentration of the additives blistering of the active materials can occur.

A combination of barium sulphate and organic additives is much more effective than the individual components since the organic additives are adsorbed at the surface of barium sulphate and remain in the working region of the negative electrode. In the absence of barium sulphate these additives are gradually transported to the positive electrode and oxidized there.

For the positive electrode the tendency to passivation depends on the crystal modification of lead dioxide. The α-PbO_2 modification has a smaller specific surface area than β-PbO_2. The crystal lattices of α-PbO_2 and $PbSO_4$ are isomorphic and a dense layer of sulphate is formed on α-PbO_2 with discharge, hindering further discharging. Therefore, the degree of utilization of α-PbO_2 is less by a factor of 1·5–3 than that of β-PbO_2. The contents of the α and β phases in the electrodes are determined by their conditions of manufacture and forming—typically, comparable amounts of both phases are formed. With cycling α-PbO_2 gradually transforms into the more stable β-PbO_2 and this process is accompanied by some increase in the capacity of the electrode.

Another cause of premature shift of the potential of the positive electrode is due to the ohmic resistance between the current collector grid and the active materials which is produced by the oxide film formed on the surface of the grid during charge. The thickness of this film increases with prolonged cycling owing to corrosion (see below).

The effect of the material of the current-collector grid. In the most widely used electrodes, the pasted ones, the active materials are applied to a cast lead current-collector grid. Since the pure lead is too soft the grid is typically manufactured from an alloy of lead and 2–12% of antimony which is harder and has better mechanical and casting properties. The composition of the alloy has a strong effect on some of the phenomena described below and, ultimately, on the parameters and cycle life of storage cells. Additional alloying additives often can be employed to obtain a required alteration of the alloy properties.

Self-discharge. Both electrodes of the lead cell are thermodynamically unstable and, in principle, can interact with the aqueous solution liberating hydrogen at the negative and oxygen at the positive electrode (see Section 4.5). Moreover, lead dioxide can react chemically with the lead grid. However, self-discharge during storage of a recently manufactured charged cell is practically insignificant, amounting to 2–3% monthly loss of capacity (at 20°C). Self-discharge increases with increasing concentration of sulphuric acid and increasing temperature.

Self-discharge rapidly increases with cycling of the cell. This is caused by dissolution of antimony owing to corrosion of the positive electrode grid. Antimony is deposited in the active mass of the negative electrode facilitating evolution of hydrogen and enhancing the rate of corrosion of lead. In practice, self-discharge of the cells with grids containing large proportions of antimony results in up to 30% loss of capacity a month. Moreover, in the second half of charging evolution of hydrogen is increased, that is the capacity of the cell is decreased. This is why wide-ranging studies are conducted aimed at using low-antimony or antimony-free alloys. An alloy containing 0·03–0·1% of calcium has good mechanical and casting properties but presents complications in manufacture owing to partial "burning-out" of calcium during melting and casting. The self-discharge of cells with grids from this alloy is low.

Evolution of hydrogen and self-discharge are also sharply enhanced by many compounds getting into the electrolyte, for instance traces of iron salts. On the other hand, the organic expanders in the negative electrode promote higher polarization during hydrogen evolution, thus reducing self-discharge. Special additives, for instance, hydroxynaphthoic acid, are even more effective in this respect.

Corrosion of the grids of positive electrodes. The forming charge of the positive electrodes gives rise to a conducting layer of PbO_2 on the grid surface which prevents further anodic oxidation of the grid. However, during cycling the difference in the specific volumes results in periodic uncovering of parts of the metal grid surface which are subjected to further corrosion, particularly owing to the electrochemical interaction between lead and PbO_2. During overcharges when oxygen is liberated on the surface of lead dioxide the grids can be corroded without any contact with the electrolyte owing to the transport of oxygen ions through the PbO_2 layer. The corrosion layer contains various lead oxides and lead sulphate.

Corrosion of the grids of positive electrodes results in decreasing cross-sectional area of their strips and increasing ohmic resistance, in shedding of the active materials and, ultimately, breakdown of the cell. Owing to variation of the specific volume, corrosion of the grids is accompanied by development of large stresses and deformation of the electrodes. Dense layers of oxides can squeeze individual grid strips giving rise to their elongation leading to "growth" of the plates.

The corrosion of the grids has, primarily, intergranular character and depends, to a considerable extent, on the structure of the lead or alloy (a large-grain structure is much more sensitive to corrosion than a fine-grain structure).

The character of corrosion depends also on the alloy composition. Corrosion of a lead–antimony alloy is higher than that of the pure lead but it is more uniformly spread over the surface. Frequently silver or arsenic (a few tenths of 1 %) is added to the alloy resulting in formation of fine-grain structures with high corrosion resistance. Various quaternary alloys have also been suggested as the grid material. The corrosion behaviour of alloys depends not only on their composition but also on the (temperature) conditions of casting of the grids which affect the character of the crystal structure of the alloy.[1]

Corrosion can be reduced also by adding cobalt sulphate to the electrolyte. The mechanism of action of the cobalt ions has not yet been fully understood. It is assumed that these ions transporting electrons decrease polarization with liberation of oxygen, that is they prevent too large a positive shift of the electrode potential. It has been suggested also that being adsorbed on the surface of lead dioxide these ions prevent penetration of oxygen into it and thus inhibit corrosion during overcharge of the cell.

Shedding of the active materials of the positive electrode. The shedding (crumbling) of the active materials of the positive electrode occurring, mostly, with charge is one of the primary factors restricting the cycle life of lead storage cells. To a considerable extent, shedding depends on the method of forming of the lead sulphate layer with discharge. If discharge is carried out at low temperature, high current density and/or high acid concentration, lead sulphate is formed as a dense fine-grain layer. When such an electrode is charged the local current density at the parts of the electrode which are not closely covered by sulphate, is sharply increased giving rise to loose lead dioxide with poor adhesion in the matrix. On the other hand, if a loose sulphate layer is formed during discharge then adhesion of the dioxide to the matrix is improved.

Shedding is markedly enhanced if barium sulphate used as an expander in the negative electrode accidentally gets into the positive electrode. A damaging effect is also produced by divalent iron ions and organic compounds since during storage of the cell they reduce lead dioxide giving rise to a dense sulphate layer which is transformed into loose lead dioxide during subsequent charge.

Shedding is somewhat greater for the lead–calcium grids than for the lead–antimony grids. Apparently, the surface layer of antimony has a favourable effect on adhesion of lead dioxide to the matrix. Moreover, under the effect of antimony electrical contact is improved and the contact resistance between the grid and the active materials is reduced.

A recently developed method for reducing shedding is to add fluoroplastic as a binder or some fibrous materials into the active materials of the positive electrode. Successful results are obtained also with a separating fibre-glass layer pressed to the active materials.

Short circuits. In operation of a lead cell, lead "bridges" can be formed between electrodes resulting in short circuits and self-discharge of the cell. The causes of short circuits can be shedding of lead dioxide which gets to the negative electrode, accumulation of a high "mud" layer, deformation of electrodes, swelling of the negative electrode under the effect of expanders, etc. Protection from short circuits is determined by the properties of separators and correct selection of their shape and size.

Sulphatation. If a lead cell is stored in the discharged state or is regularly undercharged a highly undesirable process, so-called "sulphatation", occurs on the cell electrodes (particularly on the negative electrode). This consists of gradual transformation of fine-grain lead sulphate into a hard dense layer of large-grain sulphate. A cell with sulphated electrodes is very difficult to charge since passage of the charging current produces evolution of hydrogen on the negative electrode rather than reduction of lead sulphate. The harmful effect of sulphatation is aggravated by adsorption of organic additives at the lead sulphate grains which inhibits dissolution of sulphate.

To prevent sulphatation, regular recharging of cells is recommended. To restore the cell capacity, cells with sulphated electrodes are filled with dilute sulphuric acid (in which the solubility of lead sulphate is higher), or even with deionized water, and charged at low currents, for instance currents corresponding to $j_c = 0.01$. The sulphuric acid formed in the process of charge is periodically replaced by a more dilute solution or water.

10.3. Construction and manufacture

(a) Electrodes

At present, lead cells are manufactured with electrodes (plates) of various types depending on the purpose of the cell.

Planté (surface) plates are essentially identical to the electrodes used in the early cells. A relatively thin active layer of lead dioxide is electrochemically formed on the lead surface. Thus, only a very small fraction of the lead of the plate is used and the bulk of lead actually serves as the current collector. During operation of the cell part of the lead dioxide is shed but during charge deeper layers of the plate are put into operation. This provides for large service life from Planté plates—over 15 years. At present Planté plates are made of pure lead sheets with a thickness of 10–12 mm and a large number of slits (Fig. 57). Owing to this shaping the effective surface area of the plate is larger by a factor of 8–10 than its geometric surface area.

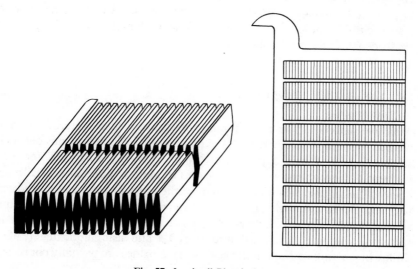

Fig. 57. Lead cell Planté plate.

Planté plates are used only as the positive electrodes in stationary cells where the power-to-weight ratio is not decisive but reliability and long service life are important.

Pasted (grid) plates are made by applying to shaped grids a paste which serves as the active material after forming of the electrode. The grids are cast of lead alloys. Figure 58 shows grids and their cross-sections. The grids for the positive plates, which are more susceptible to corrosion, are thicker. Thin pasted plates have total thickness from 1 to 5 mm and the thick plates over 6 mm.

Pasted plates have high specific capacities but not very high strength. Very high power-to-weight ratios can be obtained with the thin plates. Pasted plates are very widely used, for instance, in starter batteries and in many other types of storage cells.

Box plates differ from the pasted plates in that they have additional walls of thin perforated lead sheets which prevent the active materials from dropping out. They are about 8 mm thick. Box plates have high specific capacities and, at the same time, high strength. They are used as the negative electrodes in combination with the Planté or tubular positive electrodes.

Tubular (armoured) plates (see Fig. 59) consist of a comb cast of a lead alloy on the bars of which perforated plastic tubes (armour) or a common shaped cover is put. The tubes are packed with active material. The tubes are manufactured from ebonite, PVC, synthetic fabrics and other materials; fibre-glass lining is often used. Tubular plates are vibration-resistant and their specific capacity is 1·7–2 times that of the Planté plates. They have a long cycle life, more than 1000 cycles, and are employed as the positive electrodes in traction and stationary batteries.

(b) Technological features

Active materials. The paste for manufacturing active materials for the pasted, box and tubular electrodes is prepared by mixing lead powder with sulphuric acid. The lead powder is produced by grinding balls (or, less frequently, ingots) cast from pure lead. Grinding is done in special mills open to the air and a considerable part of the finely ground lead is oxidized to PbO. The paste for the negative electrodes contains expanders—barium sulphate and organic additives. The paste for the positive electrodes is sometimes prepared from red lead (Pb_3O_4) and lead monoxide (PbO). Following mixing with sulphuric acid the lead oxides are partially converted into lead sulphate.

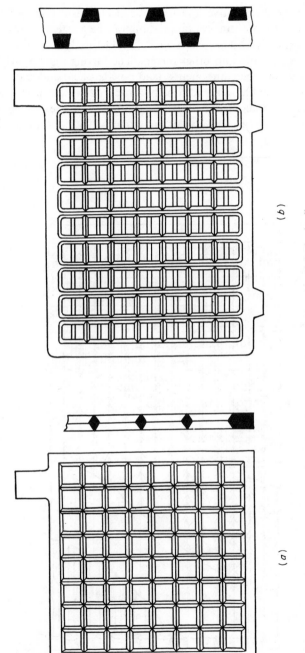

(b)

(a)

Fig. 58. (a) Single and (b) double grids of the lead cell.

The pastes for the positive and negative electrodes are manufactured at separate plants to prevent contamination of the paste for positive electrodes with expanders. The pastes are manufactured by mixing the components in auger, roller and other mixers. The mixing regime, in particular the order of feeding the components (lead powder, acid and water) into the mixer, has a noticeable influence on the properties of the pastes produced resulting from the differences between the characters of the sulphates formed in the reactions between lead oxides and concentrated and dilute sulphuric acid.

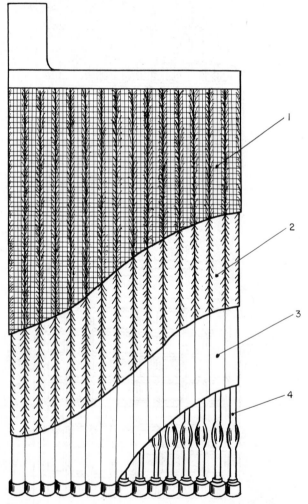

Fig. 59. Tubular plate of the lead cell. 1, tubes; 2, fibre glass; 3, active material; 4, bars.

Manufacture of the pasted plates. The paste is applied to the grids by automatic pasting machines. After pasting the plates are rolled to compact the paste and to squeeze out excessive moisture. The wet pasted plates cannot be stored because they crack and set. Drying of plates is also accompanied by cracking and setting. To prevent cracking the plates are subjected to cementation (treatment with a solution of ammonium sulphate or carbonate). This leads to formation of a dense layer of insoluble lead compounds on the plate surface. The plates are dried in chamber-, tunnel- or conveyer-type dryers at 110–140°C. The dried plates can be stored in the air for a long time.

Forming of electrodes. Prior to assembling of the cells (sometimes after assembling) the plates are subjected to forming to produce the active materials. Forming consists in charging the plates under specified conditions in electrolyte of a specified composition. Sometimes several charge–discharge cycles are carried out.

The forming is done in a sulphuric acid solution whose concentration depends on the lead sulphate content in the paste and varies between 8 % and 22 %. Immediately after impregnation of the plates and in the initial period of forming, the concentration of sulphuric acid decreases owing to its interaction with lead oxide giving rise to lead sulphate. Forming is done with low current densities ($0.001–0.01$ A/cm^2) which provide for effective "treatment" of the deeper paste layers. Frequently the current is changed in two or three steps during forming. The time of forming varies between 15 and 50 hours depending on the current density. Recently booster forming regimes have been developed which decrease the time of forming to 4–6 hours (the current densities are up to 0.04 A/cm^2). The temperature of the electrolyte during forming should not exceed 40–50°C.

After the end of the forming charge the plates are slightly discharged (for 10 min to 5 h); this gives rise to a certain amount of lead sulphate on the plate surface which cements the active layer and improves its strength. After forming the electrodes are, typically, not washed to remove sulphuric acid since this procedure is highly labour-consuming; the acid remaining in the pores is neutralized by the remaining lead oxides.

Planté plates cannot be formed in pure solution of sulphuric acid since the anodic current rapidly gives rise to a thin oxide layer on the surface leading to passivation of electrodes and evolution of oxygen at them. Therefore, the solution of sulphuric acid used for forming (10–20 %) contains additives, such as from 10 to 15 g/litre of potassium perchlorate. Forming is done at a current density of 0.001 A/cm^2 (calculated for the effective surface of the plate). Every 30–40 hours the current direction is reversed. As a result, layers of lead dioxide ("black forming") and sponge lead ("white forming") alternate on the plate surface. The process ends by formation of a lead layer whose adhesion to the surface is better and which stands up well to shipping.

Drying of electrodes. Some time ago the formed and partially discharged plates were dried in the air. During drying the dispersed lead remaining on the negative electrode was oxidized and the cells were assembled with almost fully discharged electrodes. After filling with electrolyte such cells required repeated prolonged (up to two days) charges in the forming regime which made their use more complicated. At present most lead cells are produced with dry-charged electrodes. The negative electrodes are dried under conditions ruling out possible oxidation. Typically, drying is done in special autoclaves into which superheated steam is fed preventing contact between the sponge lead and the air. Vacuum drying is also employed. Sometimes inhibitors are added into the electrolyte used for forming; being adsorbed on the electrode surface they slow down oxidation.

If the temperature of drying of the positive plates is too high, thermal passivation can occur and when the electrode is immersed into sulphuric acid it is passive since at the boundary between the grid and the active materials there has formed a poorly conducting layer of lead oxides. The electrode starts working only after short-time passage of current in the charge direction.

The first models of the dry-charged starter batteries after filling with electrolyte needed three-hour impregnation and five-hour recharge. Now an increasing number of battery models do not need recharge and have an impregnation time of only 20 min owing to improvements in the drying process.

Electrolyte. Solutions of sulphuric acid of a sufficiently high purity are employed as the electrolyte in lead storage cells. In the charged cell the solution contains from 28% to 40% of sulphuric acid. The increase of the initial concentration of the acid makes it possible to decrease its volume, that is, to improve the specific parameters. Moreover, this decreases the danger of freezing of the electrolyte at the end of discharge at low temperatures. However, excessive increase of the acid concentration is inadmissible since it results in enhancement of passivation, self-discharge and sulphatation of the electrodes and in decrease of the cycle life of the cell. Frequently, for operation at low temperatures, the acid concentration is increased by 5–8% compared with the concentration for normal temperatures. On the other hand, for operation under tropical conditions the concentration of sulphuric acid should not be higher than 30–32%.

After the first filling of the cell the concentration of sulphuric acid is typically decreased by 2–3% owing to reaction with the remaining lead monoxide, PbO, in the plates.

(c) Storage cells

Almost all the types of lead storage cell have a similar design based on the box construction (see Fig. 60). Only a small proportion of lead cells are manufactured in the form of cylindrical cells or bipolar batteries. The materials used for cell manufacture should be resistant to prolonged action of concentrated sulphuric acid. One of the few such resistent materials is lead. Therefore, all the current-carrying parts in cells are made from lead or lead alloys. Stainless steel cannot be employed since even traces of iron in the solution have a harmful effect.

The electrode assembly is placed into a tank of an insulating material (1) (Fig. 60). The end electrodes (2) are always negative. In each single electrode

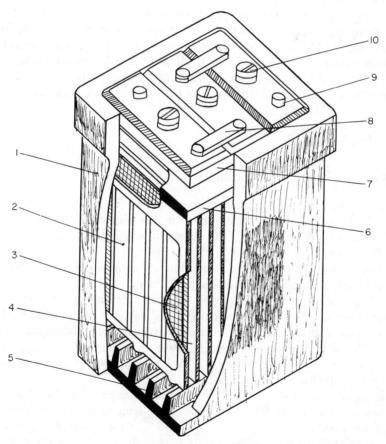

Fig. 60. Schematic of the lead cell.

group the plates are welded to a barretter (current collector) (6) with a vertical bar (9). Separators (3) are placed between the negative and positive (4) plates. The supports at the lower side of the plates rest on special prisms (5) projecting from the bottom of the tank; this provides for the mud space where the active materials shed from the electrodes are accumulated. In large stationary cells the plates are suspended from the lugs of the tank. The distance between the upper edges of the plates and the cover (7) is not less than 20–50 mm. This distance is needed to compensate for variations of the electrolyte level and to keep the cover clear of the drops of electrolyte ejected during intense gassing ("boiling") in the cell at the end of discharge. The cover has two openings for the current-collecting bars. Another opening in the cover is for the ventilation plug (10) which provides an outlet for gases during self-discharge and small overcharge and, at the same time, makes the cell spill-proof for small tilts (for instance, in a car). The opening for the ventilation plug is used also for adding electrolyte, determining its level and concentration, and for gas escape with considerable overcharge. The individual cells in the battery are connected by lead intercell connectors (8).

Starter batteries and some types of traction batteries are assembled in single cases divided by partitions into three or six compartments (according to the number of the cells respectively in the 6-V and 12-V batteries).

10.4. Performance

(a) General charge and discharge properties

Figure 61 presents typical discharge curves for starter batteries. An increase in the discharge current results in a considerable reduction of the capacity and, hence, the specific energy. The change in capacity is farly noticeable even in the range of variation of j_d from 0·05 to 0·2. This fact should be borne in mind when comparing parameters of storage cells, since cells of different types have different rated discharge modes.

As mentioned above, immediately after the beginning of discharge the voltage is not very stable. Therefore, the initial voltage is assumed to be the voltage after a small fraction of the capacity has been discharged (say, 10%). The final discharge voltage is lower than the initial voltage by approximately 0·2 V and amounts to 1·75–1·8 V for low-current drains and 1·2–1·5 V for high-current drains.

The capacity of storage cells depends on the temperature. For $j_d = 0·1$ and temperatures above 0°C a decrease of the temperature by one degree results in a decrease of capacity by 0·6–0·7%. At low temperatures the decrease of capacity is even sharper: at −25°C the capacity is about 50%

and at $-50°C$ about 20% of the capacity at room temperature (Fig. 62). The drop in capacity with temperature is even sharper with increasing discharge current. Starter batteries can work down to approximately $-30°C$ (the cell documentation gives parameters at $-18°C$). The degree of capacity decrease depends on the electrolyte concentration, the cell parameters are better at high concentrations. The main cause of capacity decrease at low temperatures is passivation of the lead electrodes.

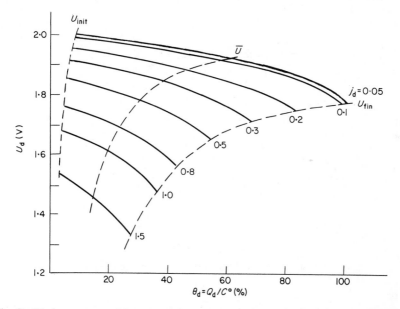

Fig. 61. Discharge curves of the starter lead battery at room temperature. Values on the curves are of j_d.

If a cell operates at a low temperature the concentration of sulphuric acid should vary in a range which should prevent freezing of the electrolyte (particularly, at the end of discharge when the solution concentration is at a minimum). Figure 19(b) shows the temperature of freezing of sulphuric acid solutions as a function of concentration. Freezing of electrolyte can result in damage of the plates.

When a cell is charged with constant current, after the negative electrode has been charged the voltage is observed to increase from $2·3$–$2·4$ V to approximately $2·7$ V (Fig. 56) and gassing starts at the same time. Intense gassing spoils the active materials of the plates, so that during gassing the charge current should be low (j_c not higher than $0·05$). Frequently storage cells are charged in a step-wise current mode. At first, the charge current is high to reduce the charge time, until a voltage of 2·4 V is reached and then low currents are used to complete charging of the positive electrode.

Lead storage cells also can be charged at a constant voltage; in this case the charge current is at first high and then decreases to a low value. When $U_c = 2·5$ V (per cell) complete charge of a cell requires 16–20 hours. Sometimes special devices are employed to limit the initial current to not too high a value. When a cell operates in a buffer mode and is discharged

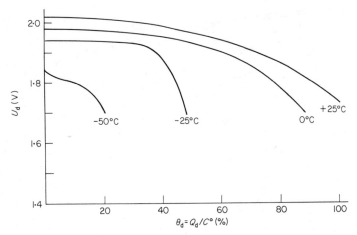

Fig. 62. Discharge curves of a lead battery at low temperatures; $j_d = 0·1$.

only partially the charge voltage can be made as low as 2·3–2·4 V to reduce gassing with overcharge. Compensation recharging can be carried out at a voltage of 2·2–2·3 V. The optimal charge voltages in the above cases strongly depend on the temperature, the state of the cells and other factors. Therefore, if the voltage of the charger is fixed and cannot be controlled, charging is frequently carried out under suboptimal conditions.

Under all conditions completion of charge is evidenced by the concentration (density) of the electrolyte being practically constant for 2–3 hours while the charge current is still being passed through the cell.

At low temperatures ($-20°$C and below) the cell chargeability sharply deteriorates since evolution of hydrogen occurs at the negative electrode instead of reduction of lead sulphate. Cell chargeability can be somewhat improved by specially selected expanders. For rated discharge currents the Ah efficiency of the lead storage cells is 80–90 % and the Wh efficiency is 70–80 %.

(b) Comparison of the properties of various cell types

Lead cells are manufactured with various design and technological features according to their purposes. The main types of lead cells are the starter, traction and stationary cells. These cell types differ, in particular, in the type of the positive electrode, which results in noticeable differences between their parameters. Table 3 compares the specific energy w_m (under the same discharge conditions with $j_d = 0.2$), the normalized effective internal resistance, the cycle life and service life for three cell types, namely starter cells with pasted plates, traction cells with tubular positive plates and stationary cells with Planté plates. All the cells have pasted negative plates.

Table 3. Characteristics of lead storage cells

Cell type	Specific energy (Wh/kg)	Normalized effective resistance (ohm Ah)	Cycle life (cycles)	Service life (years)
Starter	26–28	0·18–0·23	100–300	2–4
Traction	20–28	0·25–0·30	800–1500	4–6
Stationary	8–12	0·14–0·17	—	10–20

The table shows that each of the three cell types has a certain disadvantage along with good parameters: starter batteries have short cycle life and service life, stationary batteries have a low specific energy and, finally, traction batteries have high internal resistances. Having in mind these differences in characteristics and in the operational requirements we shall discuss the parameters of each cell type separately.

Starter storage batteries. The main requirements of starter batteries are a high specific capacity for low discharge currents (the rated current corresponds to $j_d = 0.05$) and the capability for short-time high-current discharge to start the internal combustion engine. Automobile starter batteries are required to discharge with starter currents ($j_d = 3$–5) at a temperature of $-18°C$ for no less than 30 s to the cell voltage of 1·5 V or for no less than 150 s to 1·0 V. Another characteristic of the starter properties is the current making the voltage drop to 1·2 V in 30 s.

Under normal conditions starter batteries operate in the buffer mode. After discharge of a small fraction of capacity to start the engine the batteries are recharged by the automobile generator with the output voltage stabilized in the range 2·45 ± 0·2 V or 2·37 ± 0·08 V per cell. Long drives lead to considerable overcharges accompanied by gassing. This is why water should be frequently replenished in the batteries. Starter batteries are only rarely subjected to deep discharges so that the requirement of a large number of complete cycles is not critical.

Starter batteries are assembled from thin pasted plates of thickness 1·2–5 mm (the positive plates) and 1·0–3·5 mm (the negative plates) which provide for high discharge currents. Thinner (and less strong) plates can be used in countries with good roads. The separator is typically composite consisting of a sheet of glass fibre felt pressed to the positive electrode and a sheet of a microporous material with ribs at the side of the negative electrode. The concentration of sulphuric acid in the charged battery is, typically, 35–36% and by the end of complete discharge it drops to 22–24%.

The specific energy of starter batteries in the rated regime ($j_d = 0.05$) is 30–34 Wh/kg or 60–70 Wh/dm^3. The cycle life for deep discharge is 100–300 cycles. The service life of an automobile battery is 2–4 years which corresponds to a mileage of 40–70 thousand kilometres.

Traction batteries. Lead cells are widely used in traction batteries in various countries. The work on lead traction batteries has been markedly expanded in many countries in recent years in connection with the development of electric vehicles.

In contrast to starter batteries, traction batteries are regularly subjected to deep discharges with medium currents: the rated discharge current of traction batteries corresponds to $j_d = 0.2$. In this connection, a significant problem is shedding of the active materials of the positive electrode with cycling. Traction batteries are assembled, mostly, from tubular plates which hold the particles of the active materials better than pasted plates. In some batteries thick (5–6 mm) pasted plates and strengthened (three-layer) separators are used. The adhesion of the active materials is also improved by the high antimony content in the positive grid (up to 9–10%). The resulting increased self-discharge is not very significant for traction batteries since, typically, they operate according to a regular schedule. Some time ago the negative electrodes normally comprised box-type plates but the pasted plates are used now. The acid concentration varies between 36% and 12–18% with cycling.

The specific energy of traction batteries is 20–28 Wh/kg or 55–75 Wh/dm^3. The cycle life varies between 800 and 1200 and, sometimes, 1500 cycles. To improve the cycle life it is desirable to discharge the batteries to not more than 80% of their capacity.

Stationary batteries. The typical rated discharge current of stationary batteries corresponds to $j_d = 0.1$, but often these batteries are designed for discharge with higher currents. The capacity of stationary batteries is high as a rule, but their specific energy can be relatively low.

Formerly, stationary batteries were placed in open glass jars or wooden containers lined with thin lead sheets. Now closed glass, ebonite or plastic containers are used in which evaporation of water is considerably reduced. The batteries are assembled with Planté positive plates and box-type negative plates. In the newer constructions thick pasted plates are also used, especially for the negative electrodes. The batteries are filled with relatively large amounts of electrolyte which makes it possible to use a low initial acid concentration, e.g. 30% which reduces self-discharge. With the same purpose the electrodes are manufactured from pure lead (Planté electrodes) or lead alloys with low antimony content or without antimony at all. Since for some types of stationary batteries the required number of cycles is small the lack of antimony has only a slight effect on the strength of the active materials of the positive electrode.

10.5. Maintenance of lead storage cells

The main problem in maintenance of lead cells is the monitoring and adjustment of the electrolyte level and maintenance of the charged state in storage.

During operation of the cell, water is decomposed owing to corrosion of the lead electrode and overcharge gassing. Disappearance of a part of the water lowers the level of electrolyte below the upper electrode edge and increases the acid concentration; both effects are harmful to the electrodes. Regular addition of deionized water is therefore needed during operation of the cell. Typically, water is added to the required level as often as the operational conditions dictate. In normal automobile starter batteries in daily use the electrolyte level should be checked every 2–3 weeks. Since the electrolyte is sometimes lost during operation (spilling or loss as fog with intense gassing) it is recommended also to check regularly the acid concentration. This is done by measuring the electrolyte density in the charged battery with an areometer; 2–3 hours should pass (after the end of charge or addition of water) before the checking is done to provide for levelling-out of the concentration in the bulk of the solution.

To prevent sulphatation lead cells filled with electrolyte should be stored only in the charged state. If the cells are stored for a long time they should be recharged every month to compensate for self-discharge. Moreover, it is recommended to conduct a full discharge–charge cycle with $j = 0.05$ every 3–6 months.

Conservation of lead cells is possible. First the cells are discharged at $j_d = 0.1$ to a voltage of 1.85 V. Then the electrolyte is poured off from the cells, they are washed carefully with deionized water several times (the washing period is 2–3 hours to remove the traces of sulphuric acid from the electrode pores) and then dried with warm dry air. The opening for filling up is tightly plugged.

Since charging of lead cells is accompanied by evolution of considerable amounts of hydrogen the safety rules should be complied with and the rooms where cells are charged should be carefully ventilated. Small amounts of the toxic compounds, stibine (SbH_3) or arsine (AsH_3), can be liberated during charging of cells whose grids contain antimony or arsenic.

10.6. Further development of lead storage cells

Even though the lead storage cell was invented over a hundred years ago and has been well studied, work on its development is still going on. Starting approximately from 1970 work has been underway on development of starter batteries which would need only slight maintenance or none at all ("maintenance-free" batteries). The major maintenance task with starter batteries is the regular addition of water. To reduce gassing in the new batteries the grids are manufactured from low-antimony alloys (not more than 2–3% of Sb) or lead–calcium alloys with various additives. These grids do not only reduce corrosion of lead; the voltage of the beginning of gassing for them is rather high (over 2·5 V) so that careful adjustment of the charge voltage can prevent gas liberation at the end of discharge. The starter batteries manufactured at present by some firms do not need addition of water for 1–2 years (car mileage of up to 50 thousand kilometres!). These cells are not completely gas-tight since, owing to the residual gassing, valves must be used in them. Gas liberation into the environment can be prevented to a certain extent by using special plugs with catalytic packing (see Section 6.5).

Another interesting development in the problem of simplifying cell maintenance is the system for centralizing addition of water into all the cells of a traction battery. The openings in the cells are connected to a system of pipes delivering water; each cell has a device for stopping water after the required level has been reached. Furthermore, the system of pipes can be used for centralized removal of gases during operation or charge of the cells.

Attempts have been made to manufacture hermetically sealed lead cells in which gas recombination via the oxygen cycle is employed.[2] In charge, the capacity of such cells is limited by the positive electrode and the liberated oxygen reacts with the metallic lead on the negative electrode. To improve supply of oxygen to the negative electrode the volume of the free electrolyte should be restricted which results in a decrease of the cell capacity. An additional oxygen electrode (for instance, with platinum catalyst) connected with the lead electrode is often used to intensify the reaction. As oxygen is reduced on the additional electrode, lead is oxidized to lead sulphate. It is important that the catalyst from the additional electrode should not get into the electrolyte or onto the negative electrode since this will sharply increase the rate of corrosion of lead.

Immobilized electrolytes are increasingly used in the gas-tight and leak-proof cells making possible cell operation in any position. The gelling agents used to immobilize the electrolyte are silica gel, alumogel, calcium sulphate, etc. On wetting with sulphuric acid these compounds form thixotropic gels. The typical cycle life of lead cells with immobilized electrolyte is still not more than 50–100 cycles.

Another goal in improvement of lead cells is to increase their cycle life and service life. Basically, this problem consists in decreasing the shedding of the active materials of the positive electrode and reducing the corrosion of its grid. For this purpose the effects of various additives to the electrode and electrolyte are studied. For some types of cells good results have been obtained by adding to the electrolyte small amounts of phosphoric acid (about 5–7%). This additive decreases sulphatation of the active materials, shedding and grid corrosion. A particularly marked effect has been found for the lead–calcium grids for which shedding is higher. A slight decrease in capacity and voltage has been found in the presence of phosphoric acid. The mechanism of action of phosphoric acid has not yet been fully elucidated.

The main goal in development of lead cells is to increase their specific energy, which is especially important for traction batteries designed for electric vehicles. The specific energy can be increased both by design improvements and by improving utilization of the active materials of the electrode.

The normal design of lead cells provides for high reliability but, at the same time, results in increasing weight. Given overleaf are the approximate contributions of the individual components of traction batteries to the total mass calculated for a capacity of 1 kWh; the mean discharge voltage for the rated current is assumed to be 1·92 V. The higher figures correspond to the older types of cells and the lower figures to the newer types which have only recently been produced or are still being developed (naturally, in the future the parameters can be improved even further).

The mass of the reactants and the electrolyte theoretically required for the reaction can be seen to amount to 15–24% of the total mass. An excess of electrolyte (in particular, the solvent, water) is needed to operate in the given concentration range; for instance, to operate in the concentration range from 36% to 16% a solution excess of 6·4 kg/kWh is needed. The excess of the active materials is needed owing to the low degree of their utilization. The degree of utilization can be improved by optimizing the structure of the porous electrodes and by using various additives. Of great importance for improvement of the specific parameters and the cycle life is development of new thin (0·2–0·3 mm) separator materials with high total porosity and small pore size.

Component	Contribution (kg/kWh)
Lead (theoretical consumption)	2·01
Lead dioxide	
(theoretical consumption)	2·32
Anhydrous sulphuric acid	
(theoretical consumption)	1·90
Excess active materials	3·5–6·5
Excess electrolyte	4·5–8
Current collectors (grids)	6–9
Container, cover, plugs	3·5–7
Separators	1–3
Current leads	1·3–3
Total	26–42·7

The main contribution to the mass of the structural elements of the cell is made by the current collectors of the electrodes manufactured from lead alloys. Studies are under way to attempt using lighter materials. For the grids of the negative plates use of aluminium, copper, titanium and other metals coated with a thin lead layer has been suggested and for the positive plates lead-coated titanium, lead–plastic composites and other materials. To reduce the mass of the containers the old ebonite containers are replaced with containers made from strong heat-resistant plastics, such as poly-propylene. Light-weight thin-walled containers can be manufactured from these materials (by the injection moulding technique). The cover can be readily glued onto the container so that there is no need to cement it as must be done in the ebonite containers. A considerable reduction of the cell mass is also obtained by decreasing the mass of the current leads. In new single-unit batteries the intercell connectors are passed below the cover directly through the partition between neighbouring cells, rather than above the cover.

It should be borne in mind that the measures for increasing the specific energy (thinner grids, improved utilization of the active materials, etc.) considerably decrease the cycle life of the cells. In principle, even now it is possible to manufacture traction batteries with specific energy over 40–45 Wh/kg but their cycle life would be very short. Therefore, the main goal of the research and development effort is to increase the specific energy without reducing the cycle and service life. Recently production has started of traction batteries for electric cars with specific energies of 30–35 Wh/kg and cycle life of 700–800 cycles. Optimists predict development of cells with specific energy of 40–45 Wh/kg and cycle life of more than 1000 cycles.

REFERENCES

1. M. Torralba, Present trends in lead alloys for the manufacture of battery grids. A review., *J. Power Sources* **1**, No. 4, 301–310 (1976/77).
2. B. K. Mahato, E. C. Laird, Gas recombination lead–acid batteries, *in* "Power Sources 5" (Collins, D. H., ed.), pp. 23–41. Academic Press, London and New York (1975).

BIBLIOGRAPHY

1. J. Burbank, A. C. Simon, E. W. Willihnganz, The lead–acid cell, *in* "Advances in Electrochemistry and Electrochemical Engineering" (Delahay, P., Tobias, Ch. W., eds.), Vol. 8, pp. 157–251. Interscience, New York (1971).
2. H. Bode, "Lead–Acid Batteries". Wiley, New York (1977).
3. E. Y. Weissman, Lead–acid storage batteries, *in* "Batteries", Vol. 2, Lead–acid batteries and electric vehicles (Kordesch, K. V., ed.). Dekker, New York (1977).
4. M. A. Dasoyan, I. A. Aguf, "Modern Theory of Lead Batteries". Energiya, Leningrad (1975) (in Russian).
5. P. Rüetschi, Review on the lead–acid battery science and technology, *J. Power Sources* **2**, No. 1, pp. 3–24 (1977/78).

Chapter Eleven

Nickel–Cadmium and Nickel–Iron Storage Cells

11.1. General

The alkaline nickel–cadmium (NiCd) and nickel–iron (NiFe) storage cells have many similar design features and characteristics though some of their parameters differ. Other types of alkaline nickel storage cells—the nickel–zinc and nickel–hydrogen cells—differ significantly from them. They will be discussed in Sections 12.6 and 14.3, respectively.

The NiCd and NiFe cells have specific energy of 20–30 Wh/kg and sometimes more. They have long cycle life (a few thousand charge–discharge cycles) and compact size and are relatively easy to operate. Some types of NiCd cells can be discharged with high currents, up to $j_d = 10$. Completely sealed cells can be manufactured which can operate in any position. All these advantages result in a wide use of these cells for various applications, and they occupy the second place in production of storage cells after the lead cells. The sealed cells (NiCd cells only) are manufactured with a capacity from 0·01 Ah to 160 Ah and the vented cells from 2 Ah to 1200 Ah. In high-capacity traction batteries NiFe cells are mostly used since they are cheaper than the NiCd cells and do not require scarce materials (cadmium). However, the NiFe cells have higher self-discharge owing to corrosion of iron and lower current and energy efficiency.

The first patent for the nickel–cadmium cell was granted in 1899 to the Swedish engineer W. Jungner and the patent for the nickel–iron cell to the well known American inventor T. A. Edison in 1901. The Jungner cell employed the pocket-type positive electrode. The negative electrode contained cadmium sponge produced by electrolysis of a paste of cadmium oxide and ammonium chloride. One drawback of such an electrode was rapid ageing (aggregation) of fine-grained cadmium. A few years after manufacture of the first cell models this drawback was eliminated by adding fine-grained iron into the cadmium mass. The first Edison cells also

224

employed pocket-type positive electrodes but later they were replaced with more efficient tubular electrodes. It was shown that addition of lithium hydroxide, LiOH, into the electrolyte (KOH solution) markedly increased the service life of the cell. By 1910 both cell types were considerably improved and soon plants were built for mass production of them. The designs and manufacturing technology developed at this time for cells with pocket and tubular electrodes are still in use with small modifications.

In 1928 the first experiments were made with a new electrode type based on sintered porous nickel ("sintered electrodes"). At the beginning of the thirties development of the sealed alkaline cell was reported. Both cell types were developed further after World War II. In about 1950 mass production of these new cell types was started.

11.2. Electrochemical and physicochemical processes

The charged positive electrode of these cells contain NiOOH, the hydroxide of trivalent nickel and the negative electrodes contain metallic cadmium or iron. As a rule, KOH solution serves as the electrolyte.

The main current-producing reactions on the electrodes and in the cell in general can be written as

$$(+)\ 2 \times [\text{NiOOH} + \text{H}_2\text{O} + \text{e} \underset{\text{charge}}{\overset{\text{discharge}}{\rightleftharpoons}} \text{Ni(OH)}_2 + \text{OH}^-]$$

$$(-)\ \text{M} + 2\text{OH}^- \rightleftharpoons \text{M(OH)}_2 + 2\text{e}$$

$$(\text{cell})\ 2\text{NiOOH} + 2\text{H}_2\text{O} + \text{M} \rightleftharpoons 2\text{Ni(OH)}_2 + \text{M(OH)}_2$$

where M is cadmium or iron.

Actually, the processes on the positive electrode are more complicated. There are several modifications of nickel oxides differing, in particular, in the degree of hydration. Therefore, the above equation does not give a correct description of the water balance in the reaction. The hydroxide of divalent nickel is formed as β-Ni(OH)$_2$ and has a lamellar structure with disordered crystal lattice. This disorder has a beneficial effect on the electrochemical activity. Charging typically gives rise to the hydroxide of trivalent nickel in the form of β-NiOOH. However, at high alkali concentrations or high charge currents γ-NiOOH can be formed which has a large specific volume. The resulting swelling frequently results in deterioration of contact and deformation of the electrode. The conductivity of pure Ni(OH)$_2$ is very low but even at slight oxidations it markedly increases. After discharge of the cell the active mass contains a residual 20–40% of non-reduced NiOOH which provides for a sufficiently high conductivity.

Prolonged operation results in the "ageing" of the electrodes and deterioration of their performance. One of the causes of decreasing activity is the transition of the crystal lattice into a more ordered state; another is aggregation and compressing of the particles of the active materials.

With charge higher nickel oxides are formed as well as NiOOH; the general formula of the resulting compounds can be written as $NiO_x \cdot yH_2O$ where x varies between 1·6 and 1·8 ($x = 1·5$ and $y = 0·5$ correspond to NiOOH). Opinions about the nature of these compounds differ; they may be mixtures of the oxides of trivalent nickel (NiOOH) and tetravalent nickel (NiO_2) or oxides of trivalent nickel containing an excess of oxygen.

The formation of these oxides has a number of significant consequences.

(1) The higher oxides are unstable and tend to decompose spontaneously liberating excess oxygen:

$$NiO_x \cdot yH_2O \rightarrow NiOOH + \tfrac{1}{2}(x - 1·5)O_2 + (y - 0·5)H_2O$$

The rate of this reaction increases with increasing content of oxygen. Therefore, in the first period after charging of the nickel–oxide electrodes (a few days or weeks), there occurs a noticeable self-discharge which is sharply reduced after all the higher oxides have been transformed into NiOOH. Spontaneous decomposition of NiOOH to the oxide of divalent nickel can occur, but its rate is very slow. Partial decomposition of NiO_x also occurs during charge. Owing to this the charge current, particularly by the end of the charge, is spent not only on oxidation of nickel but also on evolution of oxygen so that the capacity and energy efficiency of the cell is decreased (Fig. 12).

(2) Owing to temporary formation of the higher oxides on charge the o.c.v. of the freshly charged cell is higher and amounts to 1·45–1·7 V. With decomposition of these oxides the o.c.v. gradually decreases to the steady-state value of 1·30–1·34 V for the NiCd cells and 1·37–1·41 V for the NiFe cells.

(3) The alkali from the solution plays an essential part in formation of the higher oxides. The potassium ions are adsorbed at the oxide surface and partially penetrate the crystal lattice giving rise to mixed oxides, the composition of which is sometimes described by formulas of the type $[Ni_4O_4(OH)_4](OH)_2K$. The higher the alkali concentration the more easily are these compounds formed. Their specific volume is rather large, so that swelling of the active materials occurs. The concentration of the alkali solution decreases on charge not only because of formation of water but also

because of binding of a certain amount of KOH owing to formation of such compounds.

(4) If LiOH is added to the solution, adsorption and penetration into the lattice of Li^+ ions increases the depth of charge and prevents sintering of nickel oxides with cycling of the cell. This results in increased capacity and cycle life. However, if there is an excess of LiOH the electrochemically inactive lithium nickelate, $LiNiO_2$, can be formed and the cell performance can deteriorate. The beneficial effect on the cell capacity and cycle life is also produced by cobalt and barium which are sometimes added to the active materials in the form of oxides or salts. Iron and aluminium have a harmful effect.

On charge, oxidation of metal at the negative electrode proceeds via the intermediate formation of the ions, HMO_2^-, of the divalent metal in the solution which are then hydrolyzed giving rise to insoluble hydroxide (or oxide):

$$M + 3OH^- \underset{\text{charge}}{\overset{\text{discharge}}{\rightleftharpoons}} HMO_2^- + H_2O + 2e$$

$$HMO_2^- + H_2O \underset{\text{charge}}{\overset{\text{discharge}}{\rightleftharpoons}} M(OH)_2 + OH^-$$

Though the solubility of the HMO_2^- ions in the alkaline solution is low (about 10^{-4} mole/litre) this concentration is sufficient to enable the reaction to proceed both in the charge and discharge directions.

The equilibrium potential of the cadmium electrode is more positive by 0·02 V than the equilibrium hydrogen potential in the same solution and the equilibrium potential of the iron electrode is more negative by 0·05 V. Though the quantitative difference is small this distribution of potential has a significant consequence: the iron electrode can be corroded in the non-working state resulting in displacement of hydrogen from the solution. For cadmium this process is thermodynamically unfeasible and, hence, cadmium is corrosion-resistant.

From 5% to 30% of finely divided iron is added to the active materials of the cadmium electrode to prevent sintering of cadmium with cycling. This iron can also participate in the current-producing reaction; but in practice the contribution of iron to the operation of the cadmium electrode is small.

An important factor in stabilization of the characteristics of the iron electrode is the addition of 0·3–0·5% of iron sulphide (FeS) to the active materials. Sometimes, in the course of prolonged cell operation, sulphide is oxidized; under such circumstances the characteristics can be restored by adding a small amount of Na_2S to the solution. With deep discharge $Fe(OH)_2$ can be further oxidized to $Fe(OH)_3$; the potential of this process is more positive than the potential of the main process by 0·2–0·3 V. Thus, a characteristic second step is found on the discharge curve with a deep discharge. Though $Fe(OH)_3$ has poorer charging properties, periodic deep discharge can prove to be beneficial since it loosens the active materials.

11.3. Construction and manufacture

(a) Electrodes

There are several types of electrodes. Some of them are manufactured with two or three various thicknesses, the thinner electrodes for the cells designed for large discharge power and the thicker for the high-capacity low-power cells. Other conditions being equal, the positive electrode is thicker than the negative electrode owing to the larger volume of the active materials.

Pocket electrodes. The pressed active materials of the electrodes are placed into elongated flat boxes (pockets) manufactured from perforated strip of 0·1 mm-thick mild steel, the pocket strip. The strip for the positive electrode is always plated with nickel to prevent iron getting into the active materials. The pockets are covered with lids made from the same strip (see Fig. 63(a). The width of the pockets is 12·7 or 13·3 mm and the length is determined by the required width of the electrodes. The thickness of the pockets is 1·9–3·1 mm for the negative electrodes and 2·4–4·7 mm for the positive electrodes. In the electrode the pockets are arranged in horizontal rows with neighbouring pockets locked together by jointly beaded edges (see Fig. 63(b)). The electrode edges are press-fitted into a frame connected to a current collector. To increase rigidity and improve the contact the electrodes are additionally pressed and slightly corrugated on the surface.

The purpose of the pocket is to hold the active materials and to collect the current. The active materials contact with the electrolyte through the perforations. The ratio of the area of perforations to the total surface area is small (10–18%) resulting in shielding of the active materials and higher internal resistance. Unfortunately, this ratio cannot be increased since that would lead to washing out of the active materials. The number of perforations in the steel strip is 250–400 per cm² and the area of each perforation is 0·03–0·04 mm².

The tubular electrode is a modification of the pocket electrode. Tubes with a diameter of 4·64 or 6·35 mm and length of 80–114 mm are manufactured by spiral winding of the steel strip used for the pocket electrodes. The tubes are arranged in parallel vertical rows and, as for the pockets, fixed in a metal frame. The tubes are employed only in the positive electrodes;

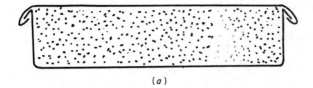

(a)

(b)

Fig. 63. Pocket electrodes of the alkaline storage battery: (a) schematic of the pocket; (b) connection between pockets.

they withstand more easily the stresses developing with swelling of the active materials. However, the tubular electrodes are difficult to manufacture.

Sintered plates (nickel oxide and cadmium only). In these electrodes the active materials are in the pores of the sintered nickel support plate. The base plate is manufactured from a high-disperse powder of nickel carbonyl, metallic nickel being produced by thermal decomposition of gaseous nickel pentacarbonyl, $Ni(CO)_5$. The mixture of nickel powder and ammonium carbonate is pressed on a coarse steel or nickel mesh or applied on a strip. The base is then heat-treated in a hydrogen atmosphere at 900–960°C so that nickel is sintered and ammonium carbonate evaporates leaving a large pore volume. The porosity of the plate can be as high as 80–85% and the pore radius varies between 5 μm and 20 μm. The finished base plates have a thickness of 1·4–1·8 mm (negative electrodes) and 1·8–2·3 mm (positive electrodes).

The base plates are filled with reactants by alternately impregnating them with concentrated solutions of the salts of the appropriate metals ($Ni(NO_3)_2$ for the positive plates and $Cd(NO_3)_2$ or $CdCl_2$ for the negative plates) and the alkali solution which produces precipitation of insoluble oxides or hydroxides. After two or four impregnation treatments the remaining NO_3^- ions are carefully washed from the electrodes and these are dried and formed by two or three charge–discharge cycles under certain conditions. In recent years efforts have been made to develop an electrochemical technique for introducing the reactants into the pores of the base plate.

The foil electrode is a modification of the sintered electrode. A thin layer of nickel powder is sprayed onto both sides of thin nickel foil (0·05 mm). An alcohol emulsion of nickel powder with binding additives is used for spraying. The total thickness of the base is 0·5–0·6 mm; it is sintered and then impregnated with reactants as for the conventional sintered electrodes.

Pressed electrodes. The active materials are pressed onto a mesh or stamped steel base under a pressure of 35–60 MPa. The electrode thickness is 0·8–1·8 mm. To improve their strength the electrodes are coated with alkali-resistant lacquer and sometimes pasted with fabric or paper. Cycling of the cells can result in partial washing out and shedding of the active materials, particularly of the positive electrode whose strength is lower owing to swelling. Therefore, the prepressed electrodes are assembled compactly so that the electrodes with the separators between them are pressed to each other. Electrodes manufactured by rolling the active materials have a somewhat better mechanical strength. Addition of binding agents to the active materials results in a marked increase in strength.[1]

The tablet electrode is a modification of the pressed electrode; the relatively small individual tablets of the active materials are arranged in a column in the cell.

Active materials of the (pocket and pressed) electrodes. Nickel hydroxide, $Ni(OH)_2$ (known as "green hydrate" in contrast to "black hydrate", $NiOOH$), for the positive electrode is obtained from a $NiSO_4$ solution by precipitation with alkali. To improve the conductivity about 20% of the natural flaky graphite is added to the active materials. The fine artificial graphite is unsuitable since it is readily oxidized during charge. Apart from graphite, other additives are 2–3% $Ba(OH)_2$ and 1–5% cobalt, typically, in the form of a solution of $CoSO_4$. About 10% of thin nickel flakes is added to the active materials of the tubular electrodes instead of graphite. The flakes form approximately three thin current-collecting layers per 1 mm of the tube length.

The cadmium material for the negative electrode is prepared from cadmium oxide produced by thermal oxidation of cadmium vapour or fine electrolytic cadmium powder. The additives are iron in the form of the oxide Fe_3O_4, and 3–4% of solar oil. Conducting additives are rarely used since conductivity is provided for by metallic cadmium which remains in the active materials even after complete discharge.

Since $Fe(OH)_2$ and fine iron powder are rapidly oxidized in the air to the oxide of trivalent iron which is rather inert electrochemically, the more active and stable magnetic iron oxide Fe_3O_4 is used for manufacture of the iron electrodes. This oxide is obtained by partial reduction with hydrogen of the naturally occurring iron oxide Fe_2O_3 (haematite, one of the most important iron ores). Use is made also of the artificial oxide Fe_2O_3 obtained from iron sulphate by roasting in the air or precipitating with alkali from solution. Sufficiently cheap and effective electrodes are produced by reducing and mixing the artificial oxide and the ore (80–85%). Another method for producing Fe_3O_4 is controlled oxidation of iron powder under strictly specified conditions. Small amounts of $NiSO_4$, FeS and graphite are added to the active material of the iron electrode.

Electrolyte. In cells operating at not too low temperatures (above $-15°C$), the electrolyte is a 20–22% KOH solution (density 1·19–1·21) to which 5–20 g/litre of LiOH are added. At lower temperatures higher concentrations of alkali are needed; for instance, at temperatures down to -30 or $-40°C$, 26–28% KOH solution is used (density 1·25–1·27). In this case LiOH is not added since this results in decreased conductivity. At higher temperatures concentrated electrolytes cannot be used in the cells with pocket electrodes since they cause considerable swelling of the active materials of the positive electrodes. There is no such danger for the sintered electrodes since, in those, nickel oxides expand in the partially filled pore space of the base plate and no swelling of the electrode occurs. Therefore, the cells with sintered electrodes can operate with higher electrolyte concentrations both at high and low temperatures. For the pocket-electrode cells operating at normal and elevated temperatures a good service life is obtained with a 15–18% NaOH solution (density 1·17–1·19) with the addition of 15 g/litre of LiOH.

(b) Storage cells

The conventional box construction is used for the vented storage cells. The current collectors of the individual electrodes are spot welded to the bridge which is a part of the terminal. Sometimes clamp contacts tightened with bolts are used. The electrodes are separated by thin ebonite rods, corrugated plastic grids or porous plastic sheets with ribbing.

The electrode block is fitted into the nickel-plated steel container, usually from the bottom which is then welded to the container. A small mud space is left between the bottom and the lower edges of the electrodes to accommodate the active materials washed out from the electrodes. The cover with the opening for pouring electrolyte and with the terminals is also tightly welded to the container. In the NiCd cells the end electrodes are positive and in the NiFe cells they are negative. The electrode half-block with the outside electrodes is frequently not insulated electrically from the container.

The distance between the upper edge of the electrodes and the cover varies between 20 mm and 70 mm. If this distance is large the size of the cell is increased but it is simpler to maintain since the electrolyte volume is large enough for there to be no need for frequent addition of electrolyte or water.

Some types of cells have the openings for filling with electrolyte fitted with special valve plugs with elongated lower bushing which prevents spilling of electrolyte when the cell is overturned. Such cells can be discharged and transported in any position and charged with tilts up to 45°.

In batteries the cells are installed in wooden containers or frames or metal racks. The individual cells should be carefully insulated from each other to prevent short circuiting. This is done by fixing the cells in certain positions with special pins on the containers so that air spaces are left between the cells. Sometimes the external surfaces of the cell containers are coated with special insulating compounds.

In recent years plastic cell containers have increasingly been used because they facilitate assembly of cells into batteries by eliminating the need for additional intercell insulation. Another convenience of the plastic containers is the fact that the electrolyte level can be seen in them, which facilitates maintenance and addition of electrolyte. Such containers can be used, in particular, for cells with sintered electrodes which do not swell and thus do not exert high pressure on the walls of the containers.

Non-sealed cells with sintered electrodes are similar in construction to the cells with pocket electrodes. Since the sintered electrodes do not swell in operation the interelectrode distance can be decreased. For this purpose thin (0·2–0·3 mm) fabrics of synthetic fibres or other materials are used as separators. By decreasing the excess of electrolyte between the electrodes and making the arrangement of the electrodes more compact the specific energy (with respect to mass or volume) of the cells with sintered electrodes can be made close to that of the cells with pocket electrodes though the sintered electrodes have a heavy metal-ceramic support.

There are three types of sealed cells, the can, disk and cylindrical cells. Operation of the sealed storage cells is described in general in Section 6.5. The construction of the can cells is similar to that of the non-sealed cells; the main difference is the fact that the volume of electrolyte is minimized to facilitate supply of oxygen to the cadmium electrode. The electrolyte is contained mainly in the electrode pores and in the separator. Both pocket and sintered electrodes can be used in such cells.

The sealed cells are often fitted with thin membranes which are ruptured when the pressure inside the cell exceeds a given level and thus act as safety devices. The sealed cells manufactured now have maximum capacity of 160 Ah. To increase the size and capacity of the cells heat removal must be improved since the intense liberation of heat at the end of charge can result in overheating of the cell.

In the disk cells (capacity from 0·01 Ah to 1·6 Ah, diameter from 11·4 mm to 51 mm, height from 5·1 mm to 15 mm) the active materials of both electrodes are pressed into tablets wrapped in thin nickel mesh (see Fig. 64). The positive tablet (1) is placed on the bottom of the steel container (7); the plastic separator (3) is placed above it and the negative electrode tablet (2) is placed on it. The cover (5) is rolled in over the packing and insulating plastic ring (6). To improve contact of the electrodes and separator with electrolyte a steel spring (4) is placed between the cover and the negative electrode tablet. The container and the cover serve as terminals for the positive and negative electrodes respectively.

Two types of the cylindrical cells are manufactured (capacity from 0·1 Ah to 6·6 Ah, diameter from 13·5 mm to 34 mm, height from 15 mm to 35 mm). In one cell type (Fig. 65(a)) pressed electrodes are used, the cylindrical positive electrode (6) is in the centre of the cell and the sectional negative electrode (2) surrounds the positive electrode, with the separator (3) placed between them. The cylindrical container (4) of the cell is in contact with the negative electrode and serves as its terminal. The ridges on the container form gas chambers (5) on the back side of the electrode through which

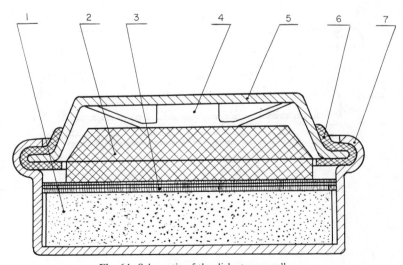

Fig. 64. Schematic of the disk storage cell.

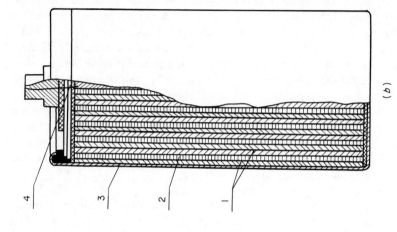

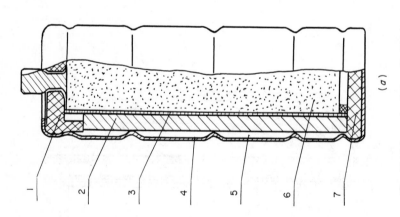

Fig. 65. Schematic of the cylindrical storage cell: (*a*) cell with pressed electrodes; (*b*) cell with coiled electrodes.

oxygen gets to the cadmium. The bottom (7) and the cover (1) are made of polyvinyl chloride plastic. Cells of another type (Fig. 65(b)) employ foil electrodes (1) and thin electrolyte-carrying separator (2) which are tightly rolled together. The outer layer contacting with the container (3) is the negative electrode; the terminal (4) from the positive electrode is welded to the cover.

11.4. Performance

(a) Common features of various types of cells

Cells with the tubular and pocket electrodes have long cycle and service lives. Their shortcoming is the restrictions on current and power; they are rarely used for normalized discharge currents j_d exceeding 0·5–1.

Cells with sintered electrodes are more expensive than cells with pocket electrodes and they consume more nickel. However, the cells with sintered electrodes have considerable advantages for high-current drain and for operation in a wide temperature range. Their higher efficiency is explained by the fact that in the sintered electrodes the path of the current from any reaction site to the current collector (here the wall of the pore) is not longer than 10–15 μm, that is, it is considerably shorter than in the pocket electrodes.

The cells with pressed electrodes have a higher specific energy with respect to mass, owing to the use of light-weight current collectors. Their manufacturing costs are lower than for other cell types. However, they have a shorter cycle life owing to the insufficient strength of the active materials.

Cells with combined electrodes are widely used. The tubular positive electrodes are combined with thinner negative pocket electrodes. Pressed negative electrodes, whose strength is much better than those of the positive ones, are often combined with positive pocket or sintered electrodes resulting in increased cycle life.

(b) Electrical characteristics

Figure 66 shows the typical discharge curves of NiCd cells in U_d–Q_d/C coordinates. The voltage for a given current increases as we go over from the pocket electrodes to the sintered and foil ones. Given below are the approximate normalized effectve internal resistance r_{eff} for various electrode types (in ohm Ah):

Electrodes $\bar{r}_{eff}$	Tubular 0·4	Pocket 0·1–0·3	Sintered 0·06	Foil 0·03

The rated regime corresponds, to discharge at $j_d = 0 \cdot 1$–$0 \cdot 2$ to the cut-off voltage of $1 \cdot 0$–$1 \cdot 1$ V at $20°C$; in this regime the mean voltage is $1 \cdot 20$–$1 \cdot 24$ V.

The discharge curves for the NiFe cells are similar but the discharge voltage is higher by $0 \cdot 02$–$0 \cdot 05$ V.

Figure 67 shows the charge curves for the rated charge regime ($j_c = 0 \cdot 2$). The shape of the curve for the vented NiCd cells is determined by the negative electrode. After cadmium oxide has been reduced by approximately 95% a distinct voltage jump from $1 \cdot 5$ V to $1 \cdot 7$ V is seen and evolution of hydrogen starts. In the sealed NiCd cells this voltage jump is not observed since cadmium is continuously oxidized by the oxygen evolved on the positive electrode and, hence, is not reduced completely. The mean charge voltage for the NiFe cell is higher by approximately $0 \cdot 25$ V; therefore, evolution of hydrogen on the iron electrode starts long before its complete charge. The voltage jump at the end of charge is not pronounced. The NiFe cells should not be charged with low currents (j_c less than $0 \cdot 05$) since the current is spent only on evolution of hydrogen and iron hydroxide is not reduced. These cells cannot therefore be charged at a constant voltage and are unsuitable for use in the buffer regime.

Owing to evolution of oxygen on the nickel oxide electrode the amount of electricity supplied on charge should be higher than the discharge capacity by not less than 40% for cells with pocket electrodes and 20% for the cells with sintered electrodes, corresponding to a capacity efficiency of 70% and 83%, respectively. According to the difference between the charge and

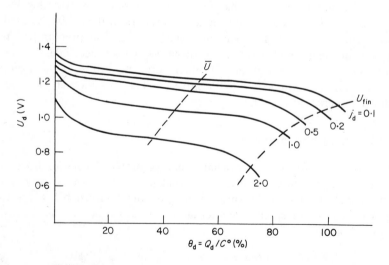

Fig. 66. Discharge curves of nickel–cadmium storage cells at $25°C$.

discharge voltages the energy efficiency in the rated regime is about 50% for NiFe cells, 55–57% for NiCd cells with pocket electrodes and 70–73% for cells with sintered and foil electrodes.

The specific energy of the conventional types of the NiCd and NiFe cells varies between 20 and 30 Wh/kg decreasing to 10–15 Wh/kg for high current drains. For the high-capacity cells in particular those with pressed electrodes, the specific energy is as high as 35–38 Wh/kg in the rated discharge mode. The specific energy with respect to volume is 40–60 Wh/dm^3 and for the high-capacity cells it is 70–80 Wh/dm^3. The specific characteristics for the small-size sealed cells (disk and cylindrical) vary about the lower margins of the above ranges.

The cell capacity and voltage are lower at low temperatures of discharge but the difference is not so large as for other widely used cell systems. NiCd cells with high-concentration electrolyte can be discharged at temperatures down to $-40°$C. The discharge capacity at this temperature for $j_d = 0 \cdot 1 – 0 \cdot 2$ is decreased to 40–60% for cells with sintered or foil electrodes and 20–30% for cells with pocket electrodes. Cells with the pressed electrodes also have good characteristics at low temperatures. The performance of the NiFe cells deteriorates more noticeably with decreasing temperature owing to faster passivation of the iron electrode. The NiCd and NiFe cells can also be charged at low temperatures, the NiCd cells down to $-40°$C and the NiFe cells down to $-10°$C, though the efficiency of charging is somewhat decreased and the charge voltage is increased.

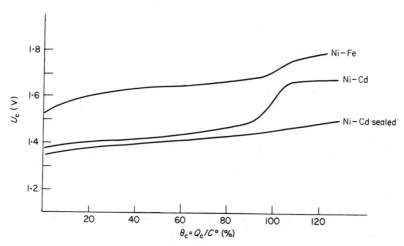

Fig. 67. Charge curves for the nickel–iron (NiFe), vented nickel–cadmium (NiCd) and sealed nickel–cadmium storage cells for $j_c = 0 \cdot 2$.

(c) Service life, cycle life and self-discharge

Cells with the tubular positive electrodes have the longest cycle life, between 2 and 4 thousand charge–discharge cycles with the discharge capacity decreasing by not more than 25%. The cells with pocket electrodes have the actual cycle life of 1000–2500 cycles. The service life for these cell types is 8–10 years but in some cases it can be 25 years or more. The cells with pressed electrodes have a cycle life of a few hundred cycles (up to 500–700). It should be remembered that all these values are strongly dependent on the operation conditions and the quality of cell maintenance.

Figure 68 shows the typical curves representing variation of the residual capacity with storage of the charged cells. The initial rapid decrease of capacity is due to decomposition of the higher nickel oxides. After the second month of storage the NiCd cells exhibit an insignificant loss of capacity (2–3% a month). For cells with sintered electrodes self-discharge is somewhat higher than for those with pocket electrodes; this is caused by chemical reaction of $NiO_x \cdot yH_2O$ with the nickel base giving rise to $Ni(OH)_2$. In the NiFe cells self-discharge is high owing to a high corrosion

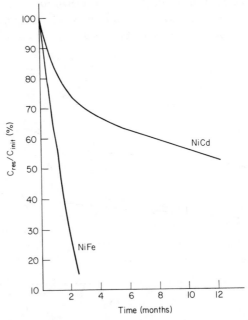

Fig. 68. Decrease in the capacity of the charged nickel–cadmium and nickel–iron storage cells during storage at room temperature.

rate of iron. After three months of storage practically the total capacity is lost. For all types of cells self-discharge depends on the temperature: at $-5°C$ self-discharge is low even for the NiFe cells.

(d) New type of nickel–iron cells[2]

A new type of the NiFe cell with improved performance is under development in Sweden, the USA and West Germany. The negative electrode is manufactured by pressing and sintering of fine pure iron powder on a thin support of mild iron. The positive electrode is also manufactured by pressing, and non-swelling materials have been developed providing for long-time stable operation of the electrode. The weight of the cell has been considerably reduced. The mass of the current collectors has been decreased from 12 kg to 2·7 kg, the mass of the structural elements from 10 kg to 2·4 kg (for instance, plastic containers are used) and the mass of electrolyte from 8·5 kg to 4·6 kg, while the total mass was decreased from 42 kg to 20 kg (all calculated for 1 kWh). Thus, the specific energy of this new cell is about 50 Wh/kg for $j = 0·5$, compared with 24 Wh/kg for the old types of cell. The specific energy can be increased in future to 55–65 Wh/kg by improving utilization of the active materials.

The charge characteristics of the new type of iron electrode are close to those of the cadmium electrode (see Fig. 67). Hence, gassing decreases and the capacity efficiency increases. Moreover, self-discharge of the iron electrode is markedly decreased making possible longer storage of charged cells. This is facilitated to a considerable extent by replacement of the nickel current collector with the iron one.

Only pilot production of the new cells is under way.

11.5. Maintenance of alkaline storage cells

The NiCd cells can be charged both with constant current and with constant voltage. In the latter case the voltage should be 1·6–1·7 V per cell. The voltage of the primary current source should be the same when these cells are used in the buffer regime. To preserve the battery charge compensation recharging is employed with normalized current j_c about 0·01 or voltage of 1·4–1·5 V per cell.

The sealed cells cannot be charged at constant voltage since the charge current does not decrease with time owing to liberation of heat and increasing temperature but, on the contrary, increases. The increase in current leads to further increasing temperature which can result in a runaway process. Therefore the sealed cells are charged only at constant current (typically, with j_c not exceeding 0·1) and prolonged recharge can be carried out with $j_c = 0·02–0·03$. The end of charge can be indicated by a temperature rise; after the end of charge, liberation of heat and the rate of temperature increase are sharply increased owing to the oxygen cycle (see Section 6.5).

Very low temperatures do not affect the operability of the cells; the alkaline solution does not freeze as a whole but forms a viscous mass so that there is no danger of container rupture. On the other hand, at elevated temperatures the positive electrode can be irreversibly damaged (owing to oxidation of graphite, dissolution of iron and iron getting into the active material of the positive electrode). Such cells therefore cannot be used at temperatures exceeding 40°C (with pocket electrodes) and 45°C (with sintered electrodes). Since charging is accompanied by heating, the maximum ambient temperature on charge should be lower by 5–7 degrees.

In contrast to the lead cells, the nickel cells can be stored for a long time without irreversible changes in both charged and discharged or partially discharged states. The capacity loss due to storage can be restored by training cycles. The nickel cells differ from the lead cells also in that in them the degree of charge cannot be determined from the density of the electrolyte solution, since, owing to the complex phenomena of hydration of the nickel oxides and absorption of the alkali, the variation of the solution concentration does not correspond to equation (11.3) and also is not reproducible.

Sealed nickel cells need no maintenance. For the vented cells periodic addition of deionized water is needed to compensate for losses due to electrolytic decomposition of water on charge. For the cells with the pocket electrodes overcharge is about 40% corresponding to water loss of 0·14 g per Ah of the capacity or to lowering of the electrolyte level by approximately 3 mm with a complete charge. Lowering of the electrolyte level below the upper edge of the electrodes is inadmissible. Hence, water should be added to the conventional cells every 3–4 complete cycles. If the gas chamber is large it accommodates a large reserve of electrolyte, and water can be added considerably less often. In the cells with sintered electrodes overcharge is about 20% and the water losses are correspondingly lower.

Though the cells are plugged, carbonates are gradually accumulated in the alkaline electrolyte (especially due to carbon dioxide dissolved in the added water). When the content of these salts becomes high (e.g., exceeding 60–70 g/litre) the electrolyte solution should be replaced. The frequency of electrolyte replacement depends on the conditions of use and maintenance; typically, it is done once every 2–3 years. Electrolyte is replaced when the cell is discharged, to prevent oxidation of metallic cadmium or iron in the air.

When vented cells are cycled a white deposit of carbonates is often found on the cover near the plug and the terminal owing to "creeping" of the alkaline solution through even the slightest leaks. The cover must be always kept clean.

REFERENCES

1. J. Jindra, J. Mrha, K. Micka, Z. Zábransky, B. Braunstein, J. Malik and V. Koudelka, Plastic bonded cadmium electrodes prepared by a rolling technique, *in* "Power Sources 6" (Collins, D. H., ed.), pp. 181–200. Academic Press, London and New York (1977).
2. J. D. Birge, J. T. Brown, W. Feduska, C. C. Hardman, W. Pollack, R. Rosey and J. Seidel, Performance characteristics of a new iron–nickel cell and battery for electric vehicles, *in* "Power Sources 6" (Collins, D. H., ed.), pp. 111–128. Academic Press, London and New York (1977).

BIBLIOGRAPHY

1. P. C. Milner and U. B. Thomas, The nickel–cadmium cell, *in* "Advances of Electrochemistry and Electrochemical Engineering" (Delahay, P. and Tobias, Ch. W., eds.), Vol. 5, pp. 1–86. Interscience, New York (1967).
2. R. Kinzelbach, "Stahlakkumulatoren". VDI, Düsseldorf (1968).

Chapter Twelve

Alkaline Cells with Zinc Anodes

12.1. Zinc anode in alkaline electrolyte

(a) Single-discharge zinc anode

Metallic zinc is a very convenient reactant for chemical power sources—it is a good reducer with a sufficiently negative potential, has a high corrosion resistance and is easy to process. The actual specific consumption of zinc is close to the theoretical value of 1·22 g/Ah. Zinc anodes were used in the earliest electric cell, Volta's pile, and are still employed in a variety of cells with salt, acid and alkaline electrolytes.

Alkaline cells with zinc anodes employ 27–40% KOH solutions (6–10 mol/litre). The 31% KOH solution has the lowest freezing temperature ($-67°C$, see Fig. 19(a)). A less frequently used electrolyte is 20% (6 mol/litre) solution of the cheaper NaOH, the freezing temperature of which is $-28°C$.

The operation of zinc anodes in alkaline solutions has the following specific features. The anodic dissolution of zinc

$$Zn + 4OH^- \rightarrow ZnO_2^{2-} + 2H_2O + 2e \qquad (12.1)$$

leads to a high consumption of the alkali since two OH^- ions are needed for each electron, and gives rise to a soluble product, zincate ions (the so-called primary process in the operation of the zinc electrode). The composition of the zincate ion is not quite definite and varies with the alkali concentration; in a wide concentration range zincate composition is close to ZnO_2^{2-} (or $Zn(OH)_4^{2-}$). The solubility of zincate in the alkaline solutions of the above concentrations is 1–2 mol/litre. When saturation has been reached zinc hydroxide starts to sediment on the zinc surface and the primary process practically stops. Thus, the capacity of the zinc electrode is limited here by the volume of the alkaline solution rather than the amount of zinc—about

242

10 ml of the solution are needed for each ampere-hour. The consumption of alkali is somewhat reduced by the marked ability of the zincate to produce super-saturated solutions.

If the current density is very low the zinc electrode also continues to operate in the saturated zincate solution giving rise to insoluble zinc hydroxide or oxide (the secondary process):

$$Zn + 2OH^- \rightarrow Zn(OH)_2 + 2e$$
$$(or \rightarrow ZnO + H_2O + 2e) \tag{12.2}$$

In the secondary process the consumption of OH^- ions is half of that in the primary process. Since discharge of the cell, as a rule, results in production of one OH^- ion for each electron at the positive electrode (see, for instance, equation (1.3)), during the secondary process in the cell there is no overall consumption of alkali (see equation (1.1)) and 1–2 ml/Ah of the solution are sufficient for operation of the cell, primarily for filling the interelectrode space and electrode pores.

Thus, there are two possible methods for utilizing zinc anodes in alkaline solutions. The first, older, method makes use only of the primary process. A comparatively thick monolithic zinc anode is immersed into a container with a large volume of electrolyte together with the cathode. The anode has a relatively smooth surface and can operate for a long time with current densities up to 5–8 mA/cm^2 until the solution is saturated with zincate. Since the density of the zincate solution is higher than the density of the starting alkali, stratification occurs in the solution. This effect is used to increase the cell capacity somewhat by placing the electrodes in the upper part of a tall can where the zincate concentration is lower than the average value.

The second method makes use of the secondary process of zinc oxidation. Such cells employ powdered anodes in which the true current density is much lower owing to their large true surface area, thus providing conditions for the secondary process. On the other hand, if the powder is sufficiently fine the total current density of the electrode can be high, up to 0·1 A/cm^2 or more. The mechanical strength of the powdered anodes is low so that these anodes are employed not in the cells with free electrolyte but in the matrix-type cells in which electrolyte is absorbed by the matrix which is pressed to the electrodes and prevents shedding of the active materials.

Three problems are encountered in operation of the zinc electrodes in alkaline solutions: zinc corrosion, zinc passivation and ageing of the solution and zinc hydroxide.

The rate of hydrogen evolution on zinc is small; hence, the rate of zinc corrosion is not high. But under certain conditions, particularly for prolonged operation or in sealed cells, corrosion has to be taken into consideration. To decrease the rate of corrosion it is necessary to employ pure reactants, that is zinc and, particularly, electrolyte. Even traces of such metals as iron sharply enhance the corrosion rate. Amalgamation of zinc is widely used to decrease the corrosion rate. The monolithic electrodes are manufactured by casting from an alloy of zinc with 0.5–2.5% mercury. Corrosion of zinc in alkaline solution is reduced with accumulation of the zincate. For this reason, in cells with powdered electrodes (for which the corrosion rate is high owing to their high true surface area), the zincate concentration of the solution is often increased by predissolving zinc oxide in it.

Operation of zinc electrodes in alkaline solutions does not produce high polarization. But if the current density is higher than a critical value this may lead to passivation of the electrode, that is to a drop in the cell voltage and the discharge current. Passivation can occur in both the primary and secondary processes. The critical current density depends to a great extent on the electrode type, electrolyte composition, etc.

Zinc passivation is caused by the fact that passage of the anodic current, simultaneously with the main reactions (12.1) or (12.2), gives rise to thin oxide layers (or just adsorbed oxygen) on the zinc surface. When these layers acquire a certain structure the main electrode reactions become inhibited. The deposits of zinc hydroxide formed in the reaction (12.2) generally have a fairly loose structure and do not serve as the direct cause of passivation. But these deposits partially shield the surface and hence increase the local current density on the working surface regions (in the deposit pores), thus possibly contributing to formation of the passivating layer. This is why passivation of the smooth electrodes occurs fairly soon after beginning of deposition of $Zn(OH)_2$ at current densities exceeding a certain (low) critical value.

Passivation effects, particularly in the secondary process of operation of the zinc electrode, are greatly influenced by the ageing processes occurring both in the supersaturated zincate solution and in the zinc hydroxide deposits. The supersaturated zincate solution most probably has a colloidal nature. At a given degree of supersaturation or under other conditions the colloid particles coagulate and the excess of zinc gradually precipitates in the form of zinc hydroxide. Additions of SiO_3^{2-} or Li^+ ions stabilize the supersaturated solution, that is they increase the maximum degree of supersaturation and inhibit precipitation of zinc hydroxide.

Zinc hydroxide has a variety of crystallographic modifications from α-$Zn(OH)_2$ to ε–$Zn(OH)_2$. The various modifications differ not only in structure but also in free energy and solubility in alkalis. The α form has the highest and ε form the lowest activity. The more active and less stable forms transform with time into the more stable forms. When the KOH solution concentration is more than 35% (8 mol/litre) the most stable form and the final product of all transformations is the non-hydrated zinc oxide ZnO. Various modifications of zinc hydroxide with different structures have different shielding effects on the surface and affect formation of the passivating layer to a varying degree.

Owing to the effect of the ageing processes, the phenomena of passivation of the zinc electrode in the alkaline solutions are fairly complex. As a rule an increase in the alkali concentration improves zinc performance and decreases the possibility of passivation. A decrease in temperature facilitates passivation and reduces the critical current density. However, in some cases an increase in temperature facilitates passivation owing to enhancement of the ageing processes giving rise to solid deposits with unfavourable structure. Under all circumstances the main method of preventing passivation is to work at very low true current densities. This is why the powdered zinc electrodes with a large true surface area have good performance at both normal and low temperatures.

(b) Rechargeable zinc electrode

Utilization of the zinc electrodes in various alkaline storage cells opens up very attractive prospects. The electrical parameters of the zinc electrodes are much better than those of the cadmium and iron electrodes employed in the widely used nickel–cadmium and nickel–iron storage cells. Development and production of the first types of cell with zinc electrodes (nickel–zinc and silver–zinc) were started in the thirties. However, there is still no solution to two problems which considerably restrict application of such storage cells: dendrite formation during charging of the zinc electrodes and displacement of the active materials (electrode shape changing) with cycling.

During electrolytic deposition of zinc from the alkaline zincate solution, zinc crystallizes in the form of thin branching crystals (dendrites) growing into the solution. During charging of the storage cell the dendrites rapidly reach the positive electrode, if special preventive measures are not taken, and the contacts thus formed result in internal short circuits and breakdown of the cell.

Storage cells with zinc electrodes became practicable only when powdered electrodes had been developed in which the secondary process is mainly used and the contribution of the primary process is reduced to a minimum. In the powdered electrodes zinc crystallizes from solid zinc hydroxide during charging. Though this process probably occurs via solution with intermediate formation of zincate ions, the morphological properties of zinc hydroxide are still retained to a certain extent; the zinc crystals are formed, mainly, in the porous bulk of zinc hydroxide and the probability of dendrite growth into the solution is lower.

The use of ultraporous swelling membranes is another method for decreasing the danger of short circuiting during charging of zinc electrodes. Although sufficiently conducting, that is permeable to the K^+ and OH^- ions of the alkaline solution, these membranes hamper passage of the zincate ions and resist the growth of zinc dendrites. Unfortunately, such membranes have low chemical resistance and break down with cycling of the cell.

Overcharge of storage cells is very dangerous from the viewpoint of dendrite formation. When all the accumulated solid zinc hydroxide has been reduced reduction of the zincate ion starts in the solution and crystallization of zinc does not remain limited to the electrode pores.

One possible method for preventing short circuiting due to dendrite growth is the use of vibrating zinc electrodes.[1] A special device vibrates the electrodes during charging of the cell, breaking the dendrites being formed. Naturally, a cell with vibrating electrode must have a fairly complicated construction. The advantages of such cell are still unclear.

The second effect which hinders the use of zinc electrodes is the gradual displacement of the active materials from one part of the electrode to another. The displacement is not caused by gravitational forces and can occur in various directions, for instance, upwards or from the electrode centre towards its edge. As a result, some parts of the electrode are bared while other parts become thicker and the density of the active materials becomes higher (shape changing). The true current density at the working electrode regions becomes higher. Moreover, higher density of the active materials leads to deterioration of the operational conditions making possible premature passivation of the electrode. In some cases the active materials are displaced downwards so that this phenomenon is often incorrectly called shedding or sliding of the zinc electrode.

The causes of displacement of the active materials with cycling have not yet been fully understood.[2] Undoubtedly, the main cause is the non-uniform distribution of the discharge and charge currents along the electrode surface. This non-uniformity may be associated with the ohmic potential drops in the electrode (current collector) or in solution and with the hindered diffusion of ions in the central part of the electrode where the porous matrix is compressed to a higher extent than at the electrode edge. At some parts the current density of discharge is higher and that of charge lower than the mean current density. As a result, a supersaturated zincate solution is produced and the zincate ions start to move towards other regions where the charge current density is higher and, hence, deposition of the excess zinc can occur. Undoubtedly, the high solubility of zincate and its tendency to form supersaturated solutions promote this phenomenon.

The non-uniformity of current distribution depends on the total current value, the higher the current the sharper the non-uniformity. Therefore, the conditions of cell operation strongly affect the character and degree of displacement of the active materials.

The displacement of the active materials leads to gradual loss of the cell capacity and after approximately 200–300 charge–discharge cycles the cell breaks down. Although this phenomenon has been extensively studied there are still no reliable methods for dealing with it. Addition of various binding or reinforcing substances to the powdered electrode produces only a small improvement in the electrode performance. Development of a stable and long-lived reversible zinc electrode for alkaline storage cells remains one of the major tasks in the work on cells with aqueous electrolytes.

12.2. Alkaline copper–zinc cells

Alkaline copper–zinc cells are manufactured mainly as high-capacity cells (250–1000 Ah). The cells are assembled in large glass or plastic containers with large volumes of liquid alkaline electrolyte and the smooth zinc electrode operates mainly via the primary process. These cells are used typically for stationary applications such as signal and communication railway devices.

The first prototype of the copper–zinc alkaline cell was developed in 1881 by F. Lalande and G. Chaperon. Commercial production of these cells started as early as 1889. The modern construction of these cells took shape by about 1910 (Fig. 69).

Construction. The active material of the positive electrode (4) is copper oxide, CuO. Technical grade copper oxide alone or with copper powder is mixed with a binder (a solution of water-glass or other substance), pressed into pellets and roasted at 700–800°C in an oxidizing atmosphere for final oxidation of the copper. A conducting grid of metallic copper is then formed on the electrode surface by means of controlled reduction (in CO atmosphere or by applying zinc powder). A steel frame placed on the end faces of the pellet serves as the external current collector.

The monolithic negative electrode (3) is cast from an alloy of zinc with mercury. The copper oxide and zinc electrodes are assembled into one electrode group with an interelectrode distance of 4–8 mm determined by spacers (2). Separators are not needed since the electrodes are rigid. The electrode group is suspended from the cover (1) in the upper part of the container (6). The capacity of the zinc electrode is higher than that of the copper oxide electrode by 20–30% to maintain the strength of the zinc electrode after dissolution of a considerable part of the zinc. The electrolyte (5) is a 20% NaOH solution (10 ml per ampere-hour of rated capacity).

The thickness of the zinc electrodes is, typically, smaller at the lower part. Special "indicator windows" with various metal thickness are made in the outside zinc electrodes to evaluate the degree of discharge of the cell which is indicated by complete dissolution of metal in the appropriate window.

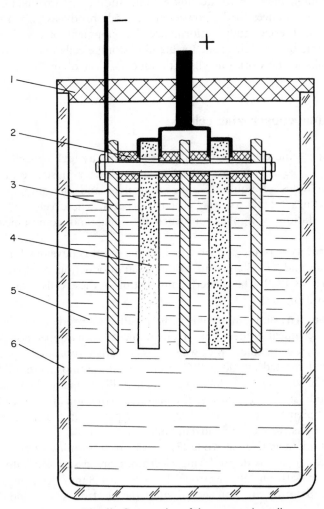

Fig. 69. Construction of the copper–zinc cell.

Performance. The final product of discharge of the copper oxide electrode is metallic copper. The reaction has a rather complicated mechanism owing to possible intermediate formation of the oxide of univalent copper:

$$2CuO \xrightarrow{+H_2O - 2OH^- + 2e} Cu_2O \xrightarrow{+H_2O - 2OH^- + 2e} 2Cu \qquad (12.3)$$

The oxide Cu_2O can also be formed in the absence of external current in chemical reaction between copper and CuO:

$$CuO + Cu \rightarrow Cu_2O \qquad (12.4)$$

The thermodynamic e.m.f. of the Zn/Cu_2O circuit (0.86 V) is lower than that of the Zn/CuO circuit (1.06 V). However, polarization with reduction of Cu_2O is considerably lower than that with reduction of CuO. Therefore, a cell with the electrode made from pure Cu_2O would have a higher discharge voltage than that with the CuO electrode (though the electrode capacity would be halved). At the beginning of discharge of a real CuO cell the voltage is elevated (0.88–0.95 V) but it rapidly drops as Cu_2O accumulated on the electrode is consumed. Further reduction of CuO (electrochemically or in the reaction (12.4)) proceeds slowly and the discharge voltage stabilizes at 0.6–0.7 V (Fig. 70).

The alkaline copper–zinc cells are designed for prolonged low-current drain—continuous discharge with current density up to 3 mA/cm^2 ($j_d = 0.003$) and intermittent discharge with the current density up to 6–10 mA/cm^2. The specific energy of these cells is 25–30 Wh/kg or 35–40 Wh/dm^3.

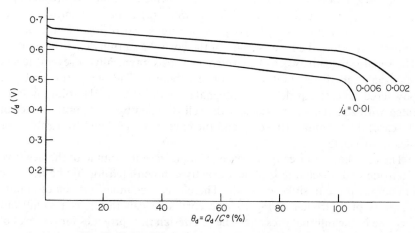

Fig. 70. Typical discharge curves for the copper–zinc cell.

Despite their low discharge voltage and poor performance, alkaline copper–zinc cells find a comparatively wide use owing to their reliability in prolonged operation, simplicity of maintenance, stability of the discharge voltage and low cost. These cells have satisfactory parameters at temperatures down to $-10°C$. When used for railway applications the cells are placed in 2–2·5 m deep pits where the temperature is acceptable even in winter. These cells exhibit negligible self-discharge and they can be operated for 10–15 years (within the limits of their capacity). To prevent carbonization and "creepage" of the alkaline solution a thin layer of unsaponifiable mineral oil is poured on top of it.

12.3. Mercury–zinc cells

Among the primary alkaline cells with zinc anodes the mercury–zinc cells are, in a sense, the opposite of the copper–zinc cells. They are manufactured as sealed cells of low capacity (0·05–15 Ah). The cells contain a limited amount of electrolyte (about 1 ml/Ah) absorbed in a porous spacer so that they operate only via the secondary process.

Modern mercury–zinc cells were developed by S. Ruben in the beginning of the forties. His "button" (disk) cell construction was so effective that large-scale production of such cells was started in the USA as early as World War II and after the war in other countries.

Construction. Figure 71(a) shows the simplest disk construction of the mercury–zinc cell. The active material of the positive electrode (1) consisting of mercuric oxide (HgO) and 5–15% fine purified graphite is pressed into the nickel-plated steel case (6). Zinc powder (2) is pressed into the steel cover (4) and amalgamated. To prevent intense corrosion of zinc at the contact with iron, the inner surface of the cover has a thin dense tin coating. The separator (3) between the electrodes consists typically of several layers of special alkali-resistant paper or cardboard. The separator and the powdered zinc electrode are impregnated with 40% KOH solution saturated with zincate. After assembly the cell is sealed by beading the edges of the case; the terminals (the case and the cover) are insulated by a rubber or plastic spacer (5).

The cells have no free (gas) space. The increase in volume of the negative electrode with discharge is almost exactly compensated for by the decrease in volume of the positive electrode. Therefore, accumulation even of minute amounts of hydrogen owing to zinc corrosion can result in a significant increase in the internal pressure. The cell design thus provides for removal of the evolved hydrogen by diffusion through the insulating spacer. The degree

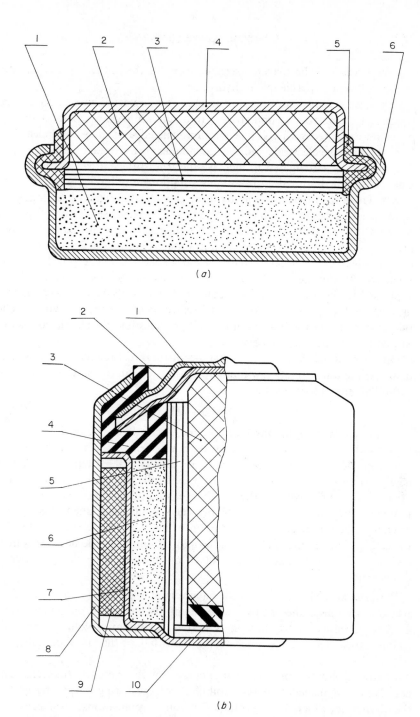

Fig. 71. Schematic of the mercury–zinc cells: (*a*) disk cell; (*b*) cylindrical cell.

of compression of the spacer is selected to make the cell leak-proof and the spacer sufficiently permeable to hydrogen.

Other cell constructions are also in use. The powdered zinc anode can be replaced by a narrow wound strip of thin zinc foil with a separator of thin paper impregnated with alkaline solution. The edge of the wound electrode faces the positive electrode and the second edge is in contact with the cover. The foil can be corrugated to increase the true surface area. The mercury–zinc cells of cylindrical type are also manufactured commercially (Fig. 71(b)). These cells typically employ more complicated types of sealing, such as the use of second containers and covers. The paper separator (9) placed between the inner (7) and outer (8) containers absorbs the electrolyte and prevents its leakage. The inner (2) and outer (1) covers are insulated from the container by a specially shaped rubber spacer (4). The powdered zinc electrode (3) is in the centre and the active mass of the positive electrode (6) is pressed to the wall of the inner container. The multi-layer diaphragm (5) impregnated with the electrolyte is pressed between the electrodes. The rubber cup (10) at the bottom of the cell provides for the reliable contact of the zinc electrode and the cover.

On the positive electrode discharge results in direct reduction of mercuric oxide to metallic mercury without formation of intermediate products or variable-composition phases:

$$HgO + H_2O + 2e \rightarrow Hg + 2OH^- \tag{12.5}$$

The final product of discharge of the zinc electrode is zinc oxide (owing to high alkali concentration and long discharge period). The coefficients of utilization of zinc and mercuric oxide are close to 100% (zinc corrosion is negligible owing to high amalgamation). The cells, typically, contain a small excess of mercuric oxide so that discharge is limited by the amount of zinc. If there were no zinc excess and the cell's circuit remained closed after discharge then after complete reduction of mercuric oxide the potential of the cell case under the effect of zinc would be shifted to the negative values and hydrogen would start to be evolved on it with the resulting danger of cell rupture.

Performance. Mercury–zinc cells have very stable o.c.v. (1.352 ± 0.002 V) which is very little affected by the degree of discharge and the temperature. Therefore, these cells are sometimes used in technology as voltage standards (though their accuracy is much worse than that of the standard cells, see Section 13.6).

Figure 72 shows typical discharge curves for the mercury–zinc cells. The cells are distinguished by good stability of voltage during the greater part of the discharge which is important for some applications. The cells are discharged to the cut-off voltage of 0·9–1·1 V (depending on the current drain) and then the voltage drops sharply. The cells employ comparatively thick electrodes with large capacity per unit surface area. Therefore, the capacity starts to decrease noticeably even at the normalized discharge currents exceeding 0.02 (the current densities exceeding $10\ mA/cm^2$). Thus, the cells are designed, primarily, for small- and medium-current drains up to $j_d = 0.1$. The normalized internal resistance varies between 1 and 8 ohm Ah depending on the cell construction.

At low temperatures the cell performance deteriorates. At $0°C$ the capacity starts to decrease at $j_d = 0.005$ and the internal resistance increases by a factor of 2–3. At $-20°C$ and $j_d = 0.002$ the cells deliver only about 20% of their capacity.

The main advantage of the mercury–zinc cells is their small size. The specific energy per unit mass is not very high (100–120 Wh/kg). But since the cells have a high mean density their specific energy per unit volume is higher than that of any other cells with aqueous electrolyte (400–500 Wh/dm^3) (all the above data are given for $j_d \leqslant 0.02$). Hence, these cells are used, primarily, in small-size devices such as electronic wristwatches, pocket calculators, etc.

Another advantage of these cells is their good storageability, the loss of capacity after storage for 3–5 years is 5–15$\%$. The cells can be stored at high temperatures, for instance for 3 months at $50°C$ and for short periods even at $70°C$.

The main drawback of mercury–zinc cells is their high cost and the scarcity of mercury sources.

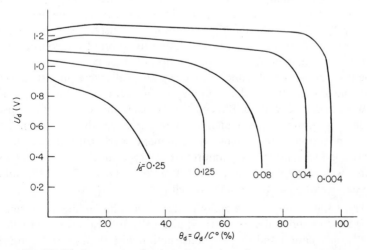

Fig. 72. Discharge curves for the mercury–zinc cell at room temperature.

Rechargeable cells. The mercury–zinc system can be utilized also for manufacturing rechargeable (storage-type) cells. The construction of the rechargeable cell differs from that of the primary cell in that a fine silver powder is added to the positive electrode mix instead of graphite. When the cell is discharged and mercury is formed the silver is amalgamated. This prevents mercury from coalescing into comparatively large beads which are difficult to oxidize during charging. The negative electrode is wound from a copper or brass strip electroplated with zinc. During discharge the anode strip is not affected and serves as a substrate for deposition of zinc during charging.

The specific energy of rechargeable cells is much lower than that of the primary cells (30 Wh/kg and 80 Wh/dm^3); their cycle life is low too.

12.4. Alkaline manganese–zinc cells

In the years after World War II transistor radios came into wide use, along with portable tape recorders and numerous other appliances requiring high-capacity small-size power sources. The conventional manganese–zinc cell with salt electrolyte (the Leclanché cells) proved to be inadequate for many of these applications. On the other hand, mercury–zinc cell cannot be widely used owing to scarcity and high cost of mercury. Hence, great attention was paid to developing the alkaline manganese–zinc cell. In comparison with the Leclanché cells these cells have a better performance at high discharge currents and low temperatures, and a better storageability. Their specific energy is also markedly higher though lower than that of the mercury–zinc cells (see Fig. 73). Although they are more expensive than the Leclanché cells their cost per unit energy is competitive and the resources of raw materials are sufficient for mass production of these cells.

The first reports on alkaline manganese–zinc cells were published as early as the end of the last and the beginning of this century. A "dry" cell of this type was suggested in 1912. But only in the beginning of the fifties did these cells become commercially available. These early cells did not permit heavy-drain applications. The intense development work carried out in many countries led to production of improved cell types in the sixties which soon became widely popular. At present storage cells based on the alkaline manganese–zinc system are being developed.

Construction. The commercially produced alkaline manganese–zinc cells mostly have cylindrical construction; disk cells are also produced while the flat cell construction and the box construction with flat electrodes are rarer. To provide for interchangeability, the cylindrical cells have the standard sizes of Leclanché cells and the disk cells have the sizes of mercury–zinc cells.

The active material of the positive electrode is electrolytic manganese dioxide (EMD) which consists mainly of the more active modification γ-MnO_2. Other components are 5–20% graphite, alkaline solution and binders. The volume of the solution in the active mass can be smaller than in the Leclanché cells owing to the higher conductivity of the alkaline solution and the absence of the danger of deposition of solid $Zn(NH_3)_2Cl_2$ in the pores. Therefore the loose carbon black (which is used in the Leclanché cells for retaining electrolyte in the pores) is not needed. Moreover, the porosity of the active materials can be decreased, they can be pressed more strongly, hence, the alkaline cell contains more MnO_2 than the Leclanché cell of the same size. For instance, the D-size cell contains 37–41 g, instead of 22–28 g.

Manganese dioxide is reduced in the alkaline solution giving rise, at first, to a variable-composition phase and then to the new phases MnOOH and $Mn(OH)_2$ (see Section 9.2). Some authors have reported possible intermediate formation of the phase Mn_3O_4 demonstrated by X-ray diffraction studies, while the existence of the reported phase MnO has not been clearly proven. In alkaline solutions the reduction reactions involve water molecules serving as proton donors; for instance reaction (9.1) has the following mechanism:

$$MnO_2 + H_2O + e \rightarrow MnOOH + OH^- \qquad (12.6)$$

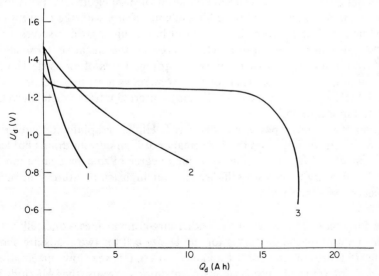

Fig. 73. Comparison of the discharge curves of the D-size cells at a current of 0·25 A: (1) manganese–zinc cell with salt electrolyte; (2) manganese–zinc cell with alkaline electrolyte; (3) mercury–zinc cell.

This leads to increasing alkali concentration in the pores. In contrast to the salt solutions, in the alkaline solutions the oxides of trivalent and divalent manganese are somewhat soluble in the form of complex anions, for instance, $[Mn(OH)_5]^{2-}$ and $[Mn(OH)_4]^{2-}$. Reaction (9.2) therefore probably proceeds via solution and is facilitated by this, in comparison with the solid state mechanism.

To provide for high performance of manganese–zinc cells, zinc electrodes with highly extended surface area should be used since the zinc can with smooth surface is ineffective as the anode. On account of this the cylindrical cells have the so-called "inside-out" construction illustrated in Fig. 74. The active material of the positive electrode (6) is pressed to the inner surface of the steel can (2). The separator (3) of unwoven plastic fabric and/or cellophane inserted into the can retains the electrolyte and prevents internal shorting. The petal-shaped brass current collector (8) is in the central part of the cell. The space between the separator and the current collector is filled with the anode paste (7), consisting of the alkaline electrolyte solution (typically, 30% KOH) gelled with starch or carboxymethylcellulose and zinc powder. The bulk concentration of the zinc powder (30–45%) provides for sufficient electron conductivity between the particles and between the current collector and the particles. This paste construction of the zinc electrode in which each particle is surrounded by a large amount of electrolyte decreases the danger of passivation at high-current discharge or at low temperatures. An additional amount of pure electrolyte (9) is inside the current collector. To provide for exchangeability with the conventional cylindrical cells the upper side of the cell has a lug (1) which serves as the positive terminal. The bottom (13) serves as the negative terminal; to improve the internal contact a pressure spring (12) is often used. The seal with the washers (10) and (11) contains sometimes a venting mechanism to relieve high pressures in the cell. The can is inserted into the metal jacket (4) with the insulator (5).

A variety of cell types is manufactured. Higher graphite content in the positive electrode improves the performance at high-current drains but leads to a somewhat lower cell capacity. A high degree of zinc amalgamation (up to 10% mercury) decreases self-discharge at high temperatures but makes the cell more expensive.

Performance. The o.c.v. of the undischarged manganese–zinc cell is 1·5–1·7 V. Typical discharge curves for the D-size cell for two loads are shown in Fig. 75 (at room temperature and at −20°C). For very low current drains and for intermittent discharge with medium currents the Ah and Wh capacities of the alkaline cells are about 1·5 times those of the Leclanché cells corresponding to the difference in the content of manganese dioxide.

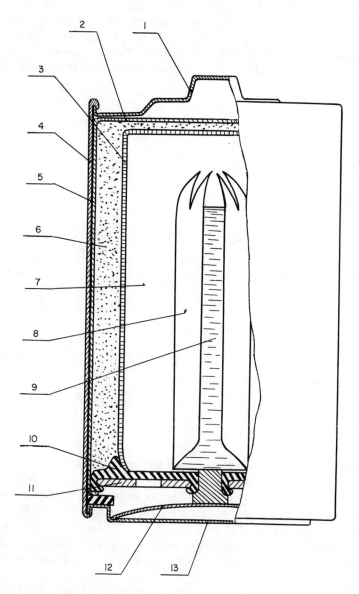

Fig. 74. Schematic of the alkaline manganese–zinc cell.

The specific energy of the alkaline cells can be as high as 80–90 Wh/kg or 190–210 Wh/dm^3. The advantages of the alkaline cells are particularly noticable for continuous discharge with medium and high current drains when the Leclanché cells have a sharply decreased capacity so that the alkaline cells have the capacity which is higher by a factor of 3–6. At $-20°$C the capacity of alkaline cells is approximately equal to that of Leclanché

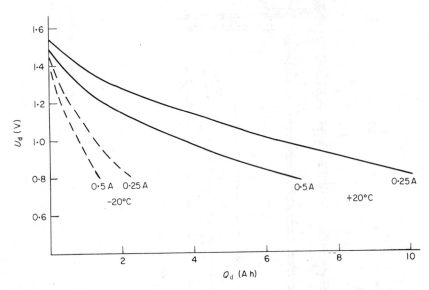

Fig. 75. Discharge curves for the D-size alkaline manganese–zinc cell at room temperature and at $-20°$C.

cells for continuous discharge at room temperature. At $-40°$C the alkaline cells can retain 7–10% of their capacity for current drains corresponding to the normalized current $j_d = 0\cdot1$–$0\cdot2$. The normalized internal resistance at the beginning of discharge is approximately $0\cdot4$, 1 and 2 ohm Ah at 20, -20 and $-40°$C, respectively.

Alkaline manganese–zinc cells have good storageability: they preserve about 90% of capacity after one year's storage at room temperature or after three months at 50°C.

The alkaline manganese–zinc cells require careful handling since in some cases alkaline electrolyte can leak owing to sudden sharp increases in the internal pressure, for instance, following short circuits, sharp temperature changes, etc. To decrease the danger of gassing and electrolyte leakage in discharged cells in closed circuits, the total amount of zinc in the cell is sometimes decreased as is done in the mercury–zinc cells (though this results in a decrease of the cell capacity).

Rechargeable cells. Alkaline manganese–zinc cells can in principle be recharged, that is used as storage cells. Such cells exhibit attractive features: they are sealed, there is no need to add electrolyte, the charged cells have very low self-discharge, the materials for the cells are freely available and cheap. However, these cells also have a number of drawbacks. The main one is that the manganese electrode retains reversibility only if not more than 25% of the primary capacity has been drained. Reversibility, apparently, depends on the variable-composition phase. When formation of the new phase, MnOOH, starts with discharge this results in immediate sharp deterioration of reversibility.

The cylindrical rechargeable cells are now commercially available. Their construction is very similar to that of the primary cells, with a few technological modifications. The positive electrode mix contains a larger amount of binding materials to improve its strength in view of periodic changes of volume during charge–discharge cycles. The content of the conducting additive (graphite or metallic powder) is also higher. The inner surface of the can is specially treated (for instance, coated with a graphite mix) to sustain electric contact with the active materials. The zinc electrode has a more effective mesh current collector (tin- or silver-plated copper) assuring better conditions for charging the electrode. The amount of mercury added to zinc is reduced to 0·5–2%. The sealed construction makes use of the oxygen cycle, charging is limited by the amount of manganese dioxide and oxygen produced reacts with metallic zinc. To facilitate permeability to oxygen, fibrous non-woven membranes are employed which, unfortunately, do not give sufficient protection from internal short circuits.

The cycle life of the rechargeable cells is about 50 cycles. The positive electrode gradually deteriorates; though only about 25% of its capacity is used in cycling, the remaining reserve capacity gradually decreases. The initial and final discharge voltages also decrease. The cells are charged at 1·7–1·75 V per cell with chargers equipped with current-limiting regulators.

These rechargeable cells are not storage cells in the proper sense of the word since their discharge capacity is limited and the cycle life is short. Work is under way to develop manganese–zinc storage cells with better performance. The main approach is to utilize thinner positive electrodes. Such electrodes make possible deeper discharge by improving uniformity of operation with thickness. Apparently, future storage cells will have a box-type construction with thin flat electrodes.

12.5. Silver–zinc cells

The silver–zinc storage cell is the first successful type of alkaline storage cell with negative zinc electrode. Despite their high cost these cells immediately attracted great attention owing to their high performance—their specific energy can be as high as 130 Wh/kg and is little dependent on the load, and their discharge currents can be high (up to $j_d = 10$).

Repeated attempts to develop storage cells using the silver–zinc system were made starting from the eighties of the last century. But a successful cell was developed only in 1941 by a French scientist, H. André, whose cell design had two essentially new features, the use of a swelling cellophane-type material for separators and a sharply limited volume of the electrolyte solution, so that the zinc electrode operated only via the secondary process. In the years of World War II twenty-five thousand such cells were manufactured in France. After the war manufacture of these cells started in many countries and they have found use in aerospace and other applications.

Construction of the silver–zinc cell. The construction of these cells is distinguished by very compact assembly of the electrodes in the plastic container (1) (see Fig. 76) in which the electrodes with the separator between them are strongly pressed to each other. The electrolyte (30–40% KOH solution saturated with zincate) is mainly in the electrode pores and in the separator (4); the electrolyte volume is about 1·5 ml/Ah. The positive electrode (2) is pressed from silver powder and sintered at about 450°C. The negative electrode (3) is pressed from fine zinc powder and zinc oxide with binders; the electrode is wrapped in thin paper. In some cells the electrode mix is pasted on paper. The zinc electrode in the free solution is not strong and the active materials are rapidly shed. Shedding is prevented by strongly pressing neighbouring electrodes together, which means that the electrodes lack any frame for applying the active materials (for instance, pockets, grids, etc.). The current collector is not used for improving the electrode strength; it consists either of a very thin silver mesh with a welded silver wire or just of wire loops pressed into the electrode. The mass of the current collector is not more than 10% of the electrode mass. The wire current lead (5) is inserted into the hollow terminal (6) and soldered to it. The cover (8) is glued or welded to the container. The vent plug (7) is used for filling the cell with electrolyte and for releasing gases evolved in the cell.

Silver oxide produced by charging the positive electrode is appreciably soluble in the alkaline solution and the concentration of the complex silver ions (for instance, of the $Ag(OH)_3^{2-}$ type) can be as high as 10^{-4} mol/litre. The film separator between the electrodes prevents not only the growth of zinc dendrites but also penetration of silver ions to the negative electrode. Contact reduction of these ions on zinc would result in growth of silver dendrites leading to internal short circuiting. The separator has several layers, typically of different materials. The layers closer to the silver electrode absorb and hold the silver ions, and the remaining layers prevent growth of the zinc dendrites. To prevent "roundabout" growth of the silver or zinc dendrites, one of the electrodes (typically, the negative one) is wrapped around with the separator (Fig. 31). Only the upper part of the electrode through which the evolved gases escape remains open but the separator wrapping edge is about 6–10 mm higher than the upper edge of the electrode.

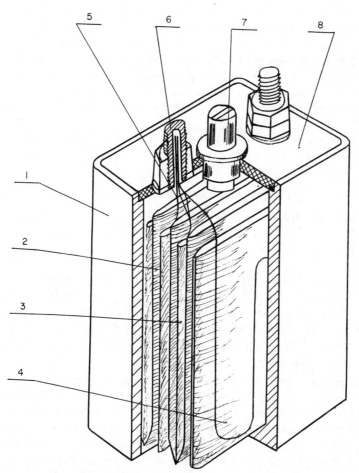

Fig. 76. Schematic of the silver–zinc storage cell.

Processes on the positive electrode. The charge–discharge curves of silver–zinc cells exhibit distinct steps (Fig. 77) typical of the operation of the silver electrode. During charging silver is at first oxidized to the univalent silver oxide:

$$2Ag + 2OH^- \rightarrow Ag_2O + H_2O + 2e \qquad (12.7)$$

At a certain moment when approximately 30–50% of the total amount of silver has been oxidized the process (12.7) suddenly stops (the electrode is passivated); the cell voltage is increased by 0·3 V and silver is oxidized to the divalent silver oxide:

$$Ag + 2OH^- \rightarrow AgO + H_2O + 2e \qquad (12.8)$$

When a certain degree of oxidation has been reached this process also stops and evolution of oxygen starts after a new increase in voltage. There is a distinct boundary between oxidation of silver and evolution of oxygen, so that, if charging is discontinued at the time of the second voltage hike 100% of the capacity can be drained from the cell. The charged electrode contains, apart from the oxides of uni- and divalent silver, a certain amount of metallic silver providing for good electronic conductivity of the active materials.

The o.c.v. of the fully charged storage cell is determined by the reaction (12·8) and is equal to 1·86 V. With low- and medium-current drain AgO is at first reduced to Ag_2O at the cell voltage of 1·7–1·8 V:

$$2AgO + H_2O + 2e \rightarrow Ag_2O + 2OH^- \qquad (12.9)$$

Later, reduction of Ag_2O to silver (this reaction is the reverse of reaction (12.7)) starts at about 1·5 V. The o.c.v. corresponding to this reaction is 1·60 V.

The considerable hysteresis seen on the charge–discharge curve where the upper step on the charge curve is longer than on the discharge curve is explained by participation of two electrons per AgO particle in the reaction (12·8) and one electron in the reaction (12·9). Moreover, the upper step of the discharge curve becomes shorter with increasing current since reduction of AgO is not proceeding uniformly in the grain's volume; at large-current drains the upper step disappears completely.

Performance. Figure 77 shows that when normalized discharge current j_d increases from 0·1 to 2 this results in a very small variation of the cell capacity. The decrease in the drained Wh capacity is due only to a decrease in the discharge voltage. If $j_d > 2$ the duration of discharge is mainly limited by overheating of the cell. The cell capacity remains high for intermittent discharge operation. At $j_d = 0·1$ noticeable decrease in capacity due to zinc passivation start only at temperatures below $-30°C$ (see Fig. 78).

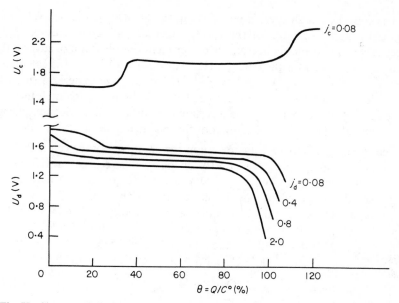

Fig. 77. Charge and discharge curves for the silver–zinc storage cell at room temperature.

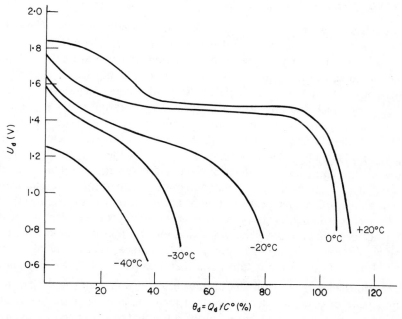

Fig. 78. Discharge curves of the silver–zinc storage cell at $j_d = 0.12$ and at various temperatures.

The cells have high specific energies (up to 130 Wh/kg and 300 Wh/dm^3) owing to high utilization of the active materials and small masses of the electrolyte and current collectors. The coefficient of utilization of silver is 0·6 to 0·7 as calculated for the reaction (12.8) or (formally) 1·2–1·4 as calculated for the reaction (12.7). A small excess of the zinc active material is used to preserve a certain amount of metallic zinc after discharge (to maintain the conductivity of the material) and a certain amount of zinc oxide after charge (to reduce the possibility of growth of the zinc dendrites).

Self-discharge. The loss of capacity in the charged storage cell is 2–4% a month at room temperature, mostly owing to zinc corrosion. To reduce corrosion a certain amount of mercury is added to the active materials. A slight self-discharge of the positive electrode can also occur. This is due to the fact that the silver ions in solution are reduced by the separator; their concentration therefore decreases and an additional amount of silver oxide (mostly Ag_2O) is dissolved. Moreover, AgO slowly decomposes giving rise to Ag_2O and oxygen, and AgO reacts with metallic silver in the electrode:

$$AgO + Ag \rightarrow Ag_2O \qquad (12.10)$$

This reaction does not lead to a decrease in the electrode Ah capacity but the upper step on the discharge curve gradually becomes shorter and finally disappears during storage of the cell.

Cycle and service life. The cycle life of silver–zinc cells varies from 30 to 200 charge–discharge cycles and the service life varies from 6 months to 2 years depending on the type of construction and operational conditions. The main causes limiting the cycle and service life are the displacement of the active material of the zinc electrode and gradual deterioration of the separator which loses its protective effect. Deterioration of the separator is due, mostly, to oxidation of the organic material by the silver ions. Deterioration starts at the layers adjacent to the silver electrode and gradually spreads to other layers. Oxidation by the oxygen evolved with overcharge contributes to deterioration of the separator. Intensive studies are under way to develop new film materials for separators. Pretreatment of cellophane with a silver salt solution has been shown to improve its stability with respect to further oxidation. Numerous attempts have been reported to replace organic film materials with oxidation-resistant inorganic separators. However, this approach has not yielded complete success owing to high brittleness of the inorganic films and their comparatively high resistance.

Maintenance. The silver–zinc storage cell can, and even should, be stored in a state of complete discharge since then the electrodes do not contain silver oxide which could be dissolved. Overcharge is dangerous owing to increased possibility of formation and growth of the zinc dendrites and oxidation of the separator with the evolved oxygen. Therefore, charging is discontinued at the final voltage of 2·0–2·05 V (at j_c about 0·1). During charging the voltage of each cell should be monitored.

Primary silver–zinc cells. There are two construction types of the primary cells—the miniature disk cells, similar to the mercury–zinc cells, and the comparatively large automatically activated reserve-type cells. The voltage of these disk cells is less sensitive to variation of the current than that of mercury cells. The silver electrode is specially prepared so that the upper step on the discharge curve is practically eliminated, that is the discharge voltage is practically constant. Such sealed cells can be recharged within certain limits; to prevent evolution of oxygen during charging it should be restricted to the lower part of the charge curve. Sealed silver–zinc cells are used for electronic watches and other miniature electronic devices.

Reserve silver–zinc cells combine high electrical parameters with good storageability in the unactivated state typical of all the reserve cells. They are employed mostly for aerospace applications. The specific parameters of these cells are about 70% of those of the silver–zinc storage cells owing to the additional volume and mass of the ampoules and auxiliary devices.

Silver–cadmium cells. A cadmium powdered electrode has been used instead of zinc in some types of storage and button cells. The charge and discharge voltages in these cells are lower by about 0·3 V. The relative variation of voltage between the steps of the charge and discharge curves is thus higher in these than in the silver–zinc storage cells. The main advantage of the cadmium electrode is the lack of self-discharge of the negative electrode and of growth of cadmium dendrites (the solubility of the cadmium ions in the alkaline solution is low, about 10^{-4} mol/litre). As a result the cadmium cells have a better cycle life than the zinc cells, about 300 cycles. The main cause of their breakdown is deterioration of the separator owing to dissolved silver ions. Their specific energy is about 70 Wh/kg or 130 Wh/dm^3.

Economic parameters. The cells with silver electrodes contain about 4–5 g/Ah of metallic silver, 75–90% of which is in the active material of the positive electrode. These cells can therefore be used only for special applications when other cell types are unsuitable and high costs are permissible.

12.6. Nickel–zinc storage cells

The high performance of silver–zinc storage cells was a stimulus for attempts to develop other types of high-capacity alkaline cells without expensive or scarce materials. Such are the nickel–zinc storage cells. These cells are extensively developed at present and they can be expected to be commercially available by the beginning of the eighties.

The first patent for the nickel–zinc cell was granted to T. Michalowski in 1899. Wide-ranging investigations were carried out by the English scientist J. Drumm in the twenties and thirties of this century. His cells were used as railway traction batteries in Ireland. However, they had poor parameters, owing to the use of monolithic zinc electrodes, and, above all, short cycle life. In 1936 Drumm was the first to suggest using the powdered zinc electrode but this did not result in any considerable improvement since large volumes of electrolyte were used. Significant advances were made only after World War II when the experience gained with development of the silver–zinc storage cells was utilized. The modern cells have essentially similar construction of the cell and the zinc electrode for both cell types.

Electrochemical processes. The reaction on the nickel oxide electrode is the same as the reaction on the positive electrode of the nickel–cadmium storage cell and the reaction on the zinc electrode is the same as the reaction on the negative electrode of the silver–zinc storage cell. The total current-producing reaction is described by the following equation:

$$2NiOOH + Zn + H_2O \rightleftharpoons 2Ni(OH)_2 + ZnO \qquad (12.11)$$

The processes in the nickel–zinc cell have some special features. The zincate ions whose concentration in the electrolyte is high affect operation of the nickel oxide electrode. Cycling of the pocket-type electrodes in the zincate solution results in a rapid decrease in their capacity (after 10–20 cycles). The decrease in capacity of the nickel oxide electrodes with a sintered support or the pressed electrodes is considerably lower owing, apparently, to a better contact between the NiOOH grains and the current collector or the conducting additive and better forming conditions. Addition of cobalt to the active materials promotes stabilization of capacity. Since there are other factors limiting the cycle life the pressed electrodes are the most suitable ones.

Another feature of the processes in the cell is that oxygen is evolved on the nickel oxide electrode at the end of charge and the capacity efficiency of this electrode is about 0·9. On the other hand, no hydrogen is evolved with charge of the zinc electrode owing to the excess of zinc oxide and the charging capacity is fully utilized. This means that the degree of charge of the zinc electrode gradually increases with cycling. To restore the starting ratio, periodic short-circuiting of discharged cells is recommended. This results in evolution of some hydrogen on the nickel electrode (under the effect of the zinc potential) and oxidation of the equivalent amount of zinc on the zinc electrode.

Performance. Immediately after charge the o.c.v. of the cell is 1·80–1·85 V and it gradually decreases to 1·74–1·78 V. The discharge and charge curves are given in Fig. 79. The difference between the charge and discharge voltages at low-current drains is smaller than in other storage cells. Since the charge curve has no distinct step at the end of charge the cells are charged to 110% of the actual discharge capacity. The cells can be drained with large currents with respective decrease in capacity and voltage. The cells can operate at low temperatures but the resulting relative decrease in capacity is higher than for the silver–zinc cells. The specific energies of the reported cell types vary between 50 and 70 Wh/kg and 100 and 150 Wh/dm³ for $j_d = 0 \cdot 1 – 0 \cdot 5$.

Cycle life. The film separator in nickel–zinc cells operates under better conditions than in silver–zinc cells. Though the amount of evolved oxygen is larger and its effect is stronger there is no deteriorating effect of the silver ions; the solubility of nickel oxides is considerably lower than the solubility of silver oxides and the effect of nickel ions on the separator is small. Therefore, the service life of the separators is longer, but nevertheless attempts are being made to find new, more stable, separator materials.

The main cause of the sharp decrease in the discharge capacity of the storage cells after 200–300 cycles and of their breakdown is the displacement of the active materials of the zinc electrode. This effect resulting in lower operational and economic parameters for nickel–zinc storage cells delayed their commercial manufacture.

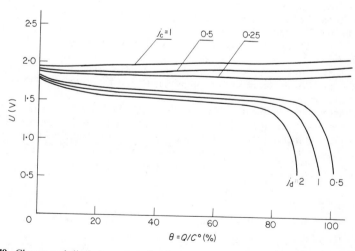

Fig. 79. Charge and discharge curves for the nickel–zinc storage cell at room temperature.

Prospects of application. Interest in nickel–zinc storage cells has been intensified mainly in connection with development of the electric cars. The specific energy of these cells is 1·5–2 times that of lead, nickel–cadmium or nickel-iron cells making possible 100–150 km runs between charges. The cost of these cells for mass production will probably be comparable with the cost of other alkaline storage cells (calculated for the same energy capacity). These cells might be widely used if their cycle life were increased at least two-fold.

Of a great interest is the work on development of sealed nickel–zinc cells utilizing the oxygen cycle. Sealed cells are not only easier to maintain but they also lack the difficulties encountered in balancing the degrees of charge of the nickel oxide and zinc electrodes discussed above.

REFERENCES

1. O. V. Krusenstierna, High-energy long-life zinc batteries for electric vehicles, *in* "Power sources 6" (Collins, D. H., ed.), pp. 303–319. Academic Press, London and New York (1977).
2. King Wai Cho, D. N. Bennion and J. Newman, Engineering analysis of changes in zinc secondary electrodes. I., *J. Electrochemical Soc.* **123**, No. 11, 1616–1627 (1976).

BIBLIOGRAPHY

1. J. McBreen and E. J. Cairns, The zinc electrode, *in* "Advances in Electrochemistry and Electrochemical Engineering" (Gerisher, H., Tobias, Ch. W., eds.), Vol. 11, pp. 275–352. Wiley, New York (1978).
2. "Zinc–Silver Oxide Batteries" (Fleisher, A., Lander, J. J., eds.). J. Wiley, New York (1971).
3. F. P. Kober and A. Charkey, Nickel–zinc: a practical high energy secondary battery, *in* "Power Sources 3" (Collins, D. H., ed.), pp. 309–326. Oriel Press, Newcastle upon Tyne (1971).
4. A. Charkey, Development of large-scale nickel–zinc cells for electric vehicles, Preprints of 4th International Electric Vehicle Symposium, Dusseldorf, 1976.
5. S. Ruben, The mercuric oxide: zinc system, *in* "Primary Batteries" (Heise, G. W. and Cahoon, N. C., eds.), Vol. 1, pp. 207–223. Wiley, New York (1971).

Chapter Thirteen

Various Systems with Aqueous Solutions

13.1. Use of magnesium and aluminium in chemical power sources

Use of magnesium and aluminium as reducers in chemical power sources would be a very attractive prospect in view of a lower specific consumption $g_Q^{(T)}$ and a more negative thermodynamic value of the standard electrode potential $E°$:

	$E°$ (V)	$g_Q^{(T)}$ (g/Ah)
Magnesium	-2.363	0.454
Aluminium	-1.662	0.336
Zinc	-0.763	1.220

The world resources of aluminium and magnesium are large. They are produced in large quantities and, in terms of Ah capacity, are cheaper than zinc. However, cells with magnesium anodes are not yet widely used and are manufactured mainly for special-purpose applications. Cells with aluminium electrodes and aqueous electrolytes are not used at all so far. This situation is explained by the peculiar features of operation of these electrodes.

There are many common features in the behaviour of magnesium and aluminium but there are also some characteristic differences. Both metals are coated with thin oxide films which prevent their further oxidation in the air. The compositions and the effects of the films in solutions depend strongly on the pH value.

Intense corrosion of magnesium occurs in acidic solutions. In alkaline solutions magnesium is completely passivated by the dense oxide layer formed on its surface. In approximately neutral and weakly alkaline solutions (roughly, in the pH range between 6 and 11) a "semipassive" state is established in which a less dense oxide layer strongly inhibits evolution of

269

hydrogen, that is, spontaneous corrosion of magnesium, but does not prevent anodic dissolution. Magnesium can be used in the cells with the electrolyte pH values lying in the above range. However, owing to a partial passivation the open-circuit steady-state potential of manganese is more positive than the thermodynamic value by 0·6–1·1 V. Thus, the gain in voltage for the manganese cell in comparison with the zinc cell decreases from 1·6 V to 0·5–1·0 V.

Depending on the pH value of the solution, anodic current leads to dissolution of magnesium giving rise to the Mg^{2+} ions or a loose precipitate of $Mg(OH)_2$:

$$pH < 9 \quad Mg \rightarrow Mg^{2+} + 2e \tag{13.1a}$$

$$pH > 10 \quad Mg + 2H_2O \rightarrow Mg(OH)_2 + 2H^+ + 2e \tag{13.1b}$$

Similar processes occur with self-dissolution (corrosion) of magnesium:

$$pH < 9 \quad Mg + 2H^+ \rightarrow Mg^{2+} + H_2 \tag{13.2a}$$

$$pH > 10 \quad Mg + 2H_2O \rightarrow Mg(OH)_2 + H_2 \tag{13.2b}$$

Magnesium and its alloys are distinguished by increasing rate of self-dissolution with increasing anodic current which is known as the negative difference effect (for other metals cathodic evolution of hydrogen typically slows down when the potential is shifted to more positive values). This effect can be caused by partial mechanical disintegration of the passivating film with anodic dissolution of magnesium and/or the stepwise mechanism of this process in which first the unstable ions Mg^+ are formed and they then interact with water liberating hydrogen. This assumption is supported by the fact that the rate of gas evolution expressed in electrical units is often close to the anodic current value, that is, the coefficient of utilization of magnesium is close to 0·5, irrespective of the current value. This means that the reactions (13.1) and (13.2) are linked to each other; practically, only one electron in the reaction (13.1) can be used (spent to form Mg^+) while the second electron is spent to liberate hydrogen. Intense gas evolution during discharge of magnesium anodes should be taken into account in design of magnesium cells.

Owing to these two factors—the decrease of the working voltage in comparison with the thermodynamic value and the concurrent evolution of hydrogen—operation of cells and batteries with magnesium anodes is accompanied by evolution of considerable amounts of heat resulting in temperature build-up in the cell. Therefore, these cells can be successfully employed at low ambient temperatures.

Another characteristic feature of magnesium anodes is their delayed-action behaviour; when the circuit is closed (or load increased) the anode potential is sharply shifted to the positive side and then gradually approaches the steady-state value. This results in a "dip" in voltage at the very beginning of the discharge curve. On the other hand, when the load is sharply decreased a small voltage peak is seen on the discharge curve. The delayed-action time is a few seconds or minutes. This is a significant drawback of cells using magnesium anodes.

Many of the above phenomena depend not only on the pH value of the electrolyte solution but also on its composition. The presence of Cl^- ions decreases the degree of passivation. This results in enhanced corrosion of magnesium but the magnesium anode can operate at much higher current densities.

The performance of magnesium anodes is greatly affected by addition of alloying agents. Addition of aluminium decreases the corrosion rate and addition of zinc decreases the delayed-action time. The optimum alloy composition includes $1-1.5\%$ zinc and $1.5-2\%$ aluminium. Addition of small amounts of calcium (0.15%) also results in better delayed-action behaviour. Even traces of iron (0.03%) lead to increased corrosion rates. In the sixties it was shown that the alloy of magnesium and 3% mercury increases the discharge voltage by $0.3-0.4$ V, particularly for high current densities (0.5 A/cm^2 and more), and decreases self-dissolution of magnesium by approximately a factor of two. The mercury in the alloy facilitates the anodic process of partial disintegration of the passivating film and simultaneously inhibits evolution of hydrogen. A disadvantage of these alloys is the volatility of mercury in the manufacturing process and the difficulties in rolling. Manganese–lead alloys may be used instead of the manganese–mercury alloys for this reason.

At present there are two types of commercially produced cells with magnesium anodes, long-life dry cells (manganese–magnesium) and reserve cells and batteries.

Corrosion of aluminium in acidic solutions is less intense than that of magnesium. In strong alkalis aluminium is intensely dissolved giving rise to soluble aluminates. Similarly to magnesium, aluminium is relatively resistant and, at the same time, electrochemically active in weakly acidic and some neutral solutions. However, noticeable corrosion of aluminium still occurs in such solutions; it often has the character of pitting and is readily enhanced by various factors. This makes the use of aluminium in chemical power sources difficult. Similarly to magnesium, the steady-state potential of aluminium is more positive than the thermodynamic value by approximately 1 V. This means that the working potential of aluminium is close to the potential of zinc or even more positive than it; thus, aluminium loses one of its advantages. Hydrogen evolution on aluminium in weakly acidic solutions is increased with anodic current but not so sharply as on magnesium; the coefficient of utilization of aluminium ranges between 0·7 and 0·9. Aluminium, like magnesium, exhibits a delayed-action effect when the circuit is closed.

Alloying agents may be added to aluminium to modify its anodic behaviour. Addition of mercury or tin shifts the potential to negative values but strongly enhances evolution of hydrogen. Some bicomponent additives (for instance, zinc plus tin) reduce evolution of gas but have a weaker effect on the potential. It should be noted that the behaviour of alloys, as is that of pure aluminium, is often unreproducible and depends on the conditions.

In recent years it has been shown[1] that addition of small amounts (less than 0·2%) of indium, gallium or thallium to aluminium markedly decreases corrosion of the electrode and, at the same time, shifts its potential by 0·4– 0·6 V to negative values. This effect opens much better prospects for future use of aluminium anodes in new types of cells.

13.2. Manganese–magnesium cells

Development of the manganese–magnesium cells was started in 1946 with the primary aim of replacing zinc in Leclanché cells since the world resources of zinc are limited. It turned out, however, that such cells can have good parameters and be of interest in themselves. At present manganese– magnesium cells are manufactured mainly in the USA and used for military communications applications.

Construction. The commercial construction of manganese–magnesium cells is similar to that of the round dry manganese–zinc cells. The can serving as the anode is manufactured by an impact extrusion process from preheated magnesium alloy. The electrolyte is a 1·2–2 M $Mg(ClO_4)_2$ solution. $MgBr_2$ solution of the same concentration is also used, but it increases the delayed-action period. A certain amount of magnesium hydroxide $Mg(OH)_2$ is added to the solution making it weakly alkaline.

It should be noted that the electrolyte does not contain Cl^- ions which partially depassivate magnesium and sharply decrease storageability of the cells. To decrease self-dissolution of magnesium, small amounts (0·2 g/litre) of Li_2CrO_4 are added to the electolyte. A higher concentration of chromate ions would result in a longer period of delayed action. To maintain the concentration of chromate in the solution slightly soluble $BaCrO_4$ is added to the positive electrode. The active materials of this electrode have the following typical composition: 10% acetylene black, 3% barium chromate, 1% magnesium hydroxide and the rest is synthetic manganese dioxide

(SMD). Up to 40% of the electrolyte is added to the active materials. The positive electrode is wrapped in a paper separator.

The cell has a vent to release hydrogen liberated during cell operation. At the same time, the top seal of the cell, including the vent, must prevent loss of moisture during prolonged storage and penetration of air into the cell.

Since the electrolyte is weakly alkaline and is alkalinized with cathode operation (reaction (9.1)), reaction (13.1b) occurs on the magnesium anode. The overall current-producing reaction in the manganese–magnesium cell is described by

$$Mg + 2MnO_2 + 2H_2O \rightarrow 2MnOOH + Mg(OH)_2 \qquad (13.3)$$

Reaction (13.2b), self-dissolution of magnesium, occurs simultaneously.

Performance. The o.c.v. of manganese–magnesium cells is about 1·9–2·0 V. In contrast to the manganese–zinc cell, discharge is not accompanied by formation of dense deposits of insoluble compounds impairing the cell performance and increasing pH. On the contrary, the loose deposit of $Mg(OH)_2$ lends buffer properties to the solution and maintains pH stability. The discharge curve of the manganese–magnesium cells is as a result somewhat less steep than that of the manganese–zinc cells (Fig. 80). Moreover, the higher total voltage results in a greater decrease of the relative voltage difference during discharge.

The specific energy of manganese–magnesium cells (for instance, of the D

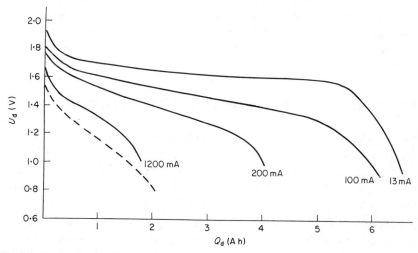

Fig. 80. Discharge curves of the D-size manganese–magnesium cell at various currents. The broken line corresponds to the discharge of a D-size Leclanché cell with a current of 100 mA.

size) with continuous discharge (corresponding to current $j_d = 0·01$–$0·04$) to the cut-off voltage of $1·25$ V is 100–120 Wh/kg or 160–200 Wh/dm^3. Since heat is evolved during cell operation the cells can work at temperatures down to $-40°C$.

The length of the delayed-action period after closing the circuit depends on various factors including the time of recuperation after the preceding operation. The delayed-action time for the freshly manufactured cells is typically longer than for the partially discharged cells. At low temperatures the delayed-action time is sharply increased.

Owing to careful selection of the electrolyte composition the manganese–magnesium cells have low self-discharge and good storageability. This advantage is particularly noticeable at elevated temperatures. The loss of capacity of manganese–magnesium cells stored at $55°C$ was 15% over two years,[3] while for the manganese–zinc cells it is 40% in 5 months. On the other hand, intermittent discharge conditions (when the cell is frequently switched on and off) result in intense corrosion of the magnesium electrodes and high self-discharge because, when the load has been disconnected, the magnesium anodes remain for some time in a more active state in which the protective effect of the passivating layer is weakened.

Prolonged discharge of manganese–magnesium cells with low-current drains ($j_d = 0·01$–$0·02$) can be accompanied by cracking of the magnesium cans. Cracking is caused by gradual increase of the internal pressure brought about, in particular, by formation of the bulky deposit of magnesium hydroxide.

Owing to the last two effects manganese–magnesium cells retain their advantage—higher Wh capacity—only under conditions of continuous discharge with not too low currents, e.g. corresponding to $j_d = 0·04$. Therefore, such cells are produced and used only in limited numbers.

13.3. Water-activated reserve cells with magnesium anodes

The reserve-type batteries with magnesium anodes are activated by the addition of water. The cathodes in them are manufactured from the barely soluble chlorides of silver, AgCl, univalent copper, CuCl, and, less commonly, lead, PbCl$_2$. The first Mg|AgCl cells were manufactured in 1943 in the USA and in 1949 production of Mg|CuCl cells was started.

Reactions. Chloride solutions are used as the electrolyte in magnesium water-activated cells. Such cells are typically filled with seawater. Cl$^-$ ions are effective depassivating agents; they penetrate the passivating film on magnesium and partially destroy it making possible easy anodic dissolution of magnesium at potentials which are more negative than in the absence of chloride ions. At the same time, the rate of magnesium corrosion in chloride solutions is high and the activated cells last only for one or two days.

Anodic dissolution in chloride solutions is described by equation (13.1a). The cathodic process consists in reduction of chlorides to the respective metals. Thus, the overall current-producing reaction, for instance, for the cell with silver chloride, is

$$2AgCl + Mg \rightarrow 2Ag + MgCl_2 \qquad (13.4)$$

The reaction of self-dissolution of magnesium (13.2a) occurs concurrently. Precipitation of $Mg(OH)_2$ starts only some time after the beginning of discharge owing to alkalinization of the solution caused by the latter reaction. Accumulation of $MgCl_2$ in the electrolyte in the first phase of discharge makes it possible to fill the cell even with fresh water. Unfortunately, the time needed to attain the working parameters in this case is much longer, since at the start of operation the concentration of chloride ions is insufficient both for depassivation of magnesium and for high enough conductivity of the solution.

In contrast to AgCl, CuCl and $PbCl_2$ are somewhat soluble in aqueous solutions. They are partially hydrolysed with dissolution and markedly decrease pH of the solution. This leads to increasing self-dissolution of magnesium.

Construction. The reserve-type magnesium batteries have, mostly, flat-cell construction. Less commonly, individual round cells are manufactured. The silver chloride electrode is made from silver foil on the surface of which a layer of AgCl is formed by passing anodic current in a NaCl solution. The electrodes designed for low-current drains are manufactured also by rolling molten AgCl. The copper chloride and lead chloride electrodes are typically manufactured by pasting a mixture of the salt and a binder (polystyrene, dextrin, etc.) to a copper screen. The electrodes can also be made by immersing copper mesh into molten CuCl. Since the discharge of the chloride electrodes produces metal deposits with sufficiently high conductivity, there is no need for conducting additives. However, several per cent of graphite is sometimes added to the CuCl electrodes. Before assembling, the electrodes are often slightly reduced to form the starting metal skeleton. In cells designed for high-current drains the anodes are manufactured from alloys of magnesium with mercury or lead.

The cells with AgCl employ separators with a small degree of shielding, for instance, plastic filaments with a diameter of 0·3–0·5 mm. The cells with CuCl require more effective separators owing to the solubility of this salt and possible diffusion of the Cu^+ ions towards the magnesium electrode.

These separators are typically manufactured from moisture-absorbing materials such as alignin, non-woven cotton sheets, etc. providing for good retention of electrolyte.

Flat-cell batteries, as well as the individual cells, do not have bottoms and/or covers, the edges of the electrodes and separator are open. When cells or batteries are immersed into water the interelectrode space is rapidly filled with water and activation occurs. After activation the cell can be removed from water and is operable up to complete discharge. The cell can also be operated immersed. To reduce the leakage currents under such conditions the edges of the electrodes are insulated with lacquer or tape.

Unactivated reserve magnesium batteries are stored in hermetically sealed packages. Penetration of moisture to the electrode surfaces during prolonged storage affects the cell operability since it enhances passivation of magnesium, increases the delayed-action time and makes possible hydrolysis of copper chloride and diffusion of copper ions to the magnesium electrode. To decrease humidity, silica gel is often placed into the package or container with the cells. Under such storage conditions the shelf-life of cells is practically unlimited.

Performance. The table below presents the thermodynamic e.m.f. $\mathscr{E}^{(T)}$, the discharge voltage U_d for $j_d = 0.2\text{--}0.5$, and the specific energy (per unit mass) w_m.

	$\mathscr{E}^{(T)}$ (V)	U_d (V)	w_m (Wh/kg)
$Mg\,|\,AgCl$	2·585	1·3–1·5	140–160
$Mg\,|\,CuCl$	2·500	1·2–1·4	60–75
$Mg\,|\,PbCl_2$	2·100	0·8–1·0	50–60

The specific energy has been calculated for the mass of the dry battery since the water for activation is added at the site of operation and is not transported. For some applications, though (for instance, for meteorological balloons), the mass of the activated batteries is significant.

The potential of the salt cathodes does not depend on the ratio between the amounts of chloride and metal and, therefore, the reserve-type magnesium batteries are distinguished by smooth discharge curves (Fig. 81). When batteries are assembled they are shorted with a thin wire to decrease the delayed-action time and more rapidly to obtain stable voltage of the magnesium anode. This wire increases the initial current after activation and accelerates accumulation of salt in the electrolyte. When stable operation has been obtained the wire burns out.

Reserve-type magnesium batteries, particularly those with silver chloride, can produce high discharge current and power. Typically, these batteries are discharged within a few hours. The total discharge time cannot exceed 24–48 hours owing to self-discharge. Such batteries used in electric torpedoes

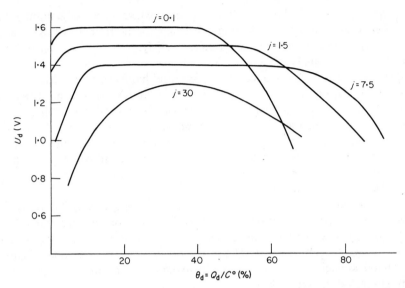

Fig. 81. Typical discharge curves for the reserve-type silver chloride–magnesium cells.

have been reported as producing powers about $1 \, kW/dm^3$ during 10 min of operation.

Reserve-type magnesium batteries are used mainly for various maritime applications such as various alarm devices, survival aids, temporary buoys, etc. because these batteries can be activated by sea water. Since silver is expensive, silver chloride–magnesium batteries are used only for critical applications demanding high-performance power sources. Magnesium batteries (mostly with CuCl) are used also for various meteorological and geophysical applications (balloons, etc.). The batteries activated at room temperature can operate at ambient temperatures down to $-70°C$ owing to intense heat generation; this is important for such applications. The main advantage of batteries with $PbCl_2$ is their better storageability in the unsealed condition.

13.4. Chemical power sources with organic reactants

At the end of the fifties development of primary cells with organic cathode materials attracted much attention. The reasons for this were, first, the decrease of the resources of high-grade manganese ores and, second, the rapid development of organic synthesis technologies making possible large-scale production of relatively cheap organic oxidants. A large number of organic compounds with oxidizing effect have been studied in this respect but only a few of them have been found to satisfy the basic requirements for the active materials (see Section 2.1).

Suitable cathode materials are the nitro- and halogen-substituted organic compounds, quinones and some other compounds. The specific capacity and the oxidation potential of organic oxidants increase with increasing number of the reducible groups in their molecules, for instance they increase when we go over from nitrobenzene to dinitrobenzene or trinitrobenzene. However, this is often accompanied with decrease in stability of these compounds (for instance, trinitrobenzene presents explosion hazards).

Organic cathode materials can be employed in long-life cells (instead of manganese dioxide) and in reserve-type cells. In the cells of the latter type more active organic oxidants may be used providing for higher voltage (2·2–2·4 V with magnesium) and higher specific energy of the cells, though the life of the activated cells is short. The group of such oxidants includes organic polyhalides such as dichlorodimethylhydantoin, tetramethylammonium tetrachloroiodide $(CH_3)_4NICl_4$, etc.

A variety of nitro and dinitro compounds have been suggested as oxidants for long-life cells. The best studied oxidant is *meta*-dinitrobenzene (*m*-DNB). Reduction of the nitro group of this compound involves eight electrons and yields hydroxylamine groups:

$$C_6H_4(NO_2)_2 + 6H_2O + 8e \rightarrow C_6H_4(NHOH)_2 + 8OH^- \qquad (13.5)$$

The specific consumption of the oxidant is 0·78 g/Ah which is lower by a factor of approximately 4 than that of manganese dioxide. The reduction potential of *m*-DNB, as is that of other nitro derivatives, is less positive than the reduction potential of manganese dioxide or other inorganic oxides. In practice, therefore, nitro compounds can be used only in combination with the magnesium electrode.

The existing prototypes of the *m*-DNB–magnesium cells have construction similar to that of the manganese–magnesium cells. The cathode is a mixture of *m*-DNB with 30% conducting carbon additive and 3% barium chromate. The content of the carbon additive is higher than in the manganese dioxide electrodes because the density of *m*-DNB is lower, that is, its specific volume is larger than that of MnO_2. The electrolyte is 1·25 M magnesium perchlorate solution with the addition of lithium chromate. The discharge curve of the *m*-DNB cells is flat and practically horizontal but their working voltage (1·0–1·2 V) is lower than that of the manganese–magnesium cells. The specific capacity of these cells is about 120 Wh/kg for $j_d = 0·04$. The current drain can be rather high, for instance, it can correspond to $j_d = 0·2$

with the relative decrease of voltage being lower than for manganese dioxide cells.

Despite these advantages and the fact that the cost of m-DNB per unit capacity is no higher than that of the high-quality grades of MnO_2, the m-DNB cells are not produced commercially. In part, this is explained by the drawbacks common to the long-life magnesium cells and primarily by higher self-discharge with intermittent discharging. Also, self-discharge is intensified owing to the fact that m-DNB is slightly soluble in aqueous solutions and diffusion of the molecules in solution towards the magnesium electrode makes possible a direct chemical reaction.

The number of studies on the use of solid organic reducers as anode materials is considerably smaller. This is understandable since metallic anode materials such as zinc, magnesium, etc., provide for good cell performance which is not easy to obtain with other materials. The liquid organic reducers are used in continuous-operation electrodes of fuel cells (see Chapter 17).

A number of attempts have been reported to develop storage cells on the basis of reversible organic systems. The electrochemical system quinone/hydroquinone is known to possess good reversibility and the so-called quinhydrone electrodes have a stable potential and can be used for measuring the pH of solutions. Many derivatives of quinone also possess reversibility. In principle, there can be developed a storage cell using the same quinone for both electrodes. After charging the positive electrode contains quinone and the negative electrode contains hydroquinone. Such storage cell can be discharged and repeatedly recharged. A significant drawback of such a concentration cell is its low discharge voltage, it can be discharged only at 0·1–0·2 V (when the solution becomes saturated with the reaction products, the voltage vanishes). The discharge voltage can be increased by using for the positive and negative electrodes different quinone derivatives with different normal redox potentials. However, this cannot increase the discharge voltage above 1 V. Other compounds can also be employed, for instance the so-called electron-exchange (redox) polymers (e.g., those based on polyanilines) which contain large numbers of redox groups. Though the specific capacity of these compounds is not too low the parameters of the storage cell employing them are not high, owing to low voltage. Cell prototypes with specific energy about 30 Wh/kg have been reported. The parameters and performance of such cells under various conditions (temperature, time of storage, etc.) are still not clear.

Though the earlier attempts have not been very successful it can be assumed that development of cells with solid organic reactants will be continued in future and will yield interesting results with the advances in organic syntheses.

13.5. Various cells with PbO$_2$ electrodes

Lead dioxide in the positive electrode of the lead storage cell is one of the strongest solid oxidants used in chemical power sources. Its potential in 20 % sulphuric acid solution is by 0·44 V more positive than the thermodynamic potential of the oxygen electrode in the same solution. Nevertheless, lead dioxide is highly resistant and does not decompose spontaneously owing to the very low rate of oxygen evolution on it. The electronic conductivity of lead dioxide is sufficient to enable high-current drain. Moreover, lead dioxide is comparatively cheap and the electrodes are simple to manufacture. In view of such a fortunate combination of advantages numerous attempts have been made to use lead dioxide in cells of other types.

(a) Lead cells with soluble electrodes

In solutions of perchloric acid, HClO$_4$, fluoroboric acid, HBF$_4$, and fluorosilicic acid, H$_2$SiF$_6$, whose lead salts are easily soluble, the electrode reactions on the lead and lead dioxide electrodes give rise to lead ions rather than insoluble lead sulphate:

$$PbO_2 + Pb + 4H^+ \rightarrow 2Pb^{2+} + 2H_2O \qquad (13.6)$$

Owing to high values of the exchange current, more compact non-porous electrodes can be used here. Since the electrodes are not porous the concentration polarization is sharply reduced and removal of the reaction products is facilitated; this results in improved coefficients of utilization of the active materials, particularly with high-density current drain. Soluble salts are formed also when hydrochloric or nitric acid is used but the process can involve side reactions (oxidation of hydrochloric acid to chlorine, oxidation of lead by nitric acid, etc.).

Various types of reserve lead cells with soluble electrodes have been developed. The positive electrode is manufactured by anodic deposition of a lead dioxide layer from a lead nitrate solution onto a thin substrate of stainless steel or special alloys, and the negative electrode is manufactured by cathodic deposition of lead. The electrodes are typically thin and the assembled batteries can have a large electrode surface per unit volume. The batteries can consist of the box-type or bipolar-type cells. Usually, the reserve batteries are automatically activated.

The o.c.v. of cells with soluble lead electrodes is about 1·8–1·9 V. They can be discharged at high current densities—about 0·05 A/cm^2 for continuous drain and up to 0·3 A/cm^2 for pulsed drain. The working voltage is 1·6–

1·8 V. High-concentration acid solutions (30–50 %) are employed in these cells making possible operation at low temperatures down to $-60°C$. The cells with $HClO_4$ have the best parameters, but concentrated solutions of this acid present explosion hazards and therefore HBF_4 and H_2SiF_6 are used more often. A drawback of these cells is their short life in the activated state (a few hours). Their life is so short because of the reaction between PbO_2 and the substrate material after activation of the cell. Although the unbroken layer of PbO_2 usually completely covers the substrate surface, slight breaks in the layer result in intense corrosion of the substrate. The cells have a comparatively low specific energy of about 30 Wh/kg owing to large consumption of the acid in the reaction (13.6). Since by the end of the discharge the acid concentration should not be too low, about 20–22 g of the 50 % acid solution are needed for each ampere-hour of the capacity.

Reserve cells of the above type are used for some specialized applications requiring short-time discharges (a few minutes to tens of minutes) with high-current drains. These cells can be stored for a long time in the unactivated state.

Attempts to develop storage cells with soluble lead electrodes have also been reported.[2] In such cells during charging the reaction (13.6) proceeds in the reverse direction, lead and lead dioxide, respectively, are deposited on the electrodes and the solution is strongly acidified. To develop such a cell corrosion-resistant materials are needed, particularly for the positive electrode. A suitable material is a conducting plastic, polypropylene with a large addition of graphite for instance. In comparison with the conventional lead cell such storage cell has a number of advantages; it tolerates deep discharge or storage in the discharge state, it has no "memory" (about incorrect handling during previous cycles for example) owing to almost complete dissolution of the reactants, and it permits high-current drain and discharge at low temperatures. However, this cell presents some difficulties. In solutions of $HClO_4$ or HBF_4 noticeable spontaneous decomposition of PbO_2 results in evolution of oxygen which gradually shifts the phases of charge of the electrodes so that the degree of charge for the negative electrode becomes higher than that for the positive electrode. Moreover, partial passivation of PbO_2 can occur with discharge resulting in a drop of voltage before complete utilization of the active materials. This makes the expected cycle life and service life of such cells still difficult to estimate.

(b) Lead–zinc and lead–cadmium cells

In the lead storage cell and in the cells with soluble lead electrodes, lead dioxide possessing a high anodic potential is used in combination with lead, which is not a very active reducer and has a low cathodic potential. Repeated attempts have therefore been made to use lead dioxide in combination with stronger reducers, for instance, with cadmium and zinc. This leads to increases in the working discharge voltage up to 2·1 V and 2·4 V, respectively. However, owing to strong corrosion of cadmium and particularly zinc in acidic solutions only reserve cells of this type can be used.

In the reported cell prototypes the construction of the positive electrode is similar to that in lead storage cells, and a sulphuric acid solution is used as the electrolyte. Owing to corrosion the shelf-life of the activated cells is not more than one hour. For 20–40 min discharge, the specific energy of the cadmium cells is about 30 Wh/kg and that of the zinc cell is about 60 Wh/kg; the specific power for high-rate discharge (2–3 min) is 300 W/kg and 600 W/kg, respectively. The zinc electrode in sulphuric acid is passivated, particularly at high current densities and low temperatures, and this considerably limits the range of applications of the lead–zinc cells.

The reserve-type lead–cadmium and lead–zinc cells have found only limited use.

13.6. Standard cells

The unit of voltage or difference of potentials (volt) is defined as the voltage producing in a circuit a current of 1 A generating a power of 1 W. This definition is not convenient for practical measurements and therefore use is made of the secondary voltage standards—the o.c.v. values of the so-called standard cells. The standard cells serve not for delivering current but for maintaining constant voltage in various measurement compensation circuits and for testing and calibrating instruments.

The o.c.v. of modern standard cells must be highly reproducible and stable for long periods (tens of years). The o.c.v.–temperature coefficient should be low. The cells should be insensitive to small current drains (a few microamperes) associated with measurements and rapidly restore the starting o.c.v. value after disconnection of the circuit (without exhibiting hysteresis).

The standard cell used nowadays is the mercury–cadmium Weston cell developed in 1892 and officially adopted in 1908. These cells are typically made in H-shaped sealed glass containers (Fig. 82) The positive electrode is mercury which is in contact with a paste of crystals of Hg_2SO_4 and $CdSO_4 \cdot \frac{8}{3}H_2O$. The negative electrode is a 6–12 % cadmium amalgam which is in contact with crystals of $CdSO_4 \cdot \frac{8}{3}H_2O$. A saturated solution of $CdSO_4$ is employed as electrolyte. The standard cells are stored at a constant

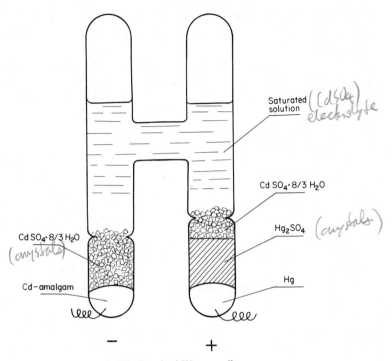

Fig. 82. Standard Weston cell.

temperature in oil or air thermostats. At 20°C the o.c.v. is 1·01864 ± 0·00002 V and its variation with temperature in the 0–40°C range is given by the following equation:

$$\mathscr{E}_t = \mathscr{E}_{20^\circ C} - 4 \times 10^{-5}(t - 20) - 10^{-6}(t - 20)^2 - 10^{-8}(t - 20)^3 \text{ V}$$

Less critical measurements in everyday laboratory or industrial work are made with the unsaturated Weston cell in which no crystals of $CdSO_4 \cdot \frac{8}{3}H_2O$ are used and the saturated $CdSO_4$ solution is replaced by a lower-concentration solution (saturated at 4°C). The o.c.v. of this cell is 1·0192 ± 0·0002 V, i.e. its reproducibility is lower. Since the solution concentration does not vary with temperature, in contrast to the saturated Weston cell, the temperature coefficient of the unsaturated cell is very small (about 1 μV/°C) and the cell does not require careful thermostatting. The unsaturated cells are placed into plastic cases with copper shields to provide for temperature equilibrium at both electrodes. In contrast to the saturated cells, the unsaturated cells are shippable. Their disadvantage is a shorter service life; while the saturated cells can be used for tens of years the ageing processes which can occur in the unsaturated cells (in particular, variation of the solution concentration) can lead to a considerable deterioration of the accuracy in 8–10 years.

The high reproducibility and stability of the o.c.v. of the standard cell is explained by the definite phase composition of the system and the lack of secondary or side reactions. The cadmium amalgam is a binary system, namely a mixture of the liquid amalgam containing about 4% cadmium and the solid amalgam containing about 14% cadmium. When the cell is stored, owing to the slight solubility of Hg_2SO_4 the mercury ions can migrate from the positive electrode to the negative electrode leading to deposition of mercury on the amalgam. Though this effect changes the general ratio of cadmium to mercury the compositions of the two amalgam phases do not change since only the ratio of their amounts is changed. Therefore, this diffusion of mercury does not affect the potential. A minute amount of sulphuric acid (less than 0·05 mol/litre) is sometimes added to the electrolyte to prevent possible hydrolysis of Hg_2SO_4.

The internal resistance of the Weston cell is 100–1000 ohms. Hence, a current of 1 μA produces a voltage change of 0·1–1 mV. Immediately after disconnection of the circuit a certain distortion of the o.c.v. (about 10–20 μV) remains. The total recovery of the starting o.c.v. takes a few minutes, sometimes an hour. After higher-current pulses the distortions can remain for longer periods and sometimes the o.c.v. cannot be reproduced at all.

REFERENCES

1. A. R. Despić, D. M. Dražić, S. K. Zečević and T. D. Grozdić, Problems in the use of high-energy-density aluminium–air batteries for traction, *in* "Power Sources 6" (Collins, D. H., ed.), pp. 361–368. Academic Press, London and New York (1977).
2. F. Beck, Lösungsakkumulatoren, *Chem.-Ing. Tech.*, Bd. **46**, No. 4, 127–131 (1974).
3. J. L. Robinson, Magnesium cells, *in* "Primary Batteries" (Cahoon, N. C. and Heise, G. W., eds.), Vol. 2, pp. 149–169. Wiley, New York (1976).

BIBLIOGRAPHY

1. J. J. Stokes and D. Belitskus, Aluminum cells, *in* "Primary Batteries" (Cahoon, N. C. and Heise, G. W., eds.), Vol. 2, pp. 171–186. Wiley, New York (1976).
2. J. S. Dereska, Organic cathodes and anodes for batteries, in "Primary Batteries" (Cahoon, N. C. and Heise, G. W., eds.), Vol. 2, pp. 187–238. Wiley, New York (1976).
3. W. J. Hamer, Standard cells, *in* "Primary Batteries" (Heise, G. W. and Cahoon, N. C., eds.), Vol. 1, pp. 433–477. Wiley, New York (1971).

Chapter Fourteen

Compound Cells

14.1. Air (oxygen) electrodes

The oxygen of the air is the most universal oxidant for all chemical reactions involving production of energy, both in nature (in living organisms) and in technological processes. Chemical power sources are, it would seem, the only energy-producing units which make use of expensive low-capacity solid oxidants instead of the freely available oxygen. The irony of the situation is that one of the most efficient ways of energy conservation, electrochemical conversion, involves use of inefficient and expensive reactants. This is why development work on cells utilizing oxygen of the air has been under way for a long time.

As early as the beginning of the nineteenth century the performance of Volta cells was found to be better in the presence of air or oxygen than in sealed conditions. Now it is known that in the presence of the air the low-efficiency cathodic process of hydrogen evolution is partially replaced by the process of oxygen reduction.

At the end of the nineteenth century carbon materials were suggested for the manufacture of the air electrodes. During World War I, owing to scarcity of manganese ore, German and French firms manufactured dry cells containing activated carbon, instead of manganese dioxide, and utilizing the aerial oxygen. These cells had very low power, and free access of air resulted in intensified zinc corrosion and drying of the electrolyte. Serious development work on air electrodes and various types of air cells started only in the twenties or thirties of this century.

The advantages of air cells are due mainly to practically unlimited availability of free air (except in such sealed environments as in submarines or spacecraft) and to the decreases in mass and cost of the cells owing to elimination of heavy oxide oxidants.

Pure oxygen has a higher activity than the air but it is rarely used because

285

special cylinders or cryogenic vessels are needed for its storage and transportation.

A variety of high performance air and oxygen electrodes have been developed for operation in concentrated alkaline solutions but the development work on air electrodes for acidic and salt solutions has not been fully successful though numerous projects are in progress in this field.

The mechanism of reduction (ionization) of oxygen

Electrochemical reduction of oxygen occurs on a number of catalytically active electrodes (platinum, silver, carbon) according to the following reaction equation:

$$O_2 + 2H_2O + 4e \rightarrow 4OH^- \qquad (14.1)$$

(the reaction is given for alkaline solutions). The equilibrium electrode potential of this reaction (as for the potentials for other reactions discussed below) varies with the pH of the solution as does the potential of hydrogen electrode. Hence it is convenient to express all potentials with respect to the reversible hydrogen electrode (r.h.e.) potential in the same solution since such values are independent of pH. The thermodynamic value of the potential of the reaction (14.1) at $p_{O_2} = 1$ atm is $1 \cdot 229$ V (r.h.e.). However, the steady-state potential of the electrode immersed in the electrolyte in an atmosphere of oxygen is more negative: $1 \cdot 0$–$1 \cdot 1$ V in alkaline solutions and $0 \cdot 9$–$1 \cdot 0$ V in acidic solutions.

It has been found that reduction of oxygen occurs via a rather complex mechanism involving possible intermediate formation of hydrogen peroxide:

$$O_2 + H_2O + 2e \rightarrow HO_2^- + OH^- \qquad (14.2)$$

In a further process hydrogen peroxide can (i) be electrochemically reduced to water, (ii) be catalytically decomposed into water and oxygen or (iii) not participate in further reactions, i.e. accumulate near the electrode and escape by diffusion. Accumulation of hydrogen peroxide is extremely undesirable since this results in shifting of the electrode potential towards negative values; the equilibrium potential of the reaction (14.2) for $c_{HO_2^-} = 1$ mol/litre is $0 \cdot 75$ V (r.h.e.). Moreover, this leads to a decrease in utilization of the active material, since every oxygen molecule reacts only with two electrons, rather than four as in the reaction (14.1). There is another possible pathway for oxygen reduction, preliminary adsorption of the oxygen molecule and its dissociation into atoms. In this case hydrogen peroxide is not formed after rupture of the –O–O– bond and the oxygen atoms adsorbed at the surface are reduced directly to OH$^-$ ions (or molecules of water).

Passage of current produces strong polarization even for platinum, owing to the low rate of oxygen reduction (Fig. 83), and the same effect takes place

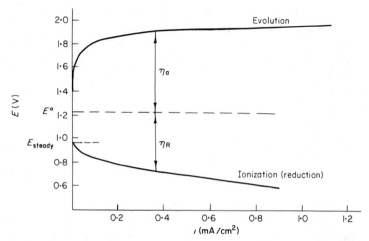

Fig. 83. Polarization curves for evolution and ionization (reduction) of oxygen on the smooth platinum electrode in sulphuric acid solution at 25°C.

with the reverse reaction of anodic evolution of oxygen from aqueous solution. The conditions of reduction of oxygen on silver (in alkaline solutions) are close to those for platinum. Polarization with reduction or evolution of oxygen for other metals is, typically, even higher.

The rate of reduction or evolution of oxygen increases and polarization decreases with increasing temperature. At temperatures higher than 170–180°C the steady-state potential in a highly concentrated alkaline solution is close to the equilibrium thermodynamic potential. This indicates a considerable improvement of reversibility of the reaction (14.1).

Construction

Development of operational air or oxygen electrodes encounters two problems: (1) selection of an effective and stable catalyst for the specific operational conditions, and (2) design of the structure of the gas-diffusion electrode. The structure is especially important for air electrodes since they must operate in direct contact with the ambient air with the gas pressure being not higher than the pressure of the liquid phase (in contrast to other types of gas-diffusion electrodes). Moreover, the reacting gas (air) contains a high concentration of an inert component (nitrogen) which hinders delivery of oxygen to the reaction zone. Only considerable hydrophobization of the porous electrode makes possible its operation in the absence of excessive gas pressure (see Section 5.4).

Air electrodes are manufactured mainly from carbon materials, such as activated carbon, graphite, carbon black, etc. Owing to its large true surface area, activated carbon can be used as a catalyst for oxygen reduction in alkaline solutions (though the specific activity of carbon (per unit area) is small). However, additional metallic or non-metallic catalysts are often used for increasing the activity. Water-repellent additives are also used.

Thick carbonaceous air electrodes for alkaline solutions developed in the thirties were produced by pressing a mixture of carbon materials and binders (sometimes with additional sintering). The block-type electrodes with thickness of up to 30 mm have channels facilitating access of air (see Fig. 84(a)). Figure 84(b) shows a more effective pocket-type electrode with 6–8 mm thick pocket walls and which has a wide opening for air. Electrodes of these types can be drained at low current densities (up to 3–6 mA/cm^2) and are employed mainly in large cells with capacity exceeding 300 Ah.

Development of thinner and more effective air electrodes started in the fifties proceeding from the experience accumulated with development of gas-diffusion electrodes for fuel cells and the theory of their operation. Generally, a thin air electrode consists of a number of layers (see Fig. 98): the active layer with the catalyst, the hydrophobic porous layer, and the metallic substrate to impart strength to the electrode. These layers can have various structure and order. The total electrode thickness is 0·5–2 mm. The hydrophobic water-tight layer is in contact with the air; it prevents electrolyte flowing out from the electrode. The active layer is in contact with the electrolyte; owing to partial hydrophobization of this layer a proportion of the pores are filled with the air and a proportion with the electrolyte. Sometimes electrodes have several active layers whose hydrophobicity

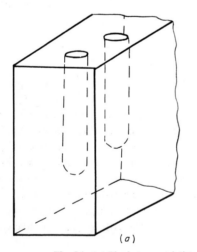
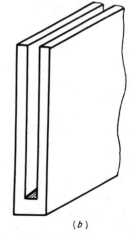

(a)   (b)

Fig. 84. (a) Block-type and (b) pocket-type carbon electrodes.

increases with the distance from the front face which is in contact with the electrolyte. Advanced manufacturing methods make possible production of the air electrodes operating in alkaline soluion with current densities up to 50–80 mA/cm^2.

Flooding; carbonization

The main difficulty arising with prolonged operation of the air electrode is the so-called flooding of the electrode, that is gradual filling of the gas pores with liquid electrolyte which hinders delivery of oxygen. This is a rather complicated effect whose mechanism is not yet fully clear. Obviously, the water-repellent properties of various materials, including PTFE, deteriorate owing to contact with the alkaline electrolyte so that they are better wetted with electrolyte. However, at the very beginning of the process the hydrophobic material is not wetted with electrolyte and, therefore, should not lose its hydrophobic properties. It is possible that the hydrophobic pores are gradually filled with liquid by evaporation and condensation of water and electroosmotic effects cannot be ruled out. During discharge, flooding of the electrodes occurs sooner than during storage of the cells. This is caused by effect of the electrode potential on the degree of wetting of the pore walls with electrolyte. Flooding of electrodes is, apparently, related to "weeping" of electrodes, the appearance of electrolyte droplets on the gas face of the electrode after a certain period of operation. All these effects result in an early breakdown of the air electrode. The thick carbon electrodes, typically, are slowly flooded throughout their bulk. The resistance of the thin electrodes to flooding is improved by careful selective hydrophobization of individual layers.

During cell operation the gas face of the electrode is always in contact with the air which contains some carbon dioxide. This leads to gradual carbonization of the alkaline electrolyte in the pores. Moreover, water can evaporate from the electrolyte through the porous electrode resulting in gradual "drying" of the electrolyte. These effects significantly reduce the efficiency of the air electrode. To prevent development of these effects during long-term storage of cells, the opening for the air in the cells is hermetically sealed with tape which is torn prior to using the cell. If the total amount of electrolyte in the cell is not large the service life of the cell after breaking the seal is limited.

A significant advantage of cells with air electrodes and acidic or neutral solutions would be the elimination of difficulties due to carbonization of electrolyte. Such cells could be drained for long periods, especially in the intermittent discharge regime. Unfortunately, no effective, stable and cheap catalysis for cells with such electrolytes are yet available.

14.2. Air–metal cells

Alkaline solutions are chiefly employed as electrolytes in the existing types of air–metal cells. The anodes are manufactured from zinc, iron and other metals. The o.c.v. is 1·4–1·5 V for air–zinc cells and 1·0–1·1 V for air–iron cells. The overall current-producing reaction is comparatively simple being associated with oxidation of the anodic metal by oxygen to the respective oxide or hydroxide. The basic advantage of air–metal cells is that they do not require expensive or scarce oxides employed in other cell types. The air electrode (cathode) of the air–metal cell in principle remains operable after discharge. Therefore, considerable efforts were made to develop renewable cells (in which the spent metal electrodes are replaced by new ones) and air–metal storage cells.

(a) Primary air–metal cells

Air–zinc cells. The air–zinc cells became commercially available in the thirties. The first cell types, still in use, had construction similar to that of the copper–zinc cells (Section 12.2). Their capacity varies between 300 and 1000 Ah and they contain large amounts of liquid electrolyte (30–40% KOH or 20% NaOH solution). The cathode is a block-type or pocket-type carbon electrode suspended in the upper part of a tall jar together with the monolithic anode of a zinc–mercury alloy. A special "breathing" opening in the cover or a protruding edge of the carbon electrode provides for free access of air to the electrode. The cells are delivered without electrolyte which is added by the user; some cell types contain solid sodium monohydrate and are sealed so that the user need only add water. The cells are designed for prolonged drain with normalized current up to 0·004 and short-term drain with normalized current up to 0·01. At low-current drain the discharge voltage is between 1·2 V and 1·3 V and its variation during discharge is small. The specific energy is about 60 Wh/kg or 80 Wh/dm^3. Self-discharge is low and the cells can be stored or discharged for 1–2 years. The cells with KOH solution can be operated at temperatures down to $-40°C$.

In similar cells of another type a mixture of lime, $Ca(OH)_2$, and a porous filler is placed at the bottom of the cell container. The zincate ions produced with discharge of the zinc electrode (reaction (12.1)) react with lime,

$$ZnO_2^{2-} + Ca(OH)_2 \rightarrow CaZnO_2 + 2OH^- \qquad (14.3)$$

giving rise to a sediment of calcium zincate and regenerating the alkali. This makes it possible to reduce the amount of the alkali solution from 8–

10 ml/Ah to about 3 ml/Ah. Owing to sedimentation of the zincate, 1 g of $Ca(OH)_2$ is equivalent to 10–12 ml of the solution. The decrease in the amount of solution and the use of plastic cell containers contributed to increasing the specific energy to 150–200 Wh/kg or 200–280 W/dm^3 (for the 1000–3000 Ah cells). However, the maximum normalized current should not exceed 0·0005–0·001 for continuous discharge.

Starting from the fifties numerous attempts were made to develop miniature "dry" air–zinc cells with improved drain characteristics and low capacity (up to 20–40 Ah). Various constructions, such as box-type, cylindrical (as for the alkaline manganese–zinc cells), etc., were suggested for these cells. To improve the cell performance more active thin carbon electrodes are employed sometimes containing additional catalysts, such as spinel-type oxides, small amounts of silver, etc. A powdered or pasted zinc electrode is used, rather than a monolithic one. Electrolyte is gelled with organic or inorganic additives or is in the pores of the membrane. Such cells have specific energies from 150 to 350 Wh/kg depending on the size and construction (120–260 Wh/kg for batteries) and can be drained at normalized currents up to 0·1–0·2.

These cells differ from other sealed alkaline cells essentially in that they have breathing openings. If the cell is designed for high drain rates the breathing opening should be large enough for the air to be delivered to the electrode only by diffusion or natural convection rather than being forced by an elevated pressure. Sometimes the carbon electrode serves as the side or edge face of the cell container. During storage the breathing opening or the electrode as a whole is sealed hermetically with tape. When the seal has been broken and the cell has started operating the contact area between the carbon electrode and the air becomes comparatively large. The ratio of this area to the electrolyte volume is tens of times that for the large cells mentioned above with liquid electrolyte. Carbonization and drying in these cells occurs much faster so that the service life of the cells after activation decreases to 1–3 weeks. This time is greatly affected by environmental factors (temperature, humidity, etc.). These processes produce deterioration of the cell performance for low-rate discharges (j_d below 0·01 or intermittent discharges). Plots of capacity or specific energy as a function of current or discharge time passes through a maximum. This is the main factor hindering practical application of these cells which in other respects possess very attractive features.

Air–iron cells. Air–iron cells employ materials that are even cheaper and less scarce than those employed in the air–zinc cells. The iron electrodes can be manufactured either by pressing iron powder with binders or by simultaneous reduction and sintering of powdered pure magnetite. The reduced iron powder exhibits pyrophorocity, that is, contact with air results in its intense oxidation ("burning"). Pyrophorocity is eliminated by treating the iron powder with benzene, water-glass solution or other material. The 400 Ah cells formerly manufactured in the USSR have low cylindrical steel cases and horizontally arranged electrodes—the iron electrode is in the lower part of the case and the air electrode is in the cell cover. The electrolyte is 40% KOH solution. Discharge of the iron electrode produces mainly $Fe(OH)_2$ and only a slight amount of ferrite ions, FeO_2^{2-}, owing to their low solubility. Consumption of alkali is small and the volume of electrolyte can be smaller than in the zinc cells operating via the primary process. For $j_d = 0.001–0.002$ the discharge voltage is $0.75–0.5$ V and the specific energy is about 65 Wh/kg and 110 Wh/dm^3. The basic drawback of the air–iron cell is its low discharge voltage so that a larger number of cells connected in series are needed to obtain a given voltage. Owing to passivation of the iron electrode the performance of these cells sharply deteriorates at temperatures below 0°C.

(b) Renewable air–zinc cells

A variety of batteries of air–zinc cells with renewable zinc electrodes ("mechanically rechargeable" batteries) have been developed for the US Army Signal Corps. The spare zinc powder electrodes have high porosity (70% or more). The electrodes are impregnated with the alkali solution and then dried so that the pores contain reserves of solid alkali. The prepared dry electrodes are mounted on plastic holders and wrapped with separator material. They are stored and shipped in sealed packets of laminated material from polyethylene film and metal foil which protect the active mass from moisture and oxidation, and the alkali from carbonization. An individual cell of the battery consists of two air electrodes serving as the side walls of the vessel and the central zinc electrode. When the battery has been discharged the used electrolyte is poured off, the zinc electrodes are replaced with new ones, and the battery is filled with water. The amount of alkali in the electrode pores is sufficient for producing electrolyte of the required concentration. To facilitate removal of the zinc electrodes they are attached to the cover and replaced together with it.

Batteries with voltage of 24 V and capacity of 20–50 Ah are designed for short-time (1–2 days) operation for radio transmissions (j_d from 0.3 to 0.5) and reception (j_d from 0.01 to 0.04). During this time the increased self-discharge and carbonization of the electrolyte do not produce significant effects. To improve air supply to the space between the air electrodes of two neighbouring cells during heavy-drain periods a small ventilator is sometimes used (the current supplied to it corresponds approximately to $j_d = 0.02$). At low ambient temperatures an electric heater is sometimes used during light-drain periods (the additional load corresponds to j_d about 0.05). The zinc electrode in these batteries can be replaced up to 30–40 times until deterioration of the air electrode becomes significant. The stability of the air electrode is improved by adding catalysts (sometimes even small amounts of platinum). The specific energy of the batteries is 130–150 Wh/kg. Leakage of electrolyte can occur during operation of the batteries owing to the difficulties encountered in sealing individual cells. Moreover, in large batteries heavy drain can lead to considerable increase in temperature resulting in concentration of solution by evaporation.

(c) Air–metal storage cells

The problem of reversibility of oxygen electrodes. Many electrodes can, in principle, be used both for cathodic reduction of oxygen and for anodic evolution of oxygen (that is, for conducting reaction (14.1) from left to right and in the reverse direction (see Fig. 83 for the platinum electrode)). However, during the anodic process evolution of oxygen is often accompanied by deep oxidation of the electrode surface which reduces its catalytic activity with respect to cathodic reduction of oxygen and sharply enhances polarization. The search for materials without this drawback continues (perhaps such materials will be found among the oxide catalysts) but so far it is practically infeasible to employ the same electrode both for evolution and reduction of oxygen. This is a significant obstacle for development of storage cells with air (oxygen) electrodes. There are a number of indirect methods to overcome these difficulties.

Three-electrode construction (Fig. 85). In the interelectrode space between the air electrode (1) and the metal electrode (2) there is placed the third electrode (3)—a rather sparse mesh of a metal which is inert under the conditions of the cell; for instance, nickel (being resistant to alkali solutions). The main air carbon electrode operates during discharge and the auxiliary electrode operates during charge. Oxygen is evolved on the auxiliary electrode during charge while metal hydroxide is reduced to metal on the negative electrode. The carbon electrode is not connected to the circuit during charge so that it does not deteriorate. This improves the reliability of this cell design. A significant drawback of this battery is that when the operational mode is changed from discharge to charge (or vice versa) the load circuits should be switched over in each cell of the in-series battery. This introduces considerable complications into operation of the battery.

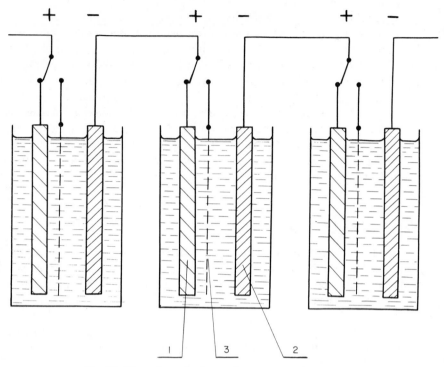

Fig. 85. Three-electrode air–metal storage cell (schematic).

Bifunctional electrodes (Fig. 86). The bifunctional electrode has a number of layers. One layer contains the active catalyst for oxygen reduction (2) and is in contact on one side with the porous hydrophobic layer (1) through which oxygen is supplied during discharge. On the other side the layer (2) is in contact with the coarse-pore non-hydrophobic nickel layer (3). During discharge the nickel layer is filled with electrolyte and produces only a slight hindering effect on operation of layer (2). In the couse of charging, oxygen is liberated mainly on the nickel layer which is closer to the electrolyte solution. The oxygen bubbles liberated in the pores partially block the active layer putting it out of operation so that it is not subjected to electrochemical oxidation during charging. The bifunctional electrode serves as a combination of the auxiliary and main electrodes and, since both parts of the electrode are electrically connected, no switching of the circuit is needed during charge–discharge cycles. However, storage cells with these electrodes have poorer reliability and shorter cycle life than the three-electrode cells.

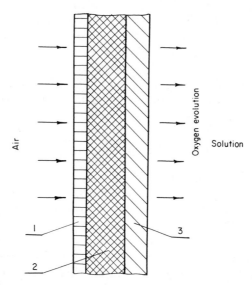

Fig. 86. Schematic of the bifunctional electrode.

Air–zinc storage cells. Though numerous attempts to develop air–zinc storage cells have been reported such cells are still not available commercially. This is apparently explained by the fact that these attempts encountered difficulties in development of both the reversible zinc electrode and the reversible air electrode. The existing cell prototypes (three-electrode and bifunctional electrode cells) have specific energy between 80 and 130 Wh/kg for $j_d = 0.1$–0.2. However, the cycle and service life of such cells is still too short for useful applications. At the same time the reported range of specific energies is very high for storage cells and therefore the research and development work on air–zinc storage cells continues.

Air–iron storage cells. The cycle life of air–iron storage cells is longer than that of air–zinc cells owing to the considerably better reversibility of the iron electrode in comparison with the zinc electrode. In recent years marked advances were made in development of the prototypes of air–iron cells in connection with the use of the modified iron electrode suggested for the new nickel–iron storage cells (Chapter 11). The prototypes of the storage cells with the new iron electrodes and air bifunctional electrodes have specific energy of 100–120 Wh/kg at $j_d = 0.05$–0.1 and highest drain rates corresponding to $j_d = 0.5$–1. A shortcoming of the air–iron storage cell is the high ratio between the charge voltage and the discharge voltage. The charge voltage is about 1·75 V and the discharge voltage is 0·65 V. This results in a

very low energy efficiency of, the cells (24–32%); the Ah efficiency is
65–85%. The cycle life is still about 200 cycles.

Removal of carbon dioxide from the air is an important prerequisite for
prolonged operation of all types of air–metal storage cells. To remove
carbon dioxide the air is passed through 40% KOH solution; though this
makes cell operation more complicated the cycle life of the cells is increased.

(d) Air–zinc cells with circulating zinc suspension[1]

An attempt to overcome difficulties encountered in development of air–zinc
cells was made in a peculiar cell construction in which zinc powder is
suspended in circulating electrolyte solution (see Fig. 87). Suspension is fed
by the pump (2) from the container (1) into the electrode chamber (3)
(which is sometimes manufactured from concentric pipes). The current-
producing electrode reaction occurs when the zinc particles are in contact
with the brass current collector (4). Zinc oxide produced in the reaction is
carried by the electrolyte flow and collected in the receptacle (6). The fine-
pore insulating coating of the air electrode (5) prevents direct contact

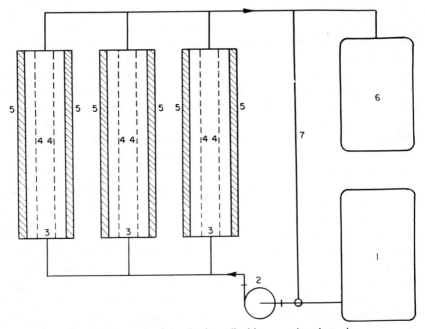

Fig. 87. Schematic of the air–zinc cell with suspension electrodes.

between the zinc particles and the electrode. If the coefficient of utilization of the zinc particles in one passage is low, a recirculation system (7) can be used.

In general, the construction of the cell is rather complicated. In the existing open-cycle cell prototypes zinc suspension is continuously fed into the cell and the discharged slurry is not utilized. Such continuous-operation devices are similar to fuel cells. In other cell types electrochemical regeneration of zinc is envisaged, that is, they operate as storage cells. The slurry is passed through the electrolytic cell whose construction is similar to that of the power source; there zinc oxide is reduced to zinc powder on the cathode and oxygen is liberated on the anode.

Such cell systems are expected to have overall specific energy of about 100 Wh/kg with a suspension stock sufficient for three-hour operation. Their drawback is the relatively large energy consumption for auxiliary operational needs. Difficulties in operation of such cell systems can be caused by "sticking" of the particles of zinc or zinc oxide to the electrode current collector resulting in deteriorating performance. For these reasons the cell prototypes have short cycle life and low energy efficiency so far. They are still not produced commercially despite intensive research and development efforts in a number of countries.

14.3. Nickel–hydrogen storage cells[2]

The nickel–hydrogen cell concept was first put forward in the Soviet Union in 1964. In recent years an intensive development effort was focused on these cells in a number of countries and small-scale commercial production of them was started.

The nickel–hydrogen storage cells employ the positive electrode from the nickel–cadmium storage cell which has good reversibility. The negative electrode is the hydrogen electrode based on platinum or other catalyst which was developed in the sixties for fuel cells. The current-producing reaction in the cell is described by

$$2NiOOH + H_2 \underset{charge}{\overset{discharge}{\rightleftharpoons}} 2Ni(OH)_2 \qquad (14.4)$$

Hydrogen liberated on charging is accumulated and cell pressure increases; during discharge hydrogen is consumed and its pressure decreases.

Construction. The breakthrough in development of the nickel–hydrogen cells was made when it was found that the rate of the direct reaction between the charged nickel oxide electrode and hydrogen is slow even at elevated hydrogen pressures. Therefore, there is no necessity for spatial separation of hydrogen and electrode, the entire electrode group can be placed into the steel cylinder in which hydrogen is accumulated (Fig. 88). The gas can be fed to the gas–liquid electrode not only from the rear face and since the electrode is not sealed at the edges hydrogen is also fed from the edges.

To facilitate feeding and removal of gas, immobilized electrolyte is employed rather than free liquid electrolyte. The 30% KOH solution is in the pores of the asbestos, or other porous matrix and in the electrode pores. Since the total volume of liquid is limited some pores in the matrix and the electrodes are not filled with electrolyte. This prevents liquid electrolyte from flowing out from the matrix and electrodes and precludes formation of liquid connections between the neighbouring cells of the battery. This makes it possible to assemble several cells connected in series within a single hydrogen cylinder.

The steel cylinders are heavy and their mass makes a considerable contribution to a decrease in the specific parameters of the cells. Some cylinders are made from special high-strength alloys making it possible to decrease the wall thickness. Sometimes the cylinder serves as a cell terminal; this somewhat simplifies the difficulties in sealing the current leads passing through the cylinder wall.

The nickel–hydrogen storage cell usually employs sintered nickel oxide electrodes. The hydrogen electrode also has a sintered nickel substrate to which a small amount of platinum catalyst is added (about 0.3 g/dm^2). Other electrode types employ a Raney nickel catalyst (without platinum). There have also been reported hydrophobized hydrogen electrodes containing a nickel mesh to which a mixture of platinum black and PTFE is applied. Such water-repellent electrodes exhibit better performance at low hydrogen pressures.

Process features. In the storage cell the hydrogen electrode operates in the discharge and charge modes, that is, alternatively as the anode and cathode. In contrast to the oxygen (air) electrode the hydrogen electrode has high reversibility. Polarization with charge and discharge is low; the anodic process produces no harmful effect on the cathodic process and vice versa. Another peculiarity of the hydrogen electrode is that it operates under variable gas pressure—the pressure is at a maximum at the beginning of discharge and at a minimum at the end.

The nickel–hydrogen storage cell can be overcharged without breaking down. When the positive electrode has been charged the anodic process of oxygen evolution starts on it. Oxygen readily gets to the hydrogen electrode through the gas pores in the matrix and its catalytic reaction with hydrogen gives rise to water. The only result produced by such overcharge is

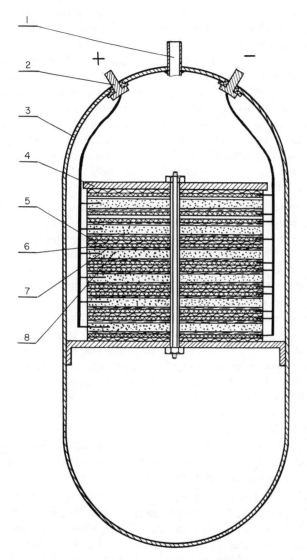

Fig. 88. Schematic of the nickel–hydrogen storage cell: (1) nozzle for feeding hydrogen and connecting manometer; (2) terminal; (3) container; (4) tie plate; (5) hydrogen gap; (6) negative electrode; (7) electrolyte-carrying separator; (8) double positive electrode.

generation of considerable amounts of heat. The hydrogen pressure reaches a certain steady-state level and 'does not increase further since the gases are consumed in the reaction with each other.

The cell is one of the few storage cells for batteries for which over-discharge and reversal are not dangerous. The discharge capacity is limited by the positive electrode; when it has been fully discharged the residual hydrogen pressure in the cell is fairly high (typically, 0·4–0·6 MPa). When the positive electrode is overdischarged, cathodic evolution of hydrogen starts on it and the cell is reversed. Hydrogen produced on the cathode is balanced out by the hydrogen entering the anodic reaction so that the gas pressure does not change. Evolution of hydrogen on the nickel oxide electrode does not result in deterioration of the electrode.

The insensitivity of these cells to overcharge and overdischarge (even at high current densities) makes operation of these cells safer and simpler and is, thus, very important.

Construction types. The nickel–hydrogen storage cell can be designed for the maximum pressure in the charged state in the range 3·5–10 MPa, depending on the need to reduce either the mass or the volume of the cells. For lower pressure the total cell volume is rather large and the mean density of the cell (the ratio of the total mass to the volume) is close to 1 kg/dm^3. For higher pressure, the cell cylinders have thicker walls and larger mass but smaller volume, and the mean density is close to 2 kg/dm^3.

In recent years a new cell construction has been developed in which a compound capable of absorbing large amounts of hydrogen, for instance $LaNi_5$, is placed into the cell so that hydrogen accumulation does not result in high internal pressures in the cell (these compounds are discussed in more detail in Section 17.3). In this cell the steady-state pressure is not higher than 0·4–0·6 MPa and the volume of the gas chamber is smaller. Since the cell contains a considerable amount of the absorbent its mean density is as high as $3–3·5 \text{ kg/dm}^3$. The absorbent is sometimes hydrophobized to increase the rate of hydrogen absorption. A drawback of this cell type is slow desorption of hydrogen at low temperatures (about 0°C or lower) decreasing the maximum discharge current.

Performance. Nickel–hydrogen cells are reported to have capacities between 10 Ah and 60 Ah. The charge and discharge voltages of these cells are higher than those of nickel–cadmium storage cells by approximately 0·02 V. The mean discharge voltage is 1·22–1·25 V and the cut-off voltage is about 1·1 V. The end-of-charge voltage is 1·55–1·58 V. The discharge current density can be as high as $0·1 \text{ A/cm}^2$ (sometimes higher) corresponding to normalized current of 2. The specific energy of these cells varies between 40 and 60–65 Wh/kg (the lower value is for the high-pressure cells and the higher value for the low-pressure cells).

A convenient feature of these cells (apart from those cells with hydrogen absorbent) is that the gas pressure in the cell indicates the degree of charge of the cell.

Though the direct reaction between hydrogen and the charged nickel-oxide electrode is inhibited it still occurs and results in a considerable self-discharge—a charged storage cell loses 6–12% of its capacity per day at room temperature.

Since both electrodes possess sufficient reversibility, nickel–hydrogen cells have long cycle life. The cycle life is decreased by corrosion of the sintered nickel substrate of the positive electrode which results in water consumption. This leads to gradual decrease of the amount of solution in the matrix and to increase of the internal resistance. Corrosion is enhanced during charging when the pH of the solution in the electrode pores drops owing to concentration polarization; overcharge produces especially high corrosion. Drying of the matrix can be made slower by using electrodes and matrix with electrolyte-filled pores of large total volume and by properly selecting the capillary structure of the porous materials to increase the "buffer capacity" of the system (see Section 5.4). Cells incorporating these and other special features can be operated for a few thousand charge–discharge cycles and even more than ten thousand cycles at 30% depth of discharge in each cycle.

The main advantages of the nickel–hydrogen storage cells are hermetic sealing, simplicity and ease of operation, and long service life and cycle life in combination with relatively high specific parameters. These cells are installed in various long-lived (3–5 years) US satellites (such as communications satellites). The use of these cells in electric vehicles is under discussion. The drawback of the cells is their high self-discharge. For some applications the feasibility of using high-pressure cells is being questioned. Under the conditions of large-scale production the cell will, apparently, be cheap, particularly if the hydrogen electrodes are manufactured without platinum catalysts.

Sealed silver–hydrogen storage cells have been developed by analogy with the nickel–hydrogen cells. Their specific energy is higher, from 75 to 90 Wh/kg but their cycle life is short (200–400 cycles) being limited by the silver electrode. These cells are manufactured for special applications.

14.4. Chlorine–zinc storage cells

Chlorine is an attractive oxidant for chemical power sources. Its activity is higher than that of oxygen; its equilibrium potential is more positive ($E = 1.36$ V) and its reduction is much easier. Unfortunately, chlorine is difficult to handle both in the gaseous and liquid state. In the presence of aqueous electrolyte, chlorine exhibits very high corrosive activity towards most metals. Therefore, the only attempts to use chlorine in chemical power sources were made primarily for systems with anhydrous electrolytes, in particular molten electrolytes (Section 16.4).

A novel type of the chlorine–zinc cell with aqueous electrolyte has been under development in the USA since the beginning of the seventies. The underlying concept of this cell is that chlorine produced with charging is accumulated in the form of a solid deposit of chlorine hydrate, $Cl_2 \cdot 6H_2O$, rather than as gas or liquid which are inconvenient to store. Solid chlorine hydrate is deposited from aqueous chlorine solution at temperatures below 9°C. The deposit is readily separated from the solution. Its corrosive activity is low and it can be stored in any vessels at low temperatures.

The system of chlorine–zinc cells shown in Fig. 89 has a circuit for recirculating the electrolyte. The system comprises the cells (1) assembled into groups or batteries, the container (2) for deposition and storage of chlorine hydrate with the cooler (3), and the pump (4) for recirculating the electrolyte.

The overall reaction in the system is

$$Cl_2 \cdot 6H_2O + Zn \underset{\text{charge}}{\overset{\text{discharge}}{\rightleftharpoons}} ZnCl_2 + 6H_2O \qquad (14.5)$$

During charging of the negative electrode the zinc ions from solution are deposited on a thin substrate (made, for instance, from copper) forming a fairly dense layer of metallic zinc. The positive electrode is manufactured

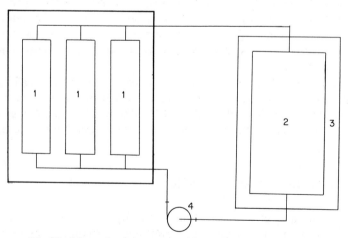

Fig. 89. Schematic of the system with chlorine–zinc storage cells.

from graphite, platinum-plated titanium or other corrosion-resistant material. Chlorine produced on the positive electrode by charging is carried by circulating electrolyte into the cooled container for deposition of chlorine hydrate. During discharge metallic zinc is dissolved while the dissolved molecular chlorine is reduced to chlorine ions. The electrolyte depleted in chlorine is recirculated through the container with chlorine hydrate where it is enriched in chlorine owing to partial dissolution of chlorine hydrate.

The concentration of zinc chloride in the solution varies from approximately 15% in the charged cell to 45–50% in the discharged cell. Deposition and decomposition of chlorine hydrate limits variation of concentration since water is formed or consumed together with zinc chloride. This makes it possible to use a smaller amount of electrolyte in the system. Chlorine hydrate also produces a thermal buffer effect in the course of discharge since a part of the heat produced is absorbed for decomposition of chlorine hydrate.

The solubility of chlorine in the electrolyte at atmospheric pressure is approximately 0·1 mol/litre (it is greatly affected by the temperature and the concentration of $ZnCl_2$). At this concentration the corrosive activity is not too high. The rate of electrolyte circulation must be sufficient to supply adequate amounts of chlorine to the surface of the positive electrode and to prevent development of too high a concentration polarization. Good results were obtained by using the technique of forcing the solution through the porous graphite electrode.

The o.c.v. of the chlorine–zinc storage cells is about 2·12 V (depending on the concentration of $ZnCl_2$). For a current density of $50 \, mA/cm^2$ the discharge voltage is about 1·7 V. Variation of the discharge voltage during discharge is slight. There have been reported bipolar battery systems with rated power of several tens of kilowatts. The system with a 3–4 hour stock of reactants has specific energy of about 70 Wh/kg and it is hoped to increase it to 100–120 Wh/kg by improving the design and suppressing side reactions. The medium-power systems have a cycle life of several hundreds of cycles.

The main peculiarity of this system is its narrow temperature range of operation. The electrolyte of the charged cell freezes below $-10°C$ while chlorine hydrate decomposes above 9°C. Cooling, with the inevitable energy consumption, is needed to operate the system at higher temperatures.

The design of the system with chlorine–zinc storage cells is similar to that of the air–zinc cell system with circulating electrolyte (Fig. 87). The reactions occurring in the chlorine–zinc cell are rather simple but they can be complicated by side reactions, for instance, by the direct reaction between dissolved chlorine and zinc resulting in loss of reactants and decrease in the current efficiency of the cell to 70%. However, such self-discharge can be minimized during breaks in cell operation, by interrupting circulation of electrolyte with dissolved chlorine. For low-current drain the electrolyte circulation rate should be decreased to prevent a rise in the relative contribution of self-discharge. Attempts have been made to reduce self-discharge by employing diaphragms with low permeability to chlorine but allowing ionic current transport.

Deposition of zinc from the solution entails the risk of dendrite formation and short-circuiting. The specific features of zinc electrocrystallization are such that the probability of dendrite formation is lower in salt solutions than in alkaline solutions (see Section 12.1). This risk is further decreased by adding chlorine-resistant surfactants, affecting zinc crystallization, to the electrolyte.

The main advantages of chlorine–zinc storage cells are their high specific parameters and the cheapness and free availability of the active materials (though, or course, the resources of zinc ores, as of other non-ferrous metals, are not infinite). Early efforts focused on using these cells in power units for electric vehicles. However, the operational features of these cells (the need for recirculationg of electrolyte and cooling) indicate that they are more suitable for use in large-scale power units. A project now underway in the USA aims at developing by 1982 a large chlorine–zinc battery with a power of 1 MW and a Wh-capacity of 10 MWh to be used as a pilot large-scale unit for levelling of loads in the power-supply system.

14.5. Lithium cells with aqueous electrolyte[3]

The very first chemical power source—the Volta pile—did not use specially added oxidants; the water molecules served in it as oxidants (donors of protons for cathodic evolution of hydrogen). The efficiency of these cells was low since the rate of hydrogen evolution on the silver electrode is low; polarization was high and voltage was low and decreased with time.

However, the use of hydrogen evolution from water as the cathodic current-producing reactions seems an attractive idea. The cell built according to this concept has metal electrodes (plates) of only two types—dissolving anodes (zinc, magnesium, etc.) and inert cathodes for hydrogen evolution. This simple cell construction does not make use of powdered reactants, separators and so on. The basic reactant (water) is freely available everywhere, especially so in marine applications.

In recent years therefore repeated attempts have been made to improve such systems. A few types of the water–magnesium cells utilizing seawater have been developed. Of particular interest are the water–lithium cells whose development started in the USA at the end of the sixties. The efficiency of these cells is greatly enhanced owing to the use of (i) an active anodic material (lithium), (ii) an iron electrode as current collector on which hydrogen is produced at a high rate, and (iii) recirculation of electrolyte to remove the reaction products (including hydrogen bubbles) and heat.

The current-producing reaction in the water–lithium cells is described by

$$2Li + 2H_2O \rightarrow 2LiOH + H_2 \qquad (14.6)$$

The process is complicated by the fact that lithium enters into intensive reaction with water so that evolution of hydrogen according to reaction (14.6) occurs not only on the iron electrode (making possible useful utilization of electric energy) but also on the surface of the lithium electrode (useless self-dissolution or corrosion of lithium).

Use of lithium electrodes in such cells became practically feasible when it had been found that at a certain concentration of LiOH in water lithium is partially passivated, that is its self-dissolution is sharply inhibited, though this results also in a certain decrease in the activity of lithium (i.e. in the maximum current density) in the primary current-producing reaction. The higher the concentration of LiOH the higher the passivation of the lithium electrode. Thus, a certain optimum concentration of LiOH corresponds to each load-current density.

In water–lithium cells bipolar electrodes are used. The cathode is a steel plate with ribs (to facilitate circulation of the solution) to which an iron mesh is welded. To increase its activity the mesh surface is roughened, for instance by sandblasting. The anode is thin lithium sheet. The passivating film on the lithium surface not only decreases self-dissolution, it also has a considerable resistance to the electron current. Hence, the lithium anode may be pressed against the cathode mesh without the danger of short-circuiting. The interelectrode gap formed by the mesh and the protrusions on the rough steel plate surface is very narrow so that the ohmic losses are small.

Water (more often seawater) is used as electrolyte. Some lithium hydroxide is added to water to initiate cell operation. When the cell has started operating the concentration of LiOH gradually increases owing to reaction (14.6). During recirculation of electrolyte some concentrated solution is removed and water is added so that the required steady-state LiOH concentration can be very accurately maintained in the electrolyte near the electrodes (by controlling the rate of electrolyte recirculation).

The o.c.v. of the water–lithium cells is 2·2 V. Figure 90 shows a typical current–voltage plot. The position of the sloping part AB depends on the LiOH concentration. The maximum efficiency (the minimum contribution of self-discharge) is obtained if the working point is near the point A where the plot starts to slope downwards. Under these conditions the discharge voltage is 1·1–1·4 V. There have been developed automatic devices which regulate the rate of electrolyte recirculation with variation of the current density, temperature, or other factors, to maintain the optimum LiOH concentration. This is done by monitoring the electrode potential, which is a good indicator of the degree of passivation of the electrode. Such systems can operate in the current density range 0·05–0·6 A/cm² with the steady-state concentration of LiOH varying from 4·5 to 3·5 mol/litre. The coefficient of lithium utilization can be made as high as 50–80% with careful control of the LiOH concentration.

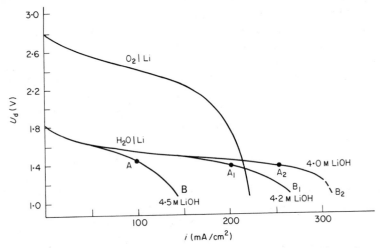

Fig. 90. Current–voltage characteristics of air–lithium and water–lithium cells.

The specific energy of the water–lithium cell (without water being taken into account) can be as high as 400 Wh/kg and the specific power can be up to 200 W/kg. Since the system has auxiliary units—recirculation pump, heat exchangers, devices for diluting the solution and controlling the recirculation rate—the above specific parameters apply to systems of comparatively high power, for instance, over 0·5–1 kW.

Water–lithium cells have been developed for various marine applications: electric torpedoes, underwater devices, buoys, etc.

This system is being adapted for land-based applications, where continuous water feeding is impractical, by making two essential modifications: (i) use of an external oxidant (aerial oxygen); and (ii) replacement of the open recirculation circuit with additions of water by a closed circuit. The reaction product, LiOH, is removed and its concentration is controlled by adding

CO_2 which reacts with LiOH giving rise to slightly soluble lithium carbonate. The overall current-producing reaction is rather simple:

$$O_2 + 4Li + 2CO_2 \rightarrow 2Li_2CO_3 \qquad (14.7)$$

Since the efficiency of the air electrode is lower the current density in the air–lithium cell is not more than 100–150 mA/cm^2; the resulting decrease in power is somewhat compensated by the higher discharge voltage which is 2·2–2·4 V. The o.c.v. of the cell is 3·4 V. The specific energy of the system (including the stock of CO_2) can be as high as 300–350 Wh/kg. Development work on these cells still lags behind the work on the water–lithium cells. The use of such system for electric vehicles is being discussed. Such systems should have enough lithium stock for 2–3 thousand kilometers and will be periodically (say, every 200–300 km) refuelled with CO_2 and water.

Other development projects on lithium cells aim at using hydrogen peroxide as oxidant.

REFERENCES

1. A. J. Appleby and M. Jacquier, The CGE circulating zinc-air battery: a practical vehicle power source, *J. Power Sources* **1**, No. 1, 17–34 (1976/77).
2. B. I. Tsenter, US Patent No. 3669744, 1971. S. Font and J. Goualard, Ni–H$_2$ performance versus Ni–Cd, *in* "Power sources 5" (Collins, D. H., ed.), pp. 331–346. Academic Press, London and New York (1975).
3. E. L. Littauer and J. J. Redlien, The lithium–water electrochemical cell as a marine power source, *in* "Proceedings of 9th Intersociety Energy Conversion Engineering Conference, San Francisco, 1974", pp. 615–619.

BIBLIOGRAPHY

1. J. P. Hoare, "The Electrochemistry of Oxygen". Interscience, New York (1968).
2. A. Damjanovic, Mechanistic analysis of oxygen electrode reactions, *in* "Modern Aspects of Electrochemistry" (Bokris, J. O'M. and Conway, B. E., eds.), No. 5, pp. 369–484. Plenum Press, New York (1969).
3. B. Siller, "Luftsauerstoffelemente". VDI Verlag, Dusseldorf (1968).
4. K. V. Kordesch, Low-temperature–low-pressure fuel cell with carbon electrodes, *in* "Handbook of Fuel Cell Technology" (Berger, C., ed.), pp. 361–424. Prentice-Hall, Englewood Cliffs (1968).
5. J. W. Cretzmeyer, H. R. Espig and R. S. Melrose, Commercial zinc–air batteries, *in* "Power Sources 6" (Collins, D. H., ed.), pp. 269–290. Academic Press, London and New York (1977).
6. K. V. Kordesch and S. J. Cieszewski, Nickel oxide–hydrogen cells, *in* "Power

Sources 6" (Collins, D. H., ed.), pp. 249–258. Academic Press, London and New York (1977).

7. W. A. Bryant and E. S. Buzzelli, A comparison of metal–air batteries for electric vehicle propulsion, *in* "Proceedings of 14th Intersociety Energy Conversion Engineering Conference, Boston, 1979", pp. 651–653.

8. Ph. C. Symons and Ch. J. Warde, Zinc–chlorine batteries for load-leveling, *in* "Proceedings of the Symposium on Load-leveling" (Yao, N. P. and Selman, J. R., eds.), Vol. 77–74, pp. 334–352. The Electrochemical Society, Princeton (1977).

Chapter Fifteen

Cells with Non-aqueous Solutions

15.1. Lithium cells with electrolytes based on aprotic solvents

(a) General description

One of the specific features of the electrolytes based on aprotic solvents (see Section 4.2) is the corrosion resistance of lithium and other alkali metals, due to the fact that cathodic evolution of hydrogen from the solution is impossible. This makes possible use of lithium for anodes in chemical power sources. Since the lithium potential is negative the o.c.v. of such cells is fairly high (3–4 V), that is, considerably higher than the o.c.v. of the cells with aqueous electrolytes. The specific consumption of lithium is as low as 0·26 g/Ah. All these factors contribute to high specific energy for lithium cells. The actual specific energy of lithium cells varies between 200 and 600 Wh/kg.

The first prototypes of the cells with lithium anodes were developed at the beginning of the sixties. Now practically all large cell-manufacturing companies carry out extensive research and development work in this field. Some lithium cells are already available commercially. These are single small-size cells with the capacity up to 15–20 Ah designed mainly for electronic and radio devices, cardiac pacemakers, electric watches, etc. Recent attempts have been reported at developing cells with capacities up to 600 Ah.

Cells with lithium anodes must be sealed, since absorption of moisture from the environment results in strong corrosion of lithium. This makes their development and manufacture more difficult but presents advantages in operation. The typical shelf-life of these cells is several years. Some types of cells show good performance at low temperatures down to about –50°C.

Table 4 describes the physical properties of some suitable solvents. Typical solvents used in the lithium cells are propylene carbonate (PC), γ-butyrolactone and tetrahydrofuran. Sometimes, mixtures of solvents are used, for instance, of PC, which has a high dielectric constant, and diethyl ether which reduces viscosity.

Some types of cells developed in recent years employ solvents which are simultaneously strong oxidants, for instance, thionyl chloride, $SOCl_2$ or liquid sulphur dioxide, SO_2. These oxidants serve not only as solvents but also as active materials in the cathodic reaction so that there is no need for use of special solid oxidants.

The essential requirement for using aprotic solvents in lithium cells is careful dehydration of them. Removal of the traces of moisture and other proton-donating contaminants from the solvents often presents a rather difficult problem.

The halogen salts of lithium, in particular LiBr, are fairly soluble in liquid sulphur dioxide but most simple salts are only slightly soluble in other aprotic solvents. Only lithium perchlorate, $LiClO_4$, and the lithium salts with complex anions, such as $LiAlCl_4$, $LiBF_4$, $LiPF_6$, have somewhat better solubilities, though lower than in water—not more than 1–2 mol/litre. The conductivity of the solutions of these salts varies between 0.5×10^{-2} and 2×10^{-2} ohm^{-1} cm^{-1}, that is, it is lower by one to two orders of magnitude than the conductivity of the aqueous solutions used as cell electrolytes.

The lithium electrodes are manufactured typically as thin rolled sheets (or strips) pressed on the current collector which is a thin plate or mesh of nickel, stainless steel or copper. There has also been a suggestion to manufacture porous lithium electrodes by applying a paste of lithium powder dispersed in mineral oil on a substrate, or by pressing lithium powder.

Though lithium does not expel hydrogen from aprotic solvents, in principle, there can occur other reactions of direct reduction of the solvent molecules due to lithium. The thermodynamic probability of such reaction is naturally high for oxidizing solvents. In the above solutions lithium is practically in a semi-passivated state and the direct reaction is strongly inhibited owing to formation of a surface film on lithium which protects it from further reaction with the solvent.

The rate of anodic reaction on lithium is sufficiently high but not as high as in aqueous solutions. The detailed mechanism of the discharge process in many aprotic solvents is still not sufficiently clear. In solutions of $LiAlCl_4$ in PC, discharge results in growth of an LiCl film on the anode which inhibits the electrode process. The electrochemical characteristics of the lithium electrode also depend on the presence of traces of moisture in the solution which not only intensify corrosion but often inhibit the anodic process.

Practically no other alkali metals are used in combination with aprotic solutions. This is explained not only by their higher equivalent mass but

Table 4. Physical parameters of various solvents at 25°C

Solvent	Formula	Dielectric constant	Viscosity (centipoise)	Density (g/cm^3)	Boiling point (°C)	Melting point (°C)
Propylene carbonate	$CH_3-CH-CH_2$ / O O / C=O	66·1	2·53	1·198	241	−49·2
γ-Butyrolactone	$CH_2-CH-CH_2$ / O —— C=O	39·1	1·75	1·125	204	−43·5
Tetrahydrofuran	CH_2-CH_2 / O / CH_2-CH_2	7·4	0·46	0·880	64	−65
Dimethylsulphoxide	$H_3CO-S-OCH_3$ =O	46·7	2·0	1·095	189	18·6
Sulphur dioxide	SO_2	12·0†	0·26†	1·491†	−10	−72·7
Thionylchloride	$SOCl_2$	9·1	0·60	1·629	77	−104·5
Water	H_2O	78·3	0·89	0·997	100	0
Ammonia	NH_3	20‡	0·26‡	0·67‡	−33	−78

† At −20°C.
‡ At −20°C and a pressure of 0·2 MPa.

also by their higher activity towards many solvents. Moreover, their easy oxidation in air makes cell manufacture more difficult.

Lithium cells with electrolytes based on aprotic solvents are manufactured as rectangular or cylindrical cells; the sizes of the cylindrical cells correspond to the standard sizes of the manganese–zinc cells.

Two constructions of the cylindrical cells have been developed. The cells with high specific energy designed for prolonged discharge at low-current drain have a steel cylindrical can into which the peripheral anode, separator and the central thick pressed cathode are inserted. There is a modification of this construction with central anode and peripheral cathode. In the second cell construction designed for higher drain power but lower specific energy, the cylindrical can contains coiled thin sheet electrodes and separator.

(b) Lithium cells with solid oxidants

Halogen salts, sulphides and oxides of various metals and numerous other compounds have been suggested for use as solid cathode materials. First to be tested were halogen salts such as NiF_2, CuF_2, $CuCl_2$, AgCl. Cell prototypes with specific energy up to 200 Wh/kg were developed for drain current densities varying between 0·2 and 1 mA/cm^2; their discharge voltages were 2–3·5 V. Such cells have poor storageability (a few months) owing to significant solubility of the halogen compounds and the possibility of their direct reaction with lithium.

Copper sulphide is considerably better in this respect. Cells with CuS were among the first lithium cells to be manufactured commercially (by the French company SAFT, 1966). The copper sulphide electrodes are manufactured by sintering powders of copper and sulphur. In contrast to halides, sulphides have a considerable electronic conductivity and are not hygroscopic. When the electrode is discharged CuS is converted first into Cu_2S and then into metallic copper. The mean discharge voltage at the first stage is 1·8 V and at the second stage 1·5 V. The specific energy can be as high as 300 Wh/kg, the shelf-life of the cell is about one year.

Much attention was paid to studying oxide electrode materials, MoO_3, for instance. Highly complicated processes are involved in discharge of such electrodes. In more recent times such electrodes have attracted greater interest in connection with the attempts to develop storage cells using aprotic solvents.

Cells employing silver chromate as the oxidant exhibit fairly good performance. SAFT have developed miniature disk cells for cardiac pacemakers with specific energy of up to 850 Wh/dm^3 for mean current drains of about 10^{-5} A. About 100 000 such cells are produced every year.

Intensive development work on cathodes based on fluorinated carbon started in 1968. This non-stiochiometric compound (berthollide) belongs to the group of intercalation compounds and has the general formula $(CF_x)_n$. It is produced in the reaction between carbon and fluorine at about 400–450°C. The value of x varies between 0·25 and 1·35 while n can have high values typical of polymers. Discharge of fluorinated carbon gives rise to carbon and fluoride ion. The theoretical consumption of fluorinated carbon of composition $(CF)_n$ is 1·16 g/Ah.

Cells with fluorinated carbon are now manufactured on a comparatively large scale (hundreds of thousands a year) by Matsushita (Japan). The cylindrical cells are assembled in nickel-plated steel cans. The positive electrode is manufactured by pressing a mixture of $(CF)_n$, acetylene black and binder onto a titanium mesh. The assembly of the 0·9 mm thick positive electrode, separator (porous polypropylene) and the 0·5 mm thick negative lithium electrode is coiled and inserted into the can. The electrolyte is a 1 M solution of $LiBF_4$ in γ-butyrolactone.

The B-size cell (see Table A.4) has a rated voltage of 2·8 V and capacity of 5 Ah for $j_d = 0·07$, corresponding to specific energy of 340 Wh/kg. Figure 91 shows the discharge curves for this cell at various temperatures (the discharge curve for the manganese–zinc cell of the same size is shown for

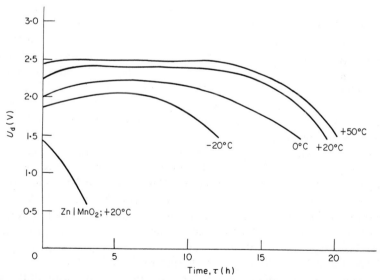

Fig. 91. Discharge curves of the B-size $(CF)_n$–Li cell at various temperatures discharged through a constant resistance of 10 ohms in comparison with the discharge curve of the manganese–zinc cell.

comparison). The cell has a high coefficient of utilization of the active materials of the positive electrode (90 % for $j_d = 0.01$).

A common feature of lithium cells with solid oxidants is the low admissible current density, not more than $0.001-0.002 A/cm^2$, corresponding to a normalized current of $0.01-0.07$ (depending on the construction) and specific power of 5–25 W/kg. The limitation of current is explained by slow electrochemical reduction of solid reactants in aprotic solutions. In the aqueous solutions such reactions are promoted owing to the fact that the resulting ions are hydrated, that is they react with water. In aprotic solutions the degree of solvation is lower.

This feature determines the main mode of operation of such lithium cells, namely prolonged low-current drain. In contrast to other cell types, self-discharge does not reduce the capacity of lithium cells even after very long periods of operation.

For solid oxidants with significant solubility (for instance, $CuCl_2$) the reaction rate increases but storageability becomes worse. Therefore, such oxidants can be used only in the reserve-type cells.

(c) Lithium cells with liquid oxidants

Use of liquid oxidants in the cells with aprotic solutions presents certain advantages. The rates of electrochemical reduction of liquids and compounds dissolved in aprotic solvents are higher than the rates of reduction of solids. While the rated current density for discharge of solid cathodes is about $10^{-3} A/cm^2$ the current densities in the cells with liquid oxidants can be higher by an order of magnitude. This makes possible development of cells not only with high specific energies but also with fairly high specific powers. Since the electrolyte in such cells is an active material this also contributes to higher specific capacity.

The liquid oxidants are in direct contact with the lithium anode in cells. A very strong passivating film of the products of reaction of lithium with the oxidant is formed on the lithium surface. There is no self-discharge of the electrode even after many years of contact with oxidant. The film has sufficient conductivity for the lithium ions and under normal conditions does not prevent anodic dissolution of lithium even at high current densities.

The discharge curves of the lithium cells with liquid oxidants (particularly, thionyl chloride) exhibit characteristic "dips" of voltage sometimes at the beginning of discharge. The magnitude and duration of the dips depend on the conditions and time of storage and the conditions of discharge. Typically, these dips appear with discharge at low temperatures after prolonged storage of the cells at higher temperatures. The nature of these dips is not yet fully clear but they are obviously associated with additional passivation of the lithium anode. It is assumed that during storage of the cells, particularly at elevated temperatures, a thicker or more dense film is formed on the lithium; this film has a noticeable inhibiting effect at the beginning of discharge. A few minutes or seconds after the beginning of discharge the film flakes off owing to anodic dissolution of the lithium beneath it, and the cell performance is restored.

Mallory & Co. (USA) manufacture cells with sulphur dioxide in large quantities. These cells employ 1·8 M KBr solution in a mixture of 28% (by volume) of acetonitrile, 8% of PC and 64% of liquid SO_2. Since the vapour pressure of liquid SO_2 at room temperature is about 0·4 MPa the cells are designed to withstand elevated internal pressure. Both electrodes are coiled and inserted into nickel-plated steel cans. Lithium foil pressed onto copper mesh is the negative electrode. The positive electrode is carbon-pasted aluminium mesh. The electrodes are separated by a membrane of porous polypropylene.

The current-producing reaction with discharge of the cell is

$$2Li + 2SO_2 \rightarrow Li_2S_2O_4$$

The insoluble reaction product (lithium dithionite) is accumulated in the pores of the carbon layer on the positive electrode resulting in a decrease in the discharge capacity of the cell.

The o.c.v. of the cells is 2·9 V. The discharge voltage is very constant; at low-current drain it is 2·7 V. The specific energy of the cells is 260–300 Wh/kg for $j_d \leqslant 0.05$–0.1. A special feature of these cells is capacity for high-current drain. They can be discharged even at $j_d = 1$ (this corresponds, for instance, to 10 A for the D-size cell). The duration of such discharges is limited by evolution of heat. To prevent a dangerous rise of the SO_2 pressure at elevated temperatures the cells are provided with safety vents.

The cells with SO_2 exhibit good performance at low temperatures owing, in particular, to the fact that the viscosity of liquid SO_2 is low and increases only slightly with decreasing temperature. Hence, the conductivity of the solution at $-50°C$ is 2.4×10^{-2} ohm^{-1} cm^{-1} which is lower by a factor of only two than the conductivity at room temperature.

These cells have good storageability. The manufacturer reports only a 20% decrease in capacity after a five-year storage at room temperature. From this, the predicted shelf-life of the cells is about 10 years.

Even higher specific energies can be obtained for cells with thionyl chloride, $SOCl_2$: for some prototypes it was as high as 600 Wh/kg. The electrolyte is typically 0·5–1 M $LiAlCl_4$ solution in pure $SOCl_2$. The advantage of $SOCl_2$ over SO_2 is lower internal pressure in the cells. The maximum current drains for such cells are somewhat lower (corresponding to j_d about 0·1) but still higher than for cells with solid oxidants.

Discharge involves a rather complicated current-producing reaction which can be approximately described by the following equation:

$$4Li + 2SOCl_2 \rightarrow 4LiCl + SO_2 + S$$

Discharge is limited by precipitation of LiCl in the pores of the carbon layer of the cathode. During discharge the internal pressure gradually increases owing to formation of SO_2. The o.c.v. of the cells is 3·65 V, the discharge voltage is fairly stable, varying between 3 V and 3·5 V depending on the current. Figure 92 shows the discharge curves of the cells with SO_2 and $SOCl_2$.

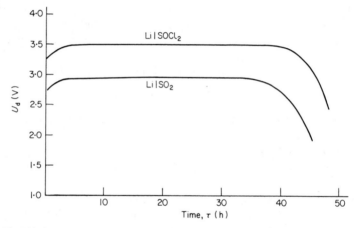

Fig. 92. Discharge curves for the B-size $SOCl_2$–Li and SO_2–Li cells for a current of 0·1 A.

Despite their attractive characteristics cells with thionyl chloride are not yet commercially available for two reasons. The first is a noticeable dip in voltage after prolonged storage of cells. It has been shown recently that some additives (e.g., SO_2) to the electrolyte can reduce the depth and duration of such dips. The second reason is the fact that in some (though rare) cases the cells can explode. Explosions occur, in particular, with overdischarge of cells assembled in batteries and during storage of discharged cells. The causes of the explosions are not yet completely clear. To eliminate the danger of explosions in cells with liquid oxidant due to overheating (lithium melts at 176°C and the passivating film disintegrates) these cells are often provided with discharge current limiters or special fuses disconnecting the cells in cases of external short circuits.

(d) Development of storage cells with aprotic solvents

Numerous attempts have been made to develop storage lithium cells with aprotic solvents. However, no commercially manufactured cells or even satisfactory prototypes have been developed so far. Difficulties are encountered in development of both the positive and negative electrodes.

During charging, lithium is in principle deposited from the solution. However, the lithium deposit has a rather loose structure and its adhesion to the substrate is inadequate. Moreover, lithium electrodes are readily passivated in the intervals between cycles. Though many studies have been conducted, this problem has not yet been solved.

The attempts to regenerate solid oxidants with charging are mostly unsuccessful owing to the extremely slow rates of these processes. Interesting properties of some intercalation compounds have been reported recently. If such compounds as TiS_2 or $NiPS_3$ are used as solid oxidants lithium penetrates them during discharge in the following reaction for instance:

$$xLi^+ + TiS_2 + xe \rightleftharpoons Li_xTiS_2$$

This reaction does not produce changes in the parameters of the crystal lattice and is fairly reversible. Further research into these materials may contribute to development of reversible positive electrodes with satisfactory performance.

15.2. Ammonia-activated cells

Liquid ammonia is an anhydrous but not an aprotic solvent. The cells employing salt solutions in liquid ammonia as the electrolyte are used exclusively as reserve-type cells, mainly for military applications, and are manufactured on a small scale.

The boiling point of ammonia is $-33°C$. At small excess pressures cells with ammonia solutions can operate in a wide temperature range; at $+20°C$ the pressure of ammonia is 0.85 MPa. The heat of vaporization of ammonia at $25°C$ is 20 kJ/mol. Similarly to water, ammonia exhibits slight dissociation into the ions NH_4^+ and NH_2^-. The ammonium ions in liquid ammonia donate protons, that is, ammonia salts can be regarded as acids.

Nitrates, perchlorates and thiocyanates have good solubilities in liquid ammonia. When ammonium salts are dissolved in liquid ammonia the volume decreases considerably owing to solvation. The pressure of ammonia vapour over solutions of ammonium salts is hence much lower than the vapour pressure over pure ammonia.

Solutions of potassium and ammonium thiocyanates are employed as electrolytes in these cells. The electrolytes have high enough conductivities, from 0·2 to 0·4 ohm^{-1} cm^{-1} at temperatures between $-60°$ and $+20°$C and concentrations between 7 and 12 mol%.

The anode materials are typically magnesium, zinc or lead; their standard potentials (with respect to the standard potential of the hydrogen electrode in ammonia solution) are $-1·74$, $-0·54$ and $+0·28$ V, respectively. Intensive corrosion of magnesium with evolution of hydrogen occurs in acidic solutions, that is, in solutions of ammonium salts. In concentrated solutions of NH_4SCN the rate of corrosion is lower; it is even lower in neutral KSCN solutions. In 30% KSCN solution the discharge current density for the magnesium anodes can be as high as 200–300 mA/cm^2 with polarizations of 0·3–0·4 V. The rate of zinc corrosion in ammonia solutions is much lower, while lead is thermodynamically stable.

Cathodes are manufactured from lead or manganese dioxides, sulphur, dinitrobenzene, etc. Lead dioxide is deposited electrochemically on a stainless steel substrate, while the construction of the MnO_2 electrodes is similar to that of Leclanché electrodes. The overall current-producing reaction in the lead dioxide–zinc cell is described by the following equation:

$$Zn + PbO_2 + 4NH_4SCN \rightarrow Zn(SCN)_2 + Pb(SCN)_2 + 2H_2O + 4NH_3$$

The actual o.c.v. values for various cells with ammonia electrolyte at $-40°$C are given below:

System	O.c.v. (V)
$Mg\,|\,NH_4SCN; NH_3\,|\,PbO_2$	2·95
$Zn\,|\,NH_4SCN; NH_3\,|\,PbO_2$	1·90
$Pb\,|\,NH_4SCN; NH_3\,|\,PbO_2$	1·10
$Mg\,|\,KSCN; NH_3\,|\,S$	2·25

Two constructions of reserve-type cells (activated with ammonia vapour and activated with liquid ammonia) have been reported. The batteries for both types of cells have the same construction. They are assembled from plates of anode material, closely pressed together, separators made of cellulose or special paper impregnated with KSCN or NH_4SCN and carefully dried, and cathodes. Liquid ammonia is stored in a special ampoule which is placed inside the sealed battery container. For activation with vapour the ampoule is broken, ammonia evaporates and condenses in the pores of the separator where the electrolyte solution is formed.

The high heat of vaporization of ammonia results in a long time of activation, particularly at low temperatures. At the same time, condensation of vapour produces considerable local heating in the battery. Therefore, a more convenient technique is activation with liquid ammonia fed into the battery under ammonia vapour pressure.

The $Mg|PbO_2$ cells are designed for short-term operation (1–2 min). Their rated discharge voltage is 2·6 V and the current density is about 50 mA/cm^2. The cells with zinc anodes can operate for longer periods at 1·5–1·7 V. The cells with lead anodes can operate for up to 2 days at 0·7–0·9 V. The temperature range of operation for all these cells is -55 to 70°C. The specific energy of the batteries varies between 15 and 30 Wh/kg for short-term discharge and between 40 and 60 Wh/kg for long-term discharge.

BIBLIOGRAPHY

1. G. J. Janz and R. P. T. Tomkins, "Nonaqueous Electrolytes Handbook", Vols. 1, 2. Academic Press, London and New York (1972, 73).
2. T. C. Waddington (ed.), "Nonaqueous Solvent Systems". Academic Press, London and New York (1965).
3. J. Quobex, J. P. Gabano, Les générateurs électrochimiques à électrolyte non aqueux, Etat de l'art, Rev. Gen. Electr. 84, No. 6, 491–496 (1974).
4. P. Bro, R. Holmes, N. Marincic and H. Taylor, The discharge characteristics of the Li–SO$_2$ battery system, in "Power Sources 5" (Collins, D. H., ed.), pp. 703–712. Academic Press, London and New York (1975).
5. A. N. Day, Primary Li/SOCl$_2$ cells, Electrochim. Acta 21, No. 11, 855–860 (1976).
6. H. S. Gleason, J. M. Freund, L. J. Minnick and W. F. Meyers, Ammonia vapour activated batteries, J. Electrochem. Soc. 106, No. 3, 157–160 (1959).
7. C. H. Elzinga and C. G. Vermeulen, A long-life multicell ammonia reserve battery, in "Power Sources 5" (Collins, D. H., ed.), pp. 803–816. Academic Press, London and New York (1975).

Chapter Sixteen

Cells with Solid and Molten Electrolytes

16.1. Solid electrolytes in chemical power sources

The use of solid electrolytes in chemical power sources opens up attractive prospects for two reasons. Firstly, they make possible the development of cells without liquids which are convenient for miniaturization since they do not have problems of sealing, current leakage (particularly, in high-voltage batteries), and so on. These cells have simpler constructions since they do not need sealing packings and separators. Secondly, in cells with liquid or gaseous reactants the solid electrolyte serves as a separator since if it does not have pores the reactants cannot penetrate it. Cells with solid electrolyte do not contain inert separator material partially shielding the ionic current.

Solid electrolytes for use in electrochemical power sources must have high ionic conductivity which must be unipolar, that is to say, due to motion of ions of one sign (see Section 4.2). Even slight electron conductivity is inadmissible since it would result in internal shortcircuiting in the cell.

Mechanism of unipolar ionic conductivity in solid electrolytes

The mobility of one ion species in the crystal lattice (we shall denote it A and the fixed ion species B) is determined by the presence of vacancies or ionic defects in the crystal lattice, i.e. unoccupied lattice sites which can in principle accommodate the ion A. Hence, an A ion near a vacancy can jump to this vacancy under the effect of thermal vibrations and electric field creating a new vacancy in the process. Successive ion jumps give rise to the current carried by the A ions.

The vacancies in crystal lattices are produced by two processes. The first is the thermal vibrations of the ions at lattice sites which can result in an ion leaving its site to become an interstitial (Frenkel defect) or to travel to the crystal surface (Schottky defect). The concentration of such thermal vacan-

320

cies is low but it rapidly increases with temperature. The conductivity associated with these defects is low (10^{-7}–10^{-4} ohm^{-1} cm^{-1}) and manifests itself mainly at elevated temperatures. The second vacancy-producing factor stems from special features of crystal structure. In some (particularly, complex) lattices the ion A has no definite site but can occupy one of several sites with almost equal probability. It is then said that the sublattice of the A ions is disordered, in contrast to the rigidly ordered sublattice B. Since an ion at a given moment can occupy only one site, other equivalent sites play the part of vacancies. The total concentration of such structural vacancies is high and the associated conductivity can be as high as 10^{-2}–1 ohm^{-1} cm^{-1}, which is similar to the conductivity of alkaline or acidic solutions of relatively high concentration. Sometimes such structures are referred to (rather inappropriately) as ionic superconductors. A more correct term is ionic conductors in contrast to ionic semiconductors with thermal excitation of defects.

Under all circumstances conductivity κ increases with temperature according to

$$\kappa = k \exp(-W_\kappa/RT)$$

which is plotted as a straight line in log κ vs $1/T$ coordinates (Fig. 93). The temperature conductivity coefficient and the parameter W_κ associated with

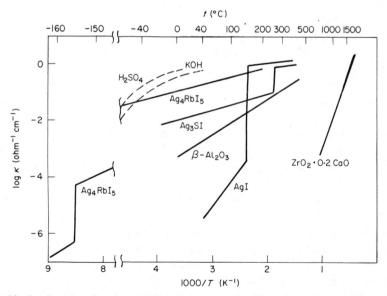

Fig. 93. Conductivity of various solid electrolytes as a function of temperature. The dashed lines correspond to 30% sulphuric acid and 35% potassium hydroxide solutions.

it are typically lower for structural vacancies than for thermal vacancies since the number of structural vacancies does not increase with temperature. Temperature rise only facilitates jumps of the A ions to the neighbouring vacancies.

Both mechanisms of vacancy generation depend on the specific properties of the crystal lattice and especially on the presence of impurities in the lattice.

Crystalline solids often have a number of crystal modifications and each of the modifications can have its own temperature range of stability. Phase transitions from one modification to another change the character of the conductivity, as well as other properties of the solid. Silver iodide serves as a typical illustration of this. Its low-temperature β- and γ-modifications have a comparatively low conductivity associated only with thermal vacancies. At temperatures over 147°C the α-modification with a highly disordered lattice of silver ions is formed and the ionic conductivity of AgI jumps upwards.

Types of solid electrolytes

The research and development work on the use of solid electrolytes in chemical power sources received a strong boost in the sixties when a number of binary salts were found to possess a disordered sublattice and high conductivity, some even at the room temperatures. A number of classes of such solid electrolytes are known at present.

(a) Binary salts based on silver iodide (and other silver halides) with the formula $nAgI \cdot mMX$ where MX has either a common cation or common anion with AgI. The best known compound of this class is $RbAg_4I_5$ ($= 4AgI \cdot RbI$) in which the silver sublattice becomes disordered at as low a temperature as $-155°C$. At 25°C its conductivity is 0.26 ohm^{-1} cm^{-1}, that is, the same as that of a 7% KOH solution. Another compound of this class is Ag_3SI whose conductivity at room temperature is 10^{-2} ohm^{-1} cm^{-1} and the conductivity of its α-phase formed at 235°C is 1 ohm^{-1} cm^{-1}. In recent years there have also been reported compounds of this type based on organic salts such as pyridine iodide ($5AgI \cdot C_5H_5NHI$) or tetramethylammonium iodide ($13AgI \cdot 2N(CH_3)_4I$). Some binary salts based on copper halides also have high conductivity.

(b) Sodium polyaluminates $Na_2O \cdot nAl_2O_3$ where $n = 3-11$ (the so-called sodium beta-aluminas). In these solids the migratory sodium ions are located in the plane between two aluminate blocks, so that in single crystals the conductivity is anisotropic being low in the direction perpendicular to this plane. In polycrystalline specimens the conductivity is about 0.005 ohm^{-1} cm^{-1} at room temperature and 0.1 ohm^{-1} cm^{-1} at 300°C.

(c) Electrolytes with the formula $ZrO_2 \cdot nMO_x$ based on zirconium dioxide, where $MO_x = CaO$, Y_2O_3, etc. When the quadruply charged zirconium ions in the crystal lattice of zirconium dioxide are partially substituted by doubly or trebly charged cations, the number of negatively charged oxygen ions should decrease to conserve electroneutrality. Thus oxygen vacancies are generated giving rise to mobility of oxygen ions. In practice, such conductivity is manifested only at high temperatures as the temperature conductivity coefficient is high. At 1000°C the solid electrolyte $ZrO_2 \cdot 0 \cdot 11Y_2O_3$ has a conductivity of about $0 \cdot 12 \, ohm^{-1} \, cm^{-1}$. Such electrolytes are used in high-temperature fuel cells (see Section 17.8).

Of considerable interest are the recently discovered solid electrolytes of binary salts based on lithium sulphate whose α-phase, formed at 586°C, has a disordered lithium sublattice. In the presence of other sulphates the temperature of transition to the high-conductivity phase is somewhat lower.

16.2. Low-temperature miniature cells with solid electrolytes

A number of cells and batteries with solid electrolytes based on simple silver halides were developed in the fifties. The feasibility of developing miniature cells of high mechanical strength, with good storageability (even at high temperatures) and without the danger of electrolyte leakage was shown. A significant drawback of these cells is the high electrolyte resistance: even with very thin electrolyte films the discharge current densities of the cells are not higher than several microamperes per cm^2. The trend to miniaturization and the need to decrease the thickness of the electrolyte films led to the use of sputtering and other technological processes developed for manufacture of solid-state devices.

When high-conductivity electrolytes, in particular $RbAg_4I_5$, were discovered they opened the way for development of cells for higher discharge current. A cell with such an electrolyte can have a silver anode and an iodine cathode. In the course of cell operation silver is dissolved and the silver ions migrate through the solid electrolyte and react with the iodine ions produced in the cathodic reaction; the final reaction product is silver iodide, AgI.

This simple cell has a number of significant drawbacks. The iodine vapour has a relatively high pressure necessitating careful sealing of the cell. Iodine can be dissolved in electrolyte and migrate towards the anode causing self-discharge and formation of a poorly conducting film of AgI on the anode. Iodine also facilitates decomposition of $RbAg_4I_5$, giving rise to separate phases of AgI and RbI_3. However, the primary drawback of these cells is the accumulation of the reaction product AgI near the cathode and the associated increase in the internal resistance of the cell.

To remove these drawbacks other cathode materials are employed, polyiodides of the type of RbI_3, $N(CH_3)_4I_3$ for instance, and others, free of such undesirable properties. Reactions with these materials produce high-conductivity compounds, for instance:

$$2Ag + RbI_3 \rightarrow 2AgI + RbI \rightarrow (RbAg_4I_5; Rb_2AgI_3; \text{etc.})$$

and the cell resistance does not increase. Since no free iodine is present in the cell the above side reactions do not occur.

Cells of this type have been developed with film, tablet and cylindrical constructions. In the first type the electrolyte film is applied onto the metal anode or cathodic current-collector by sputtering or vaporization. The tablet and cylindrical cells designed for comparatively high drains rates employ porous electrodes manufactured by pressing a mixture of the powders of the active materials (silver or polyiodide), electrolyte, and conductive additive (carbon black, etc.).

The o.c.v. of the $Ag|RbAg_4I_5|RbI_3$ cells is $0 \cdot 67$ V and their discharge current density can be as high as 50 mA/cm^2. At low-current drains the slope of their discharge curve is small and the coefficient of utilization of the active materials can be as high as 90% or more. Polarization is mainly ohmic in nature and even at high current densities it is due almost entirely to electrolyte resistance. At current densities of about $0 \cdot 1$ mA/cm^2 the cells are operable at temperatures down to $-55°C$. Owing to low o.c.v. and the large equivalent mass of the reactants the theoretical specific energy (that is, not taking into account the masses of electrolyte and structural components) is low, about 53 Wh/kg. In practice, the specific energy of the miniature cells and batteries is approximately a factor of 10 lower. The range of application of low-temperature cells with solid electrolyte is thereby limited to some special fields requiring long-time storage, operability in a wide range of temperatures, and high mechanical strength.

16.3. Sulphur–sodium storage cells

Development of a high-temperature sulphur–sodium storage cell employing solid sodium polyaluminate as electrolyte was first reported in 1966 (Ford Co., USA). This cell type immediately attracted attention and considerable research and development efforts in many countries. The main advantages of this cell are the high specific power and energy, good reversibility, the absence of side reactions, hermetic sealing and (the most important) cheapness and free availability of the main reactants: sulphur and metallic sodium. A drawback of the cell is its high working temperature which is in the range of 300–350°C. In contrast to other cell types this cell employs solid electrolyte and liquid reactants (the melting points of sulphur and sodium are 119°C and 97·5°C, respectively). The solid electrolyte also serves as the separator between the reactants of both electrodes.

Current-producing reactions. In the course of discharge sodium is anodically oxidized to sodium ions, Na^+, which penetrate the solid electrolyte and act as current carriers in it. Sulphur is reduced on the positive electrode and reacts with sodium ions from the electrolyte giving rise to various sodium polysulphides, Na_2S_m. The overall current-producing reaction can be divided into two stages:

(A)
$$5S + 2Na \underset{\text{charge}}{\overset{\text{discharge}}{\rightleftharpoons}} Na_2S_5$$

(B)
$$3Na_2S_5 + 4Na \underset{\text{charge}}{\overset{\text{discharge}}{\rightleftharpoons}} 5Na_2S_3$$

Stage A gives rise to molten Na_2S_5 which does not mix with molten sulphur so that a two-phase liquid system is formed. At stage B when free sulphur has been consumed the system consists of one phase. The melting points of polysulphides with m between 3 and 5 lie in the range of 235–285°C.

In principle, Na_2S_3 can be reduced further giving rise ultimately to the simple sulphide Na_2S. However, sodium sulphide and sodium disulphide, Na_2S_2, have higher melting points and, hence, form solid phases in the cell considerably reducing its reversibility. Therefore, the cell capacity corresponding to complete conversion of pure sulphur to pure sodium trisulphide is regarded as the limiting (theoretical) cell capacity.

The high working temperature of the sulphur-sodium cell is necessitated not only by the desire to increase the conductivity of electrolyte but also by the need to operate with molten reactants and intermediate compounds.

Electrolyte. There are two crystallographic modifications of sodium polyaluminates, $Na_2O \cdot nAl_2O_3$, with different sodium contents. One is β-alumina with $n = 9-11$ and the other is β''-alumina with $n = 3-5$. At 300°C the conductivity of the β-phase is $0.03-0.1$ ohm^{-1} cm^{-1} and the conductivity of the β''-phase (which has a higher sodium content but is less stable) is approximately twice as great. The conductivity depends on the conditions of manufacture of the electrolyte. It can be raised by a factor of 2–3 by adding magnesium or lithium oxide. However, the additions impair the stability of cell operation and give rise to hygroscopicity complicating manufacture of the cells. The existing electrolytes with optimum composition have conductivities of $0.1-0.3$ ohm^{-1} cm^{-1} at 300°C.

The stability of the electrolyte during operation presents considerable difficulties. In the early cell prototypes the electrolyte developed micro-cracks; metallic sodium leaked through them resulting in cell failure. The cracks were caused not only by mechanical stresses but also by non-uniformities of the electrolyte structure and the presence of defects. Electrolyte stability is now greatly improved owing to careful regulation of the manufacturing process and control of the microstructure of the electrolyte. The important factor in this respect is the uniformity in size of the primary electrolyte particles; their optimum size is 2–4 µm.

The electrolyte is a poreless ceramic material shaped either as a thin disk (for flat cells) or a tube with one sealed end (similar to test tube) employed in cylindrical cells. The process of electrolyte manufacture consists of the following stages: (i) preliminary calcination and grinding of the starting materials (alumina, α-Al_2O_3, and sodium carbonate); (ii) careful mixing of the components with a binder; (iii) isostatic compacting of the powder at a pressure of about 400 MPa; (iv) sintering at a temperature of about 1600°C. Sintering is the most critical stage; since sodium oxide, Na_2O, is volatile the electrolyte is sintered in closed platinum crucibles. On the whole, the technological process is rather complicated and its efficiency is so far not higher than 50–60%.

Cell construction. The cells are mostly of cylindrical construction, with the electrolyte shaped as a tube of length 20–50 cm, diameter 1·5–3·5 cm, and wall thickness about 1 mm. One reactant is inside the tube and the other is on the outside. Molten sulphur and polysulphides are typically inside the tube (Fig. 94). Molten sodium is in the gap between electrolyte and the cell container wall. A stock of sodium is stored in a container in the upper and lower part of the cell.

Since both sulphur and polysulphides lack electron conductivity a very loose felt-like graphite material is employed as the current-collector of the positive electrode; its mass is only 3–10% of the reactant masses. The graphite felt has also another function; in the course of charging at stage A sulphur is formed on the electrolyte surfaces. Since sulphur lacks ionic conductivity (in contrast to polysulphides) it can block the polysulphide melt, thus hampering further charging. The graphite felt has good wett-ability by molten sulphur but not by the polysulphide melt. When the large-pore felt layer is close to the electrolyte and the fine pore layer is farther from it capillary forces tend to draw molten sulphur away from the electrolyte surface thus reducing the blocking effect of sulphur. This ar-rangement contributes to a significant increase in chargeability of the sulphur electrode at stage A so that the coefficient of utilization of sulphur is raised to 80–90%.

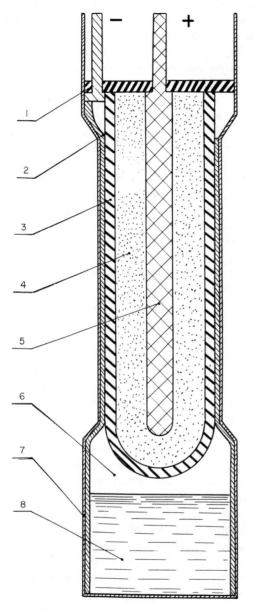

Fig. 94. Schematic of the sulphur–sodium storage cell: (1) cover; (2) wicks; (3) electrolyte; (4) active materials of the positive electrode; (5) current collector of the positive electrode; (6) gas space; (7) container; (8) sodium.

In the course of discharge the volumes of the reactants vary, the sodium volume continuously decreases as sodium is consumed and the volume of the polysulphide melt increases. Therefore, only about 60% of the poly-sulphide chamber is filled with reactants when the cell is fully charged. Owing to the capillary forces the graphite felt provides for contact between the reactants and the total electrolyte surface. Sodium can also be supplied to the working surface through a capillary system, with a good wettability by molten sodium, made, for instance, from stainless steel. Such a system delivers sodium to the whole electrolyte surface even if the residual content of sodium in the tube is small. In this way sodium can be almost fully utilized (over 90% utilization).

Selection of material for cells containing molten sodium and sulphur is not easy. Sodium is held in vessels made from stainless steel, the inside walls of which are sometimes coated with graphite, tungsten or other materials. The current collector is a tungsten wire immersed in the sodium melt. Sulphur and polysulphide melts have very high corrosive activities. Therefore, in recently suggested cell constructions the active materials of the positive electrode do not contact any materials apart from electrolyte and the graphite current collector.

Another difficult problem is sealing of the cell, especially, of the electrolyte–metal joints. A number of more or less satisfactory solutions have been suggested recently.

Owing to technological difficulties less attention is paid at present to flat cell construction than to cylindrical construction. The changes in reactant volumes with charge and discharge give rise to internal pressure differences which result in buckling and cracking of the flat electrolyte. Moreover, in cells with flat electrolytes and electrodes the size of sealing region is larger and the associated difficulties are much greater.

A battery comprising a large number of tubular storage cells must have heat insulation. An external power source is needed to preheat the battery before operation with special internal electric heaters. If the battery is large enough and has good heat insulation the temperature in the cell is maintained within the working range by the heat released during operation. The heat insulation should provide for sufficiently slow cooling of the cells after the end of operation (not more than 5–10°C per hour, say) to make possible battery operation with short-time breaks. Light-weight materials are used for heat insulation so that the use of insulation affects the specific parameters with respect to volume rather than mass.

Performance. The o.c.v. of the sulphur–sodium storage cell is 2·08 V for two-phase liquid system operation (stage A) and gradually decreases during discharge to 1·76 V for the one-phase system operation (stage B).

The charge–discharge curves exhibit a small step corresponding to transition from stage A to stage B (Fig. 95). The decrease in the cell voltage with increasing density of the discharge current is largely due to the ohmic resistance of the solid electrolyte. For 1–2 mm thick electrolyte this resistance is 1–2 ohm cm^2. Hence, the discharge current density can be higher than 500 mA/cm^2. However, to increase the cycle life of the cells the current density is now typically limited to 100 mA/cm^2.

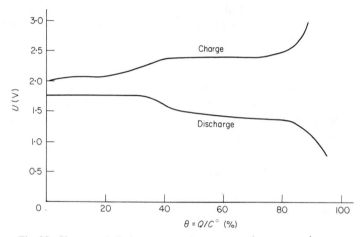

Fig. 95. Charge and discharge curves for the sulphur–sodium storage cell.

The specific parameters depend on the ratio of the mass of the reactants to the masses of the electrolyte and structural components of the cell. If the mass of the reactants is increased, the specific energy becomes higher but the specific power drops. On the other hand, if the relative mass of the reactants is small (for instance, if a large number of smaller tubes are used) the specific power is higher but the specific energy is lower. Test results for various cell prototypes predict the following cell parameters for the immediate future: specific energy 150–200 Wh/kg or 250–350 Wh/dm^3 and for the batteries (taking into account heat insulation) 100–150 Wh/kg or 70 to 100 Wh/dm^3 (for $j_d = 0.2$–0.5).

So far only prototypes of sulphur–sodium storage cells and batteries have been built. The largest battery prototype was built in Britain in 1972. The battery consisted of individual units comprising a total of 960 tubular cells. The energy capacity was 50 kWh, the rated discharge power was 15 kW and the maximum (peak) power was 29 kW. The mass of the battery together with auxiliary systems was about 800 kg and the volume was 1.5 m^3. A project is under way in the USA to develop by 1982 a large-scale battery with a power capacity of a few megawatt-hours for load levelling in utility electrical networks.

At present sulphur–sodium cells have a cycle life of a few hundred (sometimes up to two thousand) charge–discharge cycles. Apart from the stability of the electrolyte the main factors affecting the cycle life are corrosion of the vessel by the sulphur melt (for "outside" sulphur electrodes) and partial blocking of polysulphides by sulphur near the electrolyte surface (this factor is enhanced with cycling). The above factors cause a gradual decrease in the capacity of the sulphur electrode. The "thin" sulphur electrodes with sulphur melt thickness of not more than 5 mm are better in this respect.

A large number of research and development problems are still to be solved to make possible wide practical use of sulphur–sodium storage cells. However, since work in this field has gathered considerable momentum it can be expected that these problems will be largely solved and stable high-performance cells will be developed.

16.4. Cells with molten electrolyte

Considerable attention is now focused on the development of long-lived storage cells with molten electrolyte designed for high current densities and high specific powers with sufficiently high specific energies. This work is still at an early stage though a large number of studies have been reported. The major problem is that of materials since structural components in such cells are attacked by highly aggressive agents (strong oxidants and high-temperature melts). Considerable difficulties are associated also with thermal cycling—variation of the cell temperature from the ambient temperature to the working temperature and back.

Negative electrode. Most proposed cell types employ the lithium negative electrode. In such cells the electrolyte must contain a lithium salt and the electrode processes on the lithium electrode consist in simple transfer of the lithium ions from the crystal lattice of the metal to the melt and back.

At the working temperature of the cell (400–600°C) pure lithium is liquid. Two main construction types of lithium electrodes have been reported: liquid lithium in a porous matrix and a solid alloy of lithium with another metal. The matrix of the lithium electrode can be manufactured from porous plates of stainless steel from felt-like stainless steel, nickel or a similar material possessing high porosity (up to 90%), small pore diameter and the required elasticity.

Polarization of the lithium electrode is negligible both with charge and discharge; record-breaking current densities, up to 40 A/cm^2, have been obtained with lithium electrodes.

One of the main drawbacks of liquid lithium is its noticeable solubility in the salt melt and its capability to expel potassium from the melt:

$$Li + KCl \rightarrow K + LiCl$$

The vapour pressure of potassium is much higher than that of lithium and potassium can form an undesirable gas phase. Lithium dissolved in the electrolyte migrates towards the positive electrode where it is consumed in an unproductive chemical reaction. For an interelectrode distance of 1 mm the associated self-discharge is equivalent to a leakage current density of 1–10 mA/cm^2. Moreover, dissolved lithium causes disintegration of the ceramic separators. The solubility of lithium greatly increases with increasing temperature.

Lithium alloys have been suggested to replace lithium in order to reduce the activity and to decrease the solubility in electrolyte. Electrodes with lithium alloys which are liquid at the working temperature (alloys with zinc, tin, etc.) have the same construction as pure lithium electrodes. The solid lithium alloys (with silicon or aluminium) are used in the form of porous plates pressed onto current-collecting meshes.

Replacement of pure lithium with alloys results in a decrease in the o.c.v. and working voltage by 0·2–0·3 V but it is justified by a considerable reduction of self-discharge and a longer service life.

Positive electrode. The first prototypes of the cells with molten electrolyte employed gas diffusion chlorine electrodes similar to the electrodes of fuel cells. Even the early cell prototypes had discharge current densities of up to 4 A/cm^2 without significant polarization. The chlorine electrode is manufactured from porous graphite, carbides of boron and silicon, or similar materials. The difficulty encountered in the development of such electrodes is choosing the technique for storing molecular chlorine. At first, the use was suggested of activated carbon for adsorbing chlorine. A carbon with highly developed true surface can provide for an adsorption capacity for chlorine of up to 0·3–0·5 Ah/cm^3. Another suggestion was to store chlorine in vessels under elevated pressure. However, neither of the suggesions was put into practice.

Later there was a suggestion to employ sulphur and then chalcogenides, primarily sulphides, as active materials for the positive electrode. Sulphur is liquid at the working temperature of the cell. The sulphur electrode was manufactured from a mixture of sulphur and carbon or in the form of a niobium box with niobium filler packed with sulphur. The high volatility of sulphur (at 507°C the pressure of sulphur vapour is 2 atm) and its solubility in the molten electrolyte lead to self-discharge. The greatest success was obtained with iron sulphides, FeS and FeS_2. Both these sulphides have a high Ah capacity, and are cheap and non-toxic. The process of electrode manufacture is simple.

The discharge and charge processes in sulphide electrodes can be described as follows:

$$FeS + 2e \underset{\text{charge}}{\overset{\text{discharge}}{\rightleftharpoons}} Fe + S^{2-}$$

$$FeS_2 + 4e \underset{\text{charge}}{\overset{\text{discharge}}{\rightleftharpoons}} Fe + 2S^2$$

The reaction with FeS_2 involves intermediate formation of FeS which is why the charge–discharge curve of the cells with FeS_2 electrodes consists of two gently sloping parts of approximately the same length corresponding to discharge voltages of 2·05 and 1·65 V. The theoretical specific consumption of FeS_2 is somewhat lower than that of FeS (1·12 g/Ah and 1·64 g/Ah) but FeS_2 has a higher corrosive activity which can result in a shorter service life.

The sulphide electrodes are manufactured from a mixture of FeS or FeS_2 with some additives (sulphides of lithium, copper and cobalt) placed into a porous frame of molybdenum, tungsten, graphite, etc., and electrolyte is added to the positive electrode. For example, the positive electrode can have the following composition: 60% FeS_2 + 2·2% Li_2S + 29·3% LiCl + KCl eutectic + 7% carbon + 1·5% iron powder (percentages by mass). Iron is added to prevent formation of elementary sulphur in case of overcharge; it reacts with Li_2S giving rise to FeS_2.

Rather high current densities (up to 0·4 A/cm^2) are obtained with the sulphide electrodes.

Electrolyte. The maximum specific capacity of the lithium–chloride and lithium–sulphide cells could be obtained with electrolytes consisting of pure LiCl or Li_2S which are the products of the current-producing reactions. However, their melting temperatures are too high (613°C and 950°C, respectively), so that mixtures LiCl + KCl and LiCl + KCl + Li_2S with lower melting points are typically used as electrolytes to obtain working temperatures of not more than 400°C (the melting point of the LiCl + KCl eutectic is 352°C). The presence of the inert additives in the electrolyte results in a certain decrease of the specific cell capacity. Moreover, the composition of electrolyte varies in the course of operation as the content of LiCl or Li_2S increases with discharge and decreases with charge and the melting point of the electrolyte changes accordingly.

Most cell types employ immobilized or matrix electrolytes. Fine powders of boron nitride, lithium aluminate, etc., are used as immobilizing agents and matrixes are manufactured from ceramic fabrics, such as boron nitride, stabilized zirconium dioxide, and so on.

Construction and performance. Large-scale research and development projects on the Li | FeS and Li | FeS$_2$ storage cells are under way in the Argonne National Laboratory and some other US organizations. Two types of cells are being developed, one for electric vehicles and military applications and the other for load-levelling at power plants. The cells of the first type will employ positive electrodes of FeS$_2$ providing for higher performance (the rated specific energy is 150 Wh/kg) but with shorter service and cycle life (the planned cycle life is 1000 cycles and service life is 3 years) and necessitating use of more expensive materials. The cells of the second type employ FeS electrodes. The specific energy is about half that of the FeS$_2$ cells but their cost is lower by almost 50% and the planned cycle life and service life are 3000 cycles and 10 years. The cells of the second type are designed for lower rates of charge and discharge and have higher efficiency.

When the cells are not working they can be cooled to room temperature, that is, the electrolyte can be frozen. The cells can then be heated up to the working temperature without any loss of capacity or deterioration of parameters. This greatly simplifies long-term storage of the cells.

One of the cell prototypes for electric vehicles has a volume of 320 cm^3 and mass of 820 g. The positive electrode is manufactured from FeS$_2$ with addition of CoS$_2$. A few layers of the active material alternating with graphitized fabric are placed into a basket of molybdenum mesh welded to the central molybdenum current collector. The positive electrode is wrapped into a two-layer separator. The inner layer consists of ZrO$_2$ fabric and the outer layer of BN fabric. The negative electrode consists of a lithium-silicon alloy in the porous nickel matrix. The container and the cover are manufactured from stainless steel and electrically connected to the negative electrode. The prototype was drained with current up to 50 A and the specific power was as high as 53 W/kg.

A larger battery designed for a submarine had a sealed container with six positive and six negative electrodes. The negative electrodes were manufactured from a lithium–aluminium alloy. Separators made from BN fabric were inserted between the electrodes. The battery had the following dimensions: diameter, 30·5 cm; height, 21·1 cm; the mass of the battery was 43 kg, the rated discharge voltage was 1·45 V and the specific capacity was about 150 Wh/kg. The battery was designed for normalized current drain $j_d = 0.08$.

16.5. Reserve-type thermal cells and batteries

Thermal cells and batteries were developed and became commercially available at the end of the fifties. Under normal conditions the electrolyte of the thermal cells is solid and has no conductivity so that the cells can be stored for practically indefinitely long periods without self-discharge. The cells are activated by rapid heating when the electrolyte is melted and acquires ionic conductivity; when the external circuit is closed the cell starts to discharge. The working temperature is in the 400–600°C range. Thermal cells employ active reactants, similarly to other cells with molten electrolyte. These cells have high specific power and their construction provides for rapid activation. The cell volume and mass are, typically, not very large; the cells are designed for short-time operation, not more than 15 min as a rule. Thermal cells are used in small military rockets, as fuses in artillery shells, and for similar applications.

Electrolyte. Eutectic salt mixtures are used as electrolyte in thermal cells (most often $LiCl + KCl$ or $LiBr + KBr$). The molten electrolytes have conductivity of 1–2 $ohm^{-1} cm^{-1}$ at a temperature of 450°C. The high voltage of electrolyte decomposition (about 3 V) makes it possible to employ strong reducers and oxidants. Sometimes, various components are added to the electrolyte (up to 15% of chromates, bromates, perchlorates) which stabilize the discharge voltage of the cell. To increase the mechanical strength of the cell, and to reduce shrinkage with melting, immobilizing agents, such as silica or asbestos, are added to the electrolyte, or fibre glass fabric is impregnated with electrolyte and serves as a matrix.

Anode. Anodes are manufactured from calcium or magnesium which remain solid at the working temperature. The anodic process consists in transfer of the calcium or magnesium ions from the metal crystal lattice to the melt. The potential of the calcium electrode is more negative than the potential of the magnesium electrode by almost 1 V and hence the calcium cells have a correspondingly higher voltage. However, contact with electrolyte containing lithium salts can give rise to the liquid intermetallic compound $CaLi_2$ on the calcium electrode at the working temperature of the cell (the melting point of $CaLi_2$ is 230°C). This leads to a loss of capacity and instability of discharge parameters. To prevent formation of the intermetallic compound a layer of calcium acetate or other material is applied to the surface of the calcium electrode.

Cathode. Oxides of heavy metals, chromates and other oxidants can be used as the active materials for cathodes of reserve-type thermal cells. Typically, the cathode mix contains 85% calcium chromate and 15% electrolyte. In a widely used cell type the active cathodic material is completely mixed with electrolyte forming, for instance, the mixture $LiCl + KCl + CaCrO_4$ whose composition is close to eutectic. In such cells the current collector is just the cell cover.

The cathodic process has a complex mechanism. Reduction of $CaCrO_4$ gives rise to various complex salts containing chromium of lower valency. A deposit of Li_5CrO_4 can be produced in the reaction.

Construction. Two constructions of thermal cells have been developed—the tablet-type and the cup-type. In cells of the latter construction the tablets of anode, the electrolyte and the cathode mix are placed into a nickel container with a nickel cover. In the tablet-type construction the anodic material is applied directly to the surface of a nickel disk. A two-layer tablet containing electrolyte and the cathode mix is pressed between the anode disk and another nickel disk. Recently, there have been reports on sealed constructions for thermal cells which can operate at high temperatures without electrolyte leakage and can withstand high centrifugal overloads. The main type of thermal cells are small-size cells of thickness varying from 1·5 mm to 3 mm and diameter between 30 mm and 70 mm.

Batteries of thermal cells are asembled from alternating cells and heaters of approximately equal thickness. The assembled battery is placed into a sealed container with reliable thermal insulation made from asbestos, glass wool, oxides of silicon and aluminium, etc. Figure 96 shows a schematic of the reserve-type thermal battery.

The heaters of the thermal batteries consist of pyrotechnic compounds with high caloric values (thermite powders, mixtures of barium peroxide with powdered aluminium or magnesium, chromates of barium and zirconium, etc.). Electric fuses are employed to ignite these mixtures. To facilitate ignition of the heating mixtures such additives as PbO_2, $KClO_3$, $K_2Cr_2O_7$, are used. The activation time of thermal batteries is, typically, shorter than 1 s.

Performance. The rates of discharge of thermal cells can be as high as $400 \, mA/cm^2$ and more, with discharge voltages of up to 2·5–3 V. The voltage of the batteries varies from 12 V to 500 V. For short-time discharges (less than 1 min) the specific power of thermal batteries can be as high as 600 W/kg. Commercially available batteries have capacities from a fraction of a watt-hour to a few tens of watt-hours and masses from 40 g to 3 kg.

The amount of active materials in thermal cells is sufficient to operate for half an hour or longer. In practice, the batteries do not operate for more than 15 min owing to cooling and solidification of electrolyte. The specific energy of the thermal cells is therefore low, not better than 10 Wh/kg. The performance of thermal cells can be improved and the range of their application can be extended by employing better heating compounds and more efficient heat insulation making possible longer operation periods. To maintain the working temperature, heat is accumulated by inert materials possessing high specific heats or by salt mixtures with high heat of fusion.

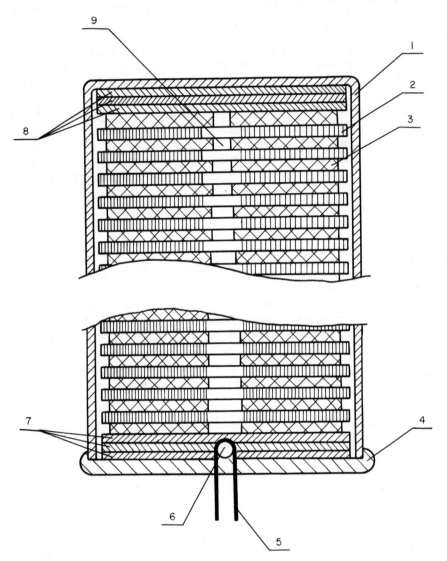

Fig. 96. Schematic of the reserve-type thermal battery: (1) container; (2) individual cells; (3) heaters; (4) cover; (5) terminals of the ignition fuse; (6) ignition fuse; (7) and (8) heat insulation; (9) ignition channel.

BIBLIOGRAPHY

1. G. Morand and J. Hladic, "Electrochimie des sels fondus", t. 1–2. Masson, Paris (1969).
2. Gool, W. van (ed.), "Fast Ion Transport in Solids. Solid State Batteries and Devices". North-Holland, Amsterdam (1973).
3. Mahan, G. D. and Rofh, W. L. (eds.), "Superionic Conductors". Plenum Press, New York and London.
4. Geller, S. (ed.), "Solid Electrodes". Springer, Berlin (1977).
5. R. O. Ivins, E. C. Gray, W. J. Walsh and A. A. Chilenskas, Design and performance of Li–Al/FeS cells, Proceedings of 27th Power Sources Symposium, Red Bank, N.J., 1976, pp. 8–13.
6. F. J. Martino, T. D. Kaun, H. Shimotake and E. C. Gay, Advances in the development of lithium–aluminium/metalsulphide cells for electric-vehicle batteries, in "Proceedings of 13th Intersociety Energy Conversion Engineering Conference, San Diego, 1978", pp. 709–716.
7. C. W. Jennings, Thermal batteries, in "Primary Batteries" (Cahoon, N. C. and Heise, G. W., eds.), Vol. 2, pp. 263–293. Wiley, New York (1976).
8. M. Lazzari and B. Scrosati, A review of silver, copper and lithium solid-state power sources, J. Power Sources 1, No. 4, 333–358 (1976/77).
9. A. T. Churmann, The sodium-sulphur battery system, Electric Vehicles 61, No. 2, 12–18 (1975).
10. G. J. May, The development of beta-alumina for use in electrochemical cells: a survey, J. Power Sources 3, No. 1, 1–22 (1978).
11. W. Fischer, W. Haar, B. Hartmann, H. Meinhold and G. Weddigen, Recent advances in Na/S cell development—a review, J. Power Sources 3, No. 4, 299–310 (1978).

Chapter Seventeen

Fuel Cells
(Electrochemical Generators)

17.1. General

Fuel cells are chemical power sources in which electrical power is generated in a chemical reaction between a reducer and an oxidizer which are continuously fed to the electrodes at a rate proportional to the current load. The products of the current-producing reaction are continuously removed from the fuel cell. Thus, the fuel cell can, in principle, work for an infinitely long time while the reactants are fed into it and the products are removed; its external functions are similar to those of combustion engine/generator sets.

Mainly liquid and gaseous reactants are used in fuel cells. In principle, even naturally-occurring fuels or their products (natural gas, water gas, etc.) can be used as reducers. This gave rise in 1889 to the name "fuel cell". In 1894 the prominent German physical chemist W. Ostwald first suggested using fuel cells for large-scale power generation as an alternative to conventional power generation in thermal power stations.[1] Since fuel cells theoretically can reach a higher efficiency of energy transformation than heat engines (see Section 1.5), fuel cells could be expected to utilize natural fuels more efficiently and to produce cheaper power. Later, a second field of use for the fuel cells was developed, namely, small-scale efficient power sources for autonomous power units. In contrast to conventional primary cells and storage cells, fuel cells employ reactants with high energy content, hydrogen for instance, which makes it possible to build units with higher specific energy characteristics. The range of use of chemical power sources can thus be considerably extended.

These are the two main fields of research and development work on fuel cells. In the first, economic factors are of decisive importance. The utility power plants must be large and long-lived (that is, the depreciation costs should be low), they should be manufactured from cheap materials and use natural fuel with minimal expenditure on its preparation. These power plants must have the requisite high efficiencies but their specific characteristics with respect to unit mass or volume need not be very high.

For the autonomous power units only the specific characteristics are of primary importance (as well as convenience of operation). Autonomous power sources may have a shorter service life than the stationary power units. The cost of the power they generate (as well as that from other chemical power sources) may be higher than the cost of bulk-generation power and is not a decisive parameter. Therefore, such units can be built from more expensive materials and use more expensive reactants, even liquid hydrogen and oxygen.

For almost 90 years fuel cells have been the subject of numerous research and development projects; many patents have been granted, working prototypes have been built and even small-scale production has been started. But nowhere in the world are fuel cells produced on a large scale. The reasons are, firstly, the considerable scientific and technological difficulties involved in development of these cells and, secondly, the fact that at various stages interest was focused on different fields of applications of the fuel cells—before World War II it was large-scale power generation, in the sixties, development of special-use cells (military, aerospace, etc.), and later, development of power units for electric vehicles and again large-scale power generation. The diversity of goals necessitated diversity of approaches. Meanwhile the lack of mass-produced fuel cells still makes it impossible to evaluate comprehensively and objectively the advantages of the fuel cells in various fields so that controversial opinions are frequently voiced regarding the prospects of development of fuel cells.

Brief historical note

In 1839 the English scientist W. Grove conducted electrolysis of sulphuric acid solution with platinum electrodes and found that after saturation of the electrodes with hydrogen and oxygen and disconnecting the external current source the electrolytic cell exhibited a measurable o.c.v. and could discharge when connected to an external load. This device was not properly a fuel cell (since fuel was not continuously fed to it) but it was a prototype of a chemical power source with gaseous electrodes. At that time this cell had no practical importance.

When the paper by W. Ostwald[1] had been published repeated attempts were made to build a fuel cell using coal as the active material of the negative electrode. However, the extreme electrochemical inertness of coal made it impossible to build cells operating at low temperatures. In the thirties of this century prototypes were developed of the high-temperature fuel cells with solid ceramic electrolytes for direct anodic oxidation of coal (by the German scientist E. Baur). At the same time various types of the oxygen–hydrogen fuel cells† were developed. However, the current densities obtained with these cells were rather low and, most importantly, the service life of the cells was too short.

Interest in fuel cells increased especially in the fifties in connection with the new developments in electrochemical power generation. An important contribution was made by the Soviet scientist O. K. Davtian who published the first monograph[2] on fuel cells in 1947. At that time there was already a clear understanding of the main problems to be solved for further development of fuel cells, namely, to develop highly effective gas-diffusion electrodes and to find methods for accelerating the electrode processes.

In Great Britain in 1958, F. Bacon built the first large power unit (5 kW) using medium-temperature oxygen–hydrogen cells. These cells employed gas-diffusion electrodes with a fine-pore gas-tight layer; to accelerate the electrode processes high temperatures (200–240°C) and elevated working gas pressures (2–4 MPa) were used making possible high current densities of $0.2–0.4$ A/cm^2 and more. Owing to the high gas pressure the battery had heavy-weight construction.

In later cell designs lower gas pressures were used (not more than 0.3 MPa). Various low-temperature units with working temperatures varying between 60°C and 110°C were developed. Starting in 1955, K. Kordesch in the USA developed prototypes of low-temperature oxygen–hydrogen fuel cells with carbon electrodes promoted with small amounts of oxide and platinum catalysts. A significant contribution to further development of fuel cells was made by E. Justi and his coworkers (West Germany) who developed non-platinum catalysts for the hydrogen and oxygen electrodes, namely, the Raney nickel and silver catalysts.

In the sixties development work on various types of fuel cells was under way in many industrial companies and research centres in all the developed countries. Various cell prototypes for exhibitions and demonstrations have been built. However, later it became clear that technologically and economically feasible fuel cells could be developed only by those organizations possessing the necessary know-how and expertise and the work was continued by several such organizations. The practical significance of fuel cells was demonstrated by the flights of the American spacecraft Gemini and Apollo in which the main on-board power supply units were

---

† It must be noted that a reverse sequence has been adopted in the terminology of fuel cells: the reducer (fuel) was mentioned first, as in "hydrogen–oxygen fuel cells". The reason is obvious: in most types of fuel cells oxygen (air) was assumed to be the only possible oxidizer. As a result, a shortened terminology also became widespread: "methanol" and "hydrazine" fuel cells instead of "air–methanol", "oxygen–hydrazine" cells, etc.

electrochemical generators using oxygen–hydrogen fuel cells (which simultaneously produced water for the crew). Various other fuel cell power units have been developed for army and navy uses; for instance, the US Army used hydrazine cell units.

Typically, the main advances in fuel cell development were made in the field of autonomous power units for special applications. These advances were made possible by large-scale financing of such projects. At the end of the sixties the scope of studies on fuel cells decreased. However, in recent years interest in fuel cells has revived (largely, for large-scale power generation applications) in view of the energy crisis.

As in the last two decades work on fuel cells is at present centred on development of oxygen–hydrogen cells.

Though fuel cells are not yet widely used the work on their development has yielded important results for development of other types of chemical power sources (in particular, compound cells) and such new branches of electrochemistry as electrocatalysis and the theory of porous electrodes.

17.2. Construction of fuel cells

In contrast to other chemical power sources, fuel cells need for their operation various auxiliary systems for feeding the reactants, removing the reaction products, maintaining the thermal regime and other functions. The battery of fuel cells in combination with the auxiliary systems is referred to as an electrochemical generator. The total power unit (see Fig. 97) comprises the electrochemical generator and a system for storage and processing of the fuel including tanks or containers with the fuel and oxidizer (if atmospheric oxygen is not used as the oxidizer) needed for operation during a given period, and the equipment for processing and purification of the fuel.

There are two types of fuel cell construction, box cells and batteries with bipolar electrodes. Thin porous diffusion electrodes are employed in the fuel cells to provide for a high activity as well as for effective feeding of the reactants to the reaction zone and removal of the reaction products.

Fuel cells can be classified by their operating temperature into low-temperature (below 130–150°C), medium-temperature (150–250°C) and high-temperature (550–1000°C) cells. In the low-temperature cells aqueous solutions of alkalies or acids or ion-exchange membranes are used as electrolytes. In the medium-temperature cells high-concentration aqueous solutions or molten crystalline hydrates are used as electrolytes. In the high-temperature cells molten salts (in particular, carbonates) or solid elec-

trolytes are employed. The solutions and melts can be used both in the free
(liquid) form or in the form of matrix or immobilized electrolytes.

In contrast to the mass-produced current sources, there are no conventional or general construction principles for fuel cells or electrochemical
generators. In this chapter we shall describe some typical units developed by
various companies.

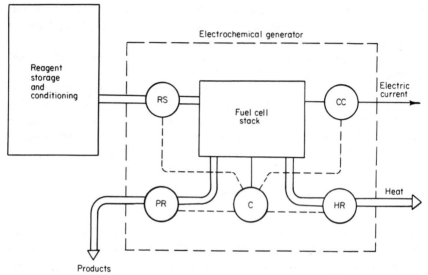

Fig. 97. Power unit with a battery of fuel cells and auxiliary systems for reactant supply (RS),
products removal (PR), heat removal (HR), current conditioning (CC) and controlling (C).

The main characteristic of the fuel cell battery or electrochemical generator in general is its specific power. To improve this characteristic the masses
of all the structural elements and the auxiliary systems should be minimized.
The specific power depends also on the structural parameters s_m and s_v and
the specific activity of the electrodes with respect to the given reaction (see
Section 3.3). Special catalysts are almost always added to the electrodes of
fuel cells (see Section 4.6).

Another important characteristic of fuel cells is their service life. Though,
in principle, operation of fuel cells should be stable as long as the reactants
are fed into them and the reaction products are removed, the performance of
real cells gradually deteriorates. This deterioration is caused by various
factors and can be both reversible (the cell performance can be restored by
various operational techniques) and irreversible. Reversible deterioration
can be caused, for instance, by variation of the electrolyte concentration
owing to unbalanced removal of the reaction products, accumulation of the

inert gases in the electrode pores, etc. Irreversible deterioration of perform-ance can be produced by corrosion, recrystallization or shedding of the catalysts, impaired contact between the catalyst and the current collector, worsened hydrophobic properties of the electrodes, etc. The relative contri-butions of various factors depend on the operational conditions. Increases in the current density and temperature, as a rule, accelerate deterioration of the cell performance. The degree of instability of the operation of a fuel cell is often described by the rate of decrease of the voltage (in $\mu V/h$) on long-time discharge with a given current density. Typically, this rate varies between 20 and 50 $\mu V/h$ (per cell). This corresponds to a cell service life of 4–10 thousand hours (with a voltage drop of no more than 0·2 V).

An important factor affecting the service life of fuel cells is the resistance of the structural materials. In contrast to other chemical power sources with aqueous electrolytes, the working temperature in the fuel cells is generally elevated, leading to difficulties in choosing suitable corrosion-resistant materials. These materials should preserve their strength, sealing, insulating and other properties for a long time; furthermore, they should not con-taminate the electrolyte with admixtures adversely affecting the operation of the electrodes and membranes.

The construction of the fuel cell battery should provide for continuous removal of the reaction products. In the oxygen–hydrogen cells the reaction product is water which either dilutes the electrolyte solution or is evap-orated from the electrode surface (at working temperatures over 60°C). According to the equations of the electrode reactions in alkaline solution, water is formed on the hydrogen electrode,

$$H_2 + 2OH^- \rightarrow 2H_2O + 2e, \qquad \tfrac{1}{2}O_2 + H_2O + 2e \rightarrow 2OH^-$$

and in acidic solution on the oxygen electrode,

$$H_2 \rightarrow 2H^+ + 2e, \qquad \tfrac{1}{2}O_2 + 2H^+ + 2e \rightarrow H_2O$$

Dilution of the solution near the electrode on which water is produced leads to increasing pressure of water vapour and evaporation of water from the back surface of this electrode. If water enters the bulk of the solution the battery should be provided with a system for circulating the electrolyte for water removal by external devices (this is discussed in more detail in Section 17.4).

17.3. Reactants for fuel cells

The performance and economic characteristics of power units with fuel cells are determined to a considerable extent by the choice of the reactants. A great variety of fuel cells using various reducers (fuels) and oxidants have been reported in the literature.

For large-scale power generation, only the oxygen of the air is suggested as the oxidant while in autonomous power units air, pure oxygen and hydrogen peroxide are variously used.

The activity of the oxygen electrode at temperatures below 150–180°C is not very high—as the reaction of electrochemical reduction of oxygen results in large polarization it is often the oxygen electrode which limits the operation of the entire fuel cell. Since the properties of the oxygen electrode have been discussed above (Section 14.1) we refer the reader to this section.

The reducers used in fuel cells can be arranged in the following series according to decreasing electrochemical activity:

hydrazine > hydrogen > alcohols >
$$\qquad\qquad\qquad\qquad\text{carbon monoxide > hydrocarbons > coal}$$

The compounds to the left of alcohols are comparatively easily oxidized in aqueous solutions at temperatures below 100°C. Moving right from alcohols the rate of oxidation decreases sharply. Hydrocarbons and especially solid carbonaceous materials cannot be oxidized at rates suitable for technological applications. In recent years, therefore, the number of reactants under study has considerably decreased.

For large-scale power plants which will utilize natural fuels, the only suitable fuel is technical-grade hydrogen produced either by conversion of hydrocarbons (natural gas or oil products) or by gasification of coal. Conversion (reforming) of hydrocarbons consists in their interaction with steam giving rise to hydrogen and carbon monoxide:

$$C_nH_m + nH_2O \rightarrow nCO + (n + \tfrac{1}{2}m)H_2$$

This reaction proceeds at temperatures of about 800°C on cheap oxide catalysts. The reaction is endothermic and requires continuous delivery of heat which is obtained by burning a part of the fuel. Thus, reforming entails certain energy losses.

Since carbon monoxide is difficult to oxidize electrochemically and is moreover a strong poison for many catalysts, the mixture of CO and H_2 is fed into another converter where at a lower temperature CO is oxidized by steam to CO_2 yielding additional hydrogen:

$$CO + H_2O \rightarrow CO_2 + H_2$$

To remove the residual CO from the gas mixture it is fed into the third reactor with a nickel chromite catalyst where the reaction is the reverse of conversion, that is, hydrogen and carbon monoxide combine giving rise to methane, small amounts of which are harmless for the fuel cells.

Gasification of coal (including underground gasification) consists in treating the pulverized coal with steam and produces water gas:

$$C + H_2O \rightarrow CO + H_2$$

Petroleum products and coal often contain sulphur in considerable concentrations. The sulphur compounds produced in conversion or gasification are very strong catalytic poisons. Since the starting fuel or its products should be processed to remove these compounds the economic characteristics of the power plants suffer greatly.

The reducers for autonomous power units should satisfy a number of requirements. They should be electrochemically active and convenient to store, handle and transport. The reaction products should be easy to remove from the cells. Unfortunately, none of the electrochemical "fuels" fully satisfies these requirements. At present, the most feasible fuels are hydrogen, hydrazine, methanol and ethylene glycol.

Liquid reactants are most convenient to store and transport. However, widespread use of hydrazine is hindered by its high cost and high toxicity. Alcohols—methanol, ethylene glycol and others—have low electrochemical activity and the reaction products, carbonates, are difficult to remove from the alkaline electrolyte solutions.

Storage of gaseous products, particularly, hydrogen, presents certain difficulties. When hydrogen is stored in conventional steel cylinders under a pressure of 15–25 MPa the mass of the cylinders is more than a hundred times that of the hydrogen. Light-weight cylinders (for instance, from reinforced glass fibre plastic) reduce this ratio by a factor of three. Cryogenic storage makes possible a greater decrease of the container mass. In the vessels available for storing liquid hydrogen the evaporation losses amount to 1–3 % per day and there are prospects of reducing this figure almost by an order of magnitude. An important drawback of cryogenic storage is its relatively high cost so that it can be used only for special applications (in particular, in aerospace technology).

Recently it has been shown that hydrogen can be stored in the absorbed state in some intermetallic compounds ($LaNi_5$, $FeTi$, etc.). These compounds can absorb considerable amounts of hydrogen (up to 15 g/kg) and then reversibly and rapidly desorb it when subjected to a slight increase in temperature and decrease in the external hydrogen pressure.[3]

Sometimes it is convenient to store reactants in the form of certain chemical substances. For instance, hydrogen can be produced by decomposition of ammonia, hydrides or boron hydrides and oxygen by decomposition of concentrated hydrogen peroxide. Ammonia is decomposed into nitrogen and hydrogen (the reverse reaction to synthesis of ammonia) at temperatures of about 800°C in reactors with ruthenium on alumogel as the catalyst. The resulting mixture can be used in fuel cells without additional purification.

The hydrides and boron hydrides of the alkali and alkali-earth metals have sufficiently high energy contents. Hydrogen is produced when these compounds are decomposed with water. The water produced during operation of the fuel cell can be used for the decomposition reaction so that the mass of water must not be taken into account in calculations of the masses of reactants. The drawbacks of hydrides and boron hydrides are their high cost and the need to remove the volatile admixtures (boron hydrides, etc.) from hydrogen.

Table 5 presents data on the specific masses of hydrogen and oxygen (taking into account the masses of containers or the starting compounds) for various methods of storage.

Hydrogen for autonomous power units can also be produced by reforming liquid hydrocarbons. Since the reforming reactors are too bulky they are employed in practice only at units with a power of no less than 10 kW (e.g., power units for the TARGET/project, USA). For units of lower power it is more convenient to use conversion of methanol, for which considerably lower temperatures are needed (300–400°C).

The gases fed into fuel cells must be sufficiently pure. The best technique for removing contaminants from hydrogen is diffusion through a thin membrane of a palladium–silver alloy which is permeable only to hydrogen. However, such membranes are expensive and cannot be widely used.

Table 5. Specific mass of hydrogen and oxygen (with containers) per 1 Ah for various methods of storage

Method of storage	Specific mass (g/Ah)		
	Hydrogen	Oxygen	Total
Without container	0·0374	0·297	0·335
In steel cylinders	4·3	2·7	7·0
In light-weight cylinders	1·2	2·0	3·2
Cryogenic storage	0·2	0·4	0·6
Absorption by intermetallic compounds	2·5–3	—	—
In compounds			
Lithium hydride	0·15†	—	—
Lithium boron hydride	0·10†	—	—
Beryllium boron hydride	0·09†	—	—
Hydrogen peroxide (80%)	—	0·8	—

† The mass of the water needed for decomposition is not taken into account.

To remove CO_2 from hydrogen and air they are passed through absorbers containing alkaline solutions. The most effective technique is to pass the gas over corrugated sheets of asbestos impregnated with alkaline solution. Another technique for CO_2 removal is absorption by monoethanolamine which is readily regenerated by heating.

17.4. Auxiliary systems

The auxiliary systems contained in the electrochemical generator consist of the working (operational) systems and the control units. The former include regulators, pumps for electrolyte, gas and cooling agent, valves, heat exchangers, etc. The control units consist of transducers monitoring appropriate parameters and the logic circuits controlling the operational systems. Some units of the auxiliary systems (for instance, the pumps) are powered by the fuel cells. The fraction of the energy produced that is consumed for operational needs varies between 2% and 10%.

Delivery of reactants. Before feeding gas into the fuel cell the working pressure of the gas should be built up and then maintained in a fairly narrow range irrespective of the load, that is, irrespective of the rate of consumption of the reactants. In oxygen–hydrogen cells the pressures of both gases must be equal to prevent bending and damage of the thin cell components.

Removal of products. The water produced by the reaction dilutes the electrolyte and/or evaporates from the electrode surfaces. To remove excess water from the liquid electrolyte, recirculation systems are used in the external loop of which there is placed a special evaporator or an electrodialytic cell for removing water.

Water can be evaporated from the electrode surface by static or dynamic techniques. In the static technique water is evaporated from the back side of the electrode and is condensed at the opposite wall of the gas chamber where a lower vapour pressure is maintained (for instance, by cooling). It can then be removed with a wick or by other means. In the dynamic technique, recirculation of the working gas is used—the gas chambers are continuously blown with a large excess of gas where it is saturated with water vapour. The gas is then passed through a separating condenser and the dry gas is returned to the fuel cell.

Removal of heat. The heat liberated in operation of the cell is removed either by circulating electrolyte to the external heat exchanger or by special built-in liquid-cooled heat exchangers. A small part of the liberated heat is removed by circulating gas. The batteries of not too high power operating in the air are blown with a ventilator. Sometimes batteries are equipped with fins, radiators, etc., to improve heat transfer to the environment.

The external and internal heat exchangers are used also for warming up the battery for start of operation.

Regulation of heat and mass exchange. Regulation of removal of the reaction products and of heat is fairly complicated, particularly at variable loads. Variation of load has a varying effect on the rates of formation of the products and liberation of heat. Fairly complicated systems for automatic regulation of the electrolyte concentration and temperature have been described in the literature. However, a more promising trend is the development of systems with self-regulation of mass and heat exchange in which a change of load automatically results in changes of the rates of removal of heat and the reaction products. In such systems the external control devices can be considerably simplified.

Regulation and monitoring of the electrical parameters. Devices used for this purpose include current and voltage regulators, DC–AC converters, etc. The system for monitoring the electrical parameters often has feedback coupling to other systems so that if the parameters go out of the specified range the rates of delivery of the reactants, removal of heat, and so on, are altered appropriately.

Auxiliary storage battery. Starting the electrochemical generator requires a certain expenditure of energy for warming-up and sometimes takes several tens of minutes. For this reason, the electrochemical generator is often combined with an auxiliary storage battery which provides the power for starting the generator and for delivery to the consumer until the generator has reached the rated operational regime. Sometimes such a battery is used as a buffer battery. The fuel cells are, typically, designed for narrow power ranges to simplify the heat and mass exchange systems. A combination of fuel cell and storage battery (a "hybrid" power unit) makes possible an almost constant loading of the fuel cell—when the total load increases the battery is discharged and when the total load decreases the battery is charged. This prevents overloading of the fuel cells and increases their service life.

17.5. Oxygen(air)–hydrogen fuel cells with alkaline electrolyte

Among all the types of fuel cells the most advanced is the oxygen–hydrogen fuel cell with alkaline electrolyte, namely, 30–40% KOH solution (though sometimes the alkali concentration is higher). Research and development projects aiming at building electrochemical generators on the basis of such cells have been carried out by companies and research centres in many countries and prototypes and small batches of cells have been manufactured. This cell type was chosen for the power units of the Apollo spacecraft.

Thermodynamic calculations yield values of 1·229 V and 1·162 V for the e.m.f. of the oxygen–hydrogen circuit in aqueous solutions at the gas pressure 0·1 MPa and the temperatures 25°C and 100°C, respectively. The e.m.f. does not depend on the pH value of the solution, that is to say it is the same in acidic and alkaline solutions. In concentrated solutions the e.m.f. increases somewhat with decreasing activity of water, for instance, in 40% KOH solution $\mathscr{E}^{(T)} = 1·243$ V at 25°C. In practice, the o.c.v. for the low-temperature oxygen–hydrogen cells is not higher than 1·08–1·10 V owing to the lack of equilibrium at the oxygen electrode.

As well as the platinum-group metals, cheaper catalysts can be used in the alkaline electrolyte both for the hydrogen electrode (Raney nickel) and the oxygen electrode (activated carbon promoted with various additives). In the Bacon cell the electrodes were manufactured from nickel powder. The material for the positive electrodes was subjected to special oxidation treatment and lithium oxide was added to it. The resulting lithiated nickel oxide has high corrosion resistance and good activity at temperatures above 180–200°C. The electrodes of the low-temperature cells (particularly, those designed for critical applications) contained platinum-group metals, sometimes in considerable amounts (up to 20–30 mg/cm²). In recent times progress has been made in decreasing the amount of these metals to 0·1–1 mg/cm². The level of 0·1 mg/cm² corresponds to platinum consumption of about 1·5 g per 1 kW of the rated power, which is quite acceptable economically.

The rated working current density of the alkaline oxygen–hydrogen cells is 50–200 mA/cm² for voltages of 0·8–0·9 V. The electrodes with platinum catalysts make possible higher current densities up to 0·4–0·8 A/cm² or even higher, for short periods. The increase in current density on hydrogen electrodes with nickel catalyst is limited by passivation of nickel with polarization over 0·2–0·3 V. The catalysts based on nickel and silver are more sensitive to variation of the working temperature than platinum catalysts and are effective only at temperatures above 70–80°C.

In the alkaline cells contamination of the reactants with carbon dioxide is inadmissible in view of possible carbonization of the electrolyte. This means that the hydrogen produced by fuel conversion and atmospheric oxygen can be fed into such cells only after careful removal of CO_2. Attention should also be paid to removing catalytic poisons (such as sulphur compounds) from the reactants. The most frequently used reactants in the alkaline cells are pure hydrogen and oxygen‡ produced by electrolysis of water.

The service life of the best oxygen–hydrogen cells with alkaline electrolyte is up to 10 000 hours. In practice, most cells work for 2–4 thousand hours. Apart from the factors limiting the service life, described in Section 17.2, gradual carbonization of the electrolyte is an important factor. Carbonization can be caused not only by the traces of carbon dioxide in the reactants but also by slow oxidation of the insufficiently resistant organic structural materials under the conditions of operation of the cell. Strong carbonization can result in deposition of solid carbonates in the pores of the positive electrode and in a sharp deterioration of its performance.

Two types of construction have been developed for oxygen–hydrogen cells with alkaline electrolyte, one with free (usually circulating) electrolyte solution and the other with matrix electrolyte. The free electrolyte design facilitates the heat and mass exchange processes and the possibility of periodic replacement or purification of the electrolyte results in less strict requirements on the purity of the reactants. In cells with matrix electrolyte the interelectrode distance can be minimized and the construction of the electrodes can be much simpler since the requirements on their mechanical strength and gas tightness are less strict (the impregnated matrix itself serves as the gas-tight layer). The electrolyte is held in porous $0.2–0.6$ mm sheets of asbestos or synthetic fibres (of potassium titanate).

Below we discuss several existing power units using oxygen(air)–hydrogen fuel cells with alkaline electrolyte. These power units, developed by various firms, exhibit some interesting and promising construction features.

Union Carbide batteries (USA)

A considerable period of development was put into carbon gas electrodes. A significant advance was made in 1963 when highly effective thin multilayer electrodes of the hydrophilic–hydrophobic type were developed. The catalytically active layers of the electrodes consist of activated carbon with the addition of oxide catalysts (spinels) for the oxygen electrode and small amounts of platinum (less than 1 mg/cm^2) for the hydrogen electrode. The subsequent layers consist of carbon material and Teflon and their hydrophobicity increases towards the rear (gas) surface of the electrode. The base of the electrode is a mesh-reinforced small-pore layer of sintered nickel powder. This layer serves simultaneously as the blocking layer. In the first cell models the blocking layer was hydrophobic (rendered so by spraying Teflon on porous nickel) and was placed at the gas side (Fig. 98(a)). In the later cell models a hydrophilic blocking layer at the electrolyte side was used

(Fig. 98(b)). The total thickness of the electrode is 0·5–2 mm. The reaction zone in such electrodes is strictly defined owing to varying hydrophobicity with thickness and the electrode characteristics are very stable.

Various constructions of batteries of the filter press type with circulating electrolyte have been developed on the basis of these electrodes with sizes of $15 \times 15 \, cm^2$ and $30 \times 35 \, cm^2$. The optimum working temperature is 65°C

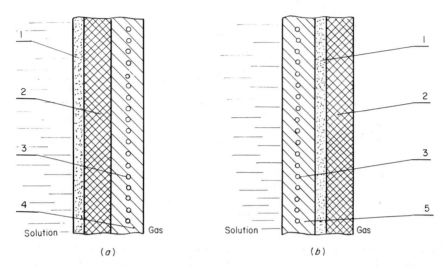

Fig. 98. Schematic of the Union Carbide air (oxygen) electrode: (1) activated carbon layer; (2) carbon layer with variable hydrophobicity; (3) nickel mesh; (4) hydrophobic porous nickel substrate; (5) hydrophilic porous nickel substrate.

since stability of operation is impaired at higher temperatures. The rated current densities are given as 50 and 100 mA/cm² for operation with aerial oxygen and pure oxygen, respectively. The voltage at this current density is 0·80–0·85 V. Short-time overloads with 3–5 times higher current densities are admissible. Figure 99 shows the current–voltage characteristics for the Union Carbide fuel cell and some other cells. The specific power of the cells with thin electrodes varies over the range 30–50 W/kg (in the rated regime with the use of pure oxygen). If the mass of the auxiliary systems is taken into account the specific power is lowered by half. In 1965 Union Carbide in cooperation with General Motors built a prototype of the electrovan with a fuel cell battery with rated power of 32 kW and peak power of 96 kW. Cryogenic storage of hydrogen and oxygen was employed in the prototype.

The reactant supply provided for the range of 160–240 km. The mass of the battery was 610 kg and the mass of the entire power unit was 1430 kg.

At the beginning of the seventies Union Carbide virtually ceased research and development work in the field of fuel cells.

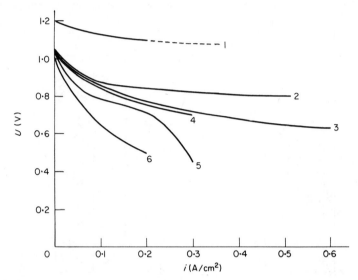

Fig. 99. Current–voltage curves for various oxygen–hydrogen fuel cells: (1) medium-temperature Pratt and Whitney cells; (2) Union Carbide cells; (3) medium-temperature Bacon cells; (4) General Electric cells with modified ion-exchange membrane; (5) air–hydrogen Union Carbide cells; (6) General Electric cells for the Gemini project.

Allis–Chalmers batteries (USA)

These are batteries of the filter press type with matrix electrolyte, using the static method of water removal. The batteries operate at 90–95°C. A single cell consists of asbestos matrix impregnated with 30% KOH solution to both sides of which there are pressed highly porous (85%) nickel plates coated with platinum–silver catalyst. At the other side of the hydrogen chamber there is placed the so-called transport membrane, an asbestos matrix wetted with KOH solution of higher concentration. The water produced on the hydrogen electrode is evaporated and then condensed in the transport membrane owing to the difference between water vapour pressures over solutions of different concentrations. The space beyond the transport membrane is periodically evacuated and the water evaporates. Figure 100 shows the cell construction and the concentration distribution of

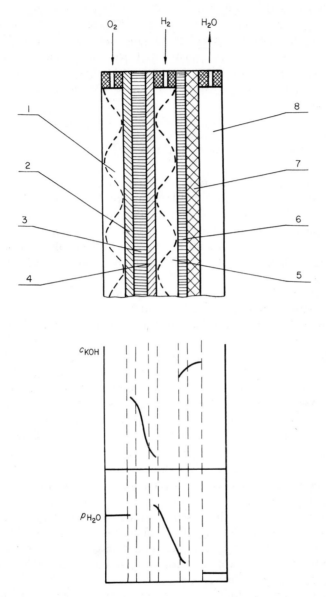

Fig. 100. Schematic of the Allis Chalmers fuel cells: (1) oxygen chamber; (2) oxygen electrode; (3) matrix electrolyte; (4) hydrogen electrode; (5) hydrogen chamber; (6) water-transport membrane; (7) porous nickel plate; (8) vacuum chamber. Given below are the profiles of alkali concentration and water vapour pressure in the cell.

the solution and the water vapour pressure during cell operation. High specific characteristics have been obtained with such cells. For instance, a battery with rated power of 5 kW has specific power of about 75 W/kg with a voltage of 0·935 V per cell. Such a high working voltage for a current density of 0·1 A/cm^2 is explained by a relatively high operational temperature. Since the current density is not very high the service life is long—about five thousand hours.

Pratt and Whitney battery (USA)

This firm acquired the patent rights to the medium-temperature Bacon cell in 1959 and began its further development. In 1962 work was begun on development of the power source for the Apollo spacecraft. The spacecraft was equipped with three identical power units. The lower part of the unit contains the fuel cell battery and the upper part the auxiliary systems. The total mass of the unit is 110 kg, the maximum diameter is 42 cm and the height is 95 cm.

The cells employ two-layer nickel electrodes with platinum catalysts (gold-containing catalysts were also used for the oxygen electrode). The electrolyte is 85 % KOH solution which is a solid crystal hydrate at room temperature and melts at about 150°C. This high-concentration electrolyte with high boiling point makes possible cell operation at 200–260°C at comparatively low gas pressures (0·3–0·4 MPa). The individual cells are sealed with respect to electrolyte and serve as examples of cells with free, but not circulating, electrolyte. Each cell has a corrugated expansion tube so that if the water removal balance is accidently disturbed and the electrolyte volume changes, the electrolyte is not driven into the electrodes. The reaction product (water) is evaporated from the electrode surface and removed by circulating hydrogen. The excess hydrogen from the battery passes through an intermediate heat exchanger and condenser; the bulk of the water vapour is condensed and separated from the hydrogen in a centrifugal separator and the gas is returned to the battery via an intermediate heat exchanger (see Fig. 101). The heat is removed from the battery (and from the condenser) by a liquid coolant circulating in the secondary loop. The heat is rejected via the thermal regulation system of the spacecraft and by special radiators into space.

The medium-temperature fuel cells have comparatively high discharge voltages (Fig. 99). A battery of 31 fuel cells has a power of 0·56–1·44 kW and voltage varying between 27 V and 31 V. When necessary the output can be increased to 2·3 kW with voltage not below 21 V. In this power range the current density varies between 25 and 135 mA/cm^2. About 10 % of the power generated is consumed by the unit for its operational needs. The service life of the power unit is about two thousand hours.

Pratt and Whitney established a new company—United Technologies Corporation—which now continues wide-ranging work on development of fuel cells, in particular cells with acid electrolytes for large-scale power units (see Section 17.6).

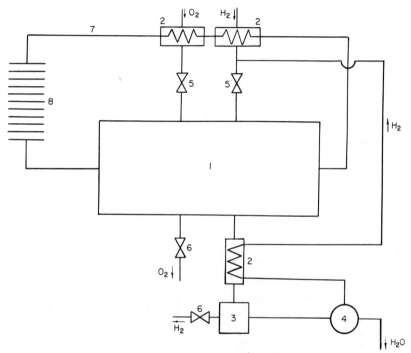

Fig. 101. Schematic of water and heat removal from the Pratt and Whitney fuel cell battery: (1) fuel cell battery; (2) heat exchangers; (3) condenser; (4) centrifugal separating pump; (5) pressure-regulating valves; (6) valves for gas blowing; (7) system for recirculating the cooling agent; (8) heat sink radiator.

Other types of alkaline oxygen–hydrogen cells

VARTA (West Germany) developed fuel cells with bilaterally operating gas diffusion electrodes. A disadvantage of the conventional gas diffusion electrodes with hydrophilic blocking layer is bending of the electrode at considerable excess gas pressure which occurs with large-diameter electrodes. To eliminate this drawback, a 5-layer electrode construction is used which looks like a back-to-back arrangement of two 2-layer electrodes with a common gas chamber between them which is a large-pore layer of sintered loose nickel powder (Fig. 102). The layer holds together both halves of the electrode and makes the electrode sufficiently strong even at high excess gas pressures. Electrodes of this type are sometimes referred to as Janus electrodes (from the double-faced god, Janus).

The cells developed by Siemens (West Germany) employ electrodes in which asbestos serves as the gas-tight layer. Catalysts applied on the asbestos sheets include Raney nickel (promoted with titanium) and silver. The catalyst is applied in such a way that the fine particles are placed on the

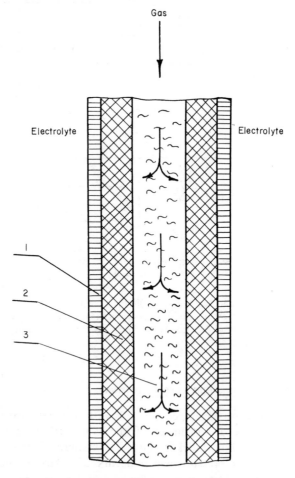

Fig. 102. Schematic of the double gas diffusion electrode: (1) gas-tight layers; (2) active layers with catalyst; (3) large-pore gas distribution layer.

asbestos and the larger-size particles are at the gas side. This improves the macrokinetic characteristics of electrodes rated for current densities of 0·4–0·5 A/cm². The electrodes have mesh current collectors and plastic frames and are assembled into filter press batteries. At 80°C these batteries have specific powers of up to 90 W/kg.

17.6. Oxygen–hydrogen cells with acidic electrolyte

The main advantage of the oxygen–hydrogen cells with acidic electrolytes, compared with the alkaline cells, is the possibility of using air and conversion-produced hydrogen without expensive purification procedures to remove carbon dioxide. These cells are highly promising for use in large power units (10 kW and more) for which good economic characteristics are important. However, considerable technological difficulties must be overcome if highly effective cells are to be developed.

Owing to lower chemical resistance the choice of catalysts for the acidic cells is smaller. Formerly, the only catalysts suitable for both the hydrogen and oxygen electrodes of acidic fuel cells were the platinum-group metals. In recent years non-platinum catalysts have been developed—tungsten carbide for the hydrogen electrode and certain complex organic compounds (porphyrins) for the oxygen electrode. Nevertheless, the platinum-group metals still remain the best catalysts for both electrodes. To decrease consumption of these catalysts, they are used on supports.

The reaction of oxygen reduction is slower in acidic electrolytes than in alkaline ones and the o.c.v. and discharge voltage of the acidic cells is typically lower by approximately 0·1 V than are those of the alkaline cells.

The maximum possible working temperature is chosen for the acidic cells. The first prototypes operated at 80°C; subsequently the temperature was raised and now the working temperature is 190–200°C. This is done to increase the effectiveness of the platinum catalysts and to reduce the amounts needed. Moreover, at elevated temperatures adsorption and poisoning effects of carbon monoxide are markedly decreased and, hence, the cell operation is more stable.

High-concentration acids are employed for operation at temperatures over 80°C. At first sulphuric acid was used and later 85–98% phosphoric acid. The important advantage of phosphoric acid over sulphuric acid is that it makes possible self-regulation of water removal from the cells. In concentrated phosphoric acid solutions the pressure of water vapour drops sharply with increasing concentration. Therefore, a matrix impregnated with a certain amount of the acid cannot be overdried by excessive gas blowing, so that blowing can be set at the maximum rate and there is no need to regulate it with decreasing current load. This makes possible the use of matrix electrolytes and considerable simplification of the construction of the cell and electrochemical generator. Sulphuric acid does not possess this property and it needs to be circulated to remove heat and water (giving rise to problems associated with sealing and corrosion resistance).

Two large-scale projects for development of oxygen–hydrogen cells with acidic electrolyte are being carried out by numerous American firms (the main participant is United Technologies Corporation).

The TARGET Programme aims at development of 12·5 kW power units utilizing natural gas supplied by pipes. These power units are designed for supplying electric power and heat for small buildings. The power unit comprises a device for conversion and purification of gas, a fuel cell battery and control systems. A few tens of such power units have been built and are now being tested in various towns.

The second programme is concerned with development of large-scale power plants; this will be discussed in Section 17.11.

Considerable amounts of platinum catalysts were used in the first prototypes of the fuel cells for these plants (for instance, 12·5 mg/cm^2 on each electrode) but now this amount has been reduced to 0·5–0·7 mg/cm^2. The platinum catalyst is applied to the carbon support. During operation at 190°C some recrystallization of the platinum particles occurs resulting in gradual deterioration of the cell performance. After 1000 hours of operation, a fuel cell with these electrodes reaches a stable voltage of 0·65 V at a current density of 200 mA/cm^2.

17.7. Oxygen–hydrogen cells with ion-exchange membranes (solid polymer electrolyte)

Thin ion-exchange membranes used as electrolyte make it possible to build very compact and simple fuel cell batteries, which is particularly important for low-power units. In such cells, moreover, it is easier to separate the reaction product (water) from the electrolyte. At not too high temperatures (below 80°C) the drops of water formed on the membrane surface covered with electrolyte can simply be removed by wicks.

In practice, the only types of membrane used so far in fuel cells are cation-exchange membranes in the hydrogen form in which current is carried by hydrogen ions. Such membranes have the properties of acidic electrolytes which means that ion-exchange membrane cells have the same electrochemical features as the acidic cells. A significant advantage of membrane electrolyte is the possibility of minimizing its contact with the structural materials, thus considerably relieving corrosion problems. The catalysts, naturally, still remain in contact with the electrolyte so that the main type of catalyst for the electrodes of such cells is still corrosion-resistant metals of the platinum group.

One of the main problems in development of ion-exchange membrane cells is obtaining an ion-exchange membrane with sufficient conductivity and sufficient chemical resistance under the conditions of operation of the cell.

A power unit with fuel cells was employed for the first time on board the Gemini spacecraft in 1965. The unit built by General Electric comprised two fuel cell batteries with ion-exchange membranes, tanks for cryogenic storage of oxygen and hydrogen and auxiliary systems. The fuel cells contained ion-exchange membranes of sulphonated polystyrene bonded with a Teflon binder. The catalysts (platinum or palladium) were applied directly onto the membrane and a mesh current collector was pressed against the membrane. The working temperature varied from 25°C to 60°C. Special thin-tube coolers were used for heat removal. Water which formed on the surface of the oxygen electrode was removed by wicks.

The mass of each battery was 31 kg. The battery comprised three groups of cells connected in parallel. The cell size was 18×20 cm^2. The battery had a rated power of 0·6 kW and a peak power of 1 kW, corresponding to current densities of 20 and 37 mA/cm^2 respectively. The battery voltage varied between 22 V and 30 V. The service life was about 1000 hours. A significant drawback of this battery was considerable consumption of platinum catalysts—about 30 mg/cm^2 or more than 2 kg for the total unit.

The ion-exchange membrane cells were considerably improved later. A significant advance was the development of a new perfluorosulphonic membrane (Nafion) which is distinguished by a high chemical resistance as well as by thermal resistance. This made it possible to increase the working temperature to 80–120°C and the current density to 100–250 mA/cm^2. The service life also increased considerably (one battery was tested for 34 thousand hours). Figure 99 shows the discharge curves of the ion-exchange membrane cells (for old and modified cell types).

17.8. Hydrazine fuel cells

Hydrazine, N_2H_4, is a liquid reducer with one of the highest activities and energy contents. This is why considerable attention has been paid to development of the hydrazine fuel cell despite such drawbacks as the high toxicity and high cost of hydrazine. The air–hydrazine cell is one of the few fuel cell types which has been commercially manufactured (though in very small numbers).

Alkaline electrolyte is employed in hydrazine cells. The reaction on the hydrazine electrode is described by the following equation:

$$N_2H_4 + 4OH^- \rightarrow N_2 + 4H_2O + 4e$$

The equilibrium potential for this reaction is by 0·33 V more negative than the potential of the reversible hydrogen electrode in the same solution. Hence the associated processes of oxidation of hydrazine and evolution of hydrogen on metals with a small hydrogen overpotential can occur in the absence of external current. Moreover, hydrazine can catalytically decompose on various materials giving rise to nitrogen and hydrogen.

The actual potential established on many metals in alkaline hydrazine solutions is close to the potential of the reversible hydrogen electrode. Thus, the voltage of the oxygen–hydrazine cell is close to the voltage of the oxygen–hydrogen cell. The rate of anodic oxidation of hydrazine depends to a considerable extent on the electrode material. Polarization curves for oxidation of hydrazine on various metals are not identical to the corresponding polarization curves for oxidation of hydrogen, showing that the process of hydrazine oxidation does not consist simply in catalytic decomposition of hydrazine and oxidation of the produced hydrogen. Hydrazine electrodes are manufactured from nickel, nickel boride or palladium-promoted nickel. A temperature rise increases the rate of anodic oxidation of hydrazine, but results in considerable enhancement of its spontaneous decomposition leading to large losses. Most existing hydrazine cells operate at ambient temperatures (the working temperature in the cell is somewhat higher owing to liberation of heat by the reaction). The rated current density of hydrazine cells lies in the range of 50–100 mA/cm^2.

In contrast to the hydrogen and oxygen electrodes, the starting active material of the hydrazine electrode is dissolved in the electrolyte. If hydrazine gets to the oxygen electrode its performance deteriorates considerably owing to the shifting of the potential to the negative values; this leads, moreover, to unproductive expenditure of hydrazine and oxygen. In the first prototypes of oxygen–hydrazine cells of the box-type, diffusion electrodes were used, the fuel (in the form of hydrazine–alkaline solution) was fed into and gas was evolved at the rear side. Since the porous electrode is not a sufficiently effective barrier to penetration of the dissolved hydrazine into the electrolyte and to the oxygen electrode, rather low hydrazine concentrations (about 1·5%) were used in these cells and the feed rate of hydrazine was automatically kept proportional to the current load.

Filter press constructions of the oxygen–hydrazine and air–hydrazine cells with asbestos separating diaphragms have also been developed. For instance, Monsanto (USA) has developed such batteries with output from 0·06 kW to 60 kW using hydrazine solutions of a higher concentration (3–6%) in 5M KOH solution. The anodes were made from porous nickel with palladium catalyst, the cathodes were made from carbon rendered water-repellent and carrying a platinum catalyst. The fuel cells generated a voltage of 0·8 V with current density of 100 mA/cm^2 at the working temperature of 65–70°C. The alkaline hydrazine solution was recirculated at the rear face of the anode and then through systems for heat removal, separation of gaseous nitrogen, removal of water, and resaturation with hydrazine. The service life of such batteries was over a thousand hours.

Union Carbide (USA) has developed renewable air–hydrazine cells in which the alkaline solution of hydrazine is replaced periodically. This construction minimizes the number of auxiliary systems and is very convenient for low-power units. The water produced in the reaction is removed by blowing air (a small amount of alkali is consumed here in decarbonization of the air before feeding it into the battery). To reduce penetration of hydrazine to the oxygen electrode the battery is filled with hydrazine–alkaline mixture (containing about 30% hydrazine) just before operation. A separating diaphragm is placed between electrodes and somewhat inhibits diffusion of hydrazine.

The relatively high electrochemical activity of hydrazine enabled the French firm Alstom (later in cooperation with Exxon, USA) to develop novel hydrogen peroxide–hydrazine batteries which employed comparatively smooth electrodes washed with reactant solutions, rather than porous diffusion electrodes. In fact, these were the only working batteries using liquid oxidant. They have a filter press construction assembled from thin bipolar profiled plates between which there are pressed thin electrodes from profiled foil and semipermeable membranes. The catalysts are nickel or cobalt for the anode and silver for the cathode. The battery is very compact—the total thickness of one fuel cell is 0·4 mm! The structural parameter s_V is about 120 dm^2/dm^3 and hence, the specific power of the battery is high (up to 0·4–0·5 kW/dm^3), despite a relatively low current density (50 mA/cm^2). When the battery is working, solutions of hydrazine and hydrogen peroxide in 1 M KOH are recirculated through the anodic and cathodic spaces. The recirculation rate is chosen to provide for complete consumption of the active materials in one passage through the battery. Thus, the battery rejects only foam consisting of nitrogen and dilute electrolyte. After separation of the gas, doses of alkali and hydrazine or hydrogen peroxide are added to the electrolyte and it is again fed into the cell. This battery is very simple and reliable in operation. It can be regarded as a promising prototype for development.

17.9. Low-temperature fuel cells with organic fuels

From the viewpoints of energy content, convenience of storage and transportation—and cost—liquid organic substances are ideal fuels for fuel cells. Efforts have been made for some time to develop fuel cells utilizing such fuels for this reason. However, an important obstacle stood in the way of these efforts, in the form of the low electrochemical activity of the organic compounds. At present the work on direct oxidation of the cheapest fuels, hydrocarbons, in low-temperature cells has practically ceased. Meanwhile, attempts are still continuing to develop fuel cells using alcohols and other oxygen-containing organic compounds, chiefly methanol, ethylene glycol, formaldehyde and formic acid.

Complete oxidation of methanol in acidic and alkaline solutions is described by the following equations:

$$CH_3OH + H_2O \rightarrow CO_2 + 6H^+ + 6e$$

$$CH_3OH + 8OH^- \rightarrow CO_3^{2-} + 6H_2O + 6e$$

The high energy content of the fuel is explained by the production of six electrons for the oxidation of one methanol molecule. The theoretical specific consumption of methanol is 0.2 g/Ah. Oxidation of organic fuel in the alkaline medium necessitates large consumption of the alkali. The consumption of alkali is 2 moles per mole of methanol (taking into account the reaction on the oxygen electrode) corresponding to a specific consumption of KOH of 0.7 g/Ah.

The equilibrium potential of the reaction of methanol oxidation in acidic solution is $+0.02$ V (r.h.e.). For other organic compounds mentioned the thermodynamic value of the potential is also close to the potential of the reversible hydrogen electrode in the same solution.

Anodic oxidation of organic fuels in acidic solutions can be catalysed only by the platinum-group metals (applied on supports). Though methanol is an easily oxidizable organic fuel its rate of anodic oxidation is low, so that the equilibrium electrode potential cannot be realized in practice on any electrode material in acidic or alkaline solutions. The zero-current potential obtained with the best catalysts (mixed blacks of platinum-group metals) is approximately by 0.2 V more positive than the equilibrium potential. The polarization of the electrodes with anodic current is also high. The same is true for other types of organic fuel.

To increase the reaction rate concentrated methanol solutions and elevated temperatures may be used effectively. However, at temperatures exceeding 60–70°C the pressure of the methanol vapour over the concentrated solutions is fairly high. In this respect ethylene glycol has an advantage since its boiling point (200°C) is much higher than that of methanol while the electrochemical properties of ethylene glycol are close to those of methanol.

In alkaline solutions oxidation of organic compounds occurs with a higher rate than in acidic solutions. Smaller amounts of platinum catalysts can be used in alkaline solutions and under certain conditions even non-platinum catalysts such as Raney nickel may be employed. This explains the great interest shown in alkaline fuel cells with organic fuels despite their major drawback of large consumption of alkali. Continuous removal of the reaction product (carbonate) from alkaline fuel cells presents considerable difficulties which is why most cells of this type are designed for one-time operation or as chemically rechargeable cells with periodic replacement of the fuel–alkaline mixture after it has been spent. The construction of cells of this type is similar to that of air–zinc cell but the active material of the anode is an alkaline solution of an organic compound, rather than zinc.

In the cells with acidic electrolyte the anodic reaction gives rise to gaseous carbon dioxide which is easily removed. The construction of such cells may be similar to that of hydrazine cells in which a liquid reactant also gives rise to a gaseous reaction product (nitrogen). Despite a number of obvious operational advantages, fuel cells with acidic electrolyte and organic fuel have still not been developed since considerable amounts of platinum catalysts are needed for them.

17.10. High-temperature fuel cells

For many years two types of high-temperature fuel cells have been under development: (a) cells with molten carbonate electrolyte operating at 550–700°C; and (b) cells with solid electrolyte operating at 850–1000°C.

The primary advantage of the high-temperature cells is that their reaction heat is evolved at a high temperature, and, hence, is highly efficient and can be used either for maintaining the endothermic reaction of conversion of hydrocarbon fuel or for operating a heat engine producing additional power. Since about 30–40% of the energy is converted into heat during operation, the resulting increase in economic efficiency is quite substantial. Moreover, at high temperatures the rates of the electrode processes are higher, polarization and ohmic losses in the electrolyte are lower and the electrical parameters of cells are improved. The danger of catalyst poisoning with various additives is considerably decreased and the electrode stability is improved.

At first, it was assumed that the high-temperature fuel cells could utilize hydrocarbon fuel such as natural gas without previous processing since hydrocarbons fed into the cell with steam can easily be directly converted to oxidizable products (hydrogen and carbon monoxide) in the cell. Furthermore, it was suggested that carbon monoxide and mixtures rich in it which are produced at the first stage of hydrocarbon conversion or with gasification of coal, could be directly oxidized in such cells. All these reactions were shown to be feasible in principle. However, in practice difficulties are presented by spontaneous decomposition of carbon monoxide (the Boudouard reaction)

$$2CO \rightarrow CO_2 + C$$

which occurs under certain conditions and results in deposition of carbon black on the electrodes and in the piping, blocking the gas flows and inhibiting the electrode process. Though these difficulties could be overcome, to date the development work is mainly aimed at cells utilizing technical-grade hydrogen which does not contain large amounts of carbon monoxide.

The peculiar features of operation of the high-temperature cells, namely, the necessity of maintaining the temperature conditions, the difficulties involved in starting and discontinuing operation and the long periods needed for this, dictate development of such cells primarily for large-scale stationary units which work for long periods, and for which a high total efficiency of fuel utilization is important.

Fuel cells with molten electrolyte. Only carbonates can be used as molten electrolyte in fuel cells. Melts of other salts are unstable, either gradually evaporating, decomposing giving rise to solids, or reacting with CO_2 admixtures in the technical-grade hydrogen or air and giving rise to carbonates. The melts used in the cells include eutectic binary and ternary mixtures of carbonates of alkali metals with melting points of about 500°C.

Molten carbonate electrolyte has high corrosive activity; therefore, matrix or immobilized electrolyte is typically used. The matrices are manufactured from sintered magnesium oxide or other pure oxides. Powders of magnesium oxide or lithium aluminate are added to the electrolyte for immobilization. The immobilized electrolytes which are practically solids proved to be more convenient since they do not crack.

In fuel cells with molten electrolyte the anodic process consumes carbonate ions:

$$H_2 + CO_3^{2-} \rightarrow H_2O + CO_2 + 2e$$

and so carbon dioxide from the exhaust gas is added to the oxygen for cathodic reaction and carbonate ions are regenerated in the reaction

$$2CO_2 + O_2 + 4e \rightarrow 2CO_3^{2-}$$

Thus, the current-producing reaction is accompanied by transfer of an additional CO_2 molecule for each two electrons. CO_2 is transferred through the electrolyte from the cathodic space to the anodic space, to be returned to the cathodic space through the gaseous phase (through the external loop).

The oxygen electrodes are manufactured from silver, lithia-containing (lithiated) nickel oxide and some other oxides. The fuel electrodes are manufactured from platinum, palladium, silver, nickel or palladium-coated graphite.

The main difficulties in development of fuel cells with molten electrolytes are caused by the instability of the electrolyte carriers, corrosion of the electrodes and structural materials, and by sealing problems. At the end of the sixties most laboratories discontinued research and development in this field but recently there has been some renewal of interest in such cells. For instance, American scientists have designed a cell with oxygen electrodes made from lithiated nickel oxide, hydrogen electrodes made from a nickel–cobalt alloy and electrolyte immobilized with lithium aluminate. Such a cell with a working electrode surface area of 9 dm^2 produces a current density exceeding 300 mA/cm^2 at a voltage of 0·6 V. The service life is estimated as seven thousand hours but it is expected to be increased to 40 thousand.

Fuel cells with solid oxide electrolyte. The electrolyte in such cells is a ceramic based on zirconium dioxide with the following typical composition: $ZrO_2 \cdot 0 \cdot 17CaO$ or $ZrO_2 \cdot 0 \cdot 11Y_2O_3$ (see Section 16.1). At 1000°C the conductivity of the first electrolyte is about 0·02 $ohm^{-1} cm^{-1}$ and that of the second is about 0·12 $ohm^{-1} cm^{-1}$. When the temperature is reduced the conductivity decreases sharply. The current carriers in the oxide electrolyte are oxygen ions (O^{2-}).

Anodes for cells with solid electrolyte can be manufactured from platinum as well as from the iron-group metals and some oxide compositions with semiconductor properties (based, for instance, on oxides of cerium or lanthanum). The cathodes can be made from platinum, silver or semiconducting oxides (e.g. based on oxides of zinc and zirconium).

Two basic constructions for fuel cells with solid electrolyte are being developed. One is based on a ceramic electrolyte onto which the electrode materials are applied by sputtering or pasting and to which the current collectors are pressed. In the other, one of the electrodes is used as a support on which electrolyte is applied by plasma sputtering and the other electrode is then deposited. This makes it possible to reduce the electrolyte thickness to 0·1 mm.

Westinghouse (USA) has developed complex-shaped cylindrical cells which can be easily assembled into a tubular in-series battery. Hydrogen or other fuel gas is passed through the tube whose outer wall is in contact with hot air. Electrolyte thickness is 0·4 mm. Figure 103 shows the construction of this cell. Voltage losses in the high-temperature solid-electrolyte cells result mainly from the ohmic resistance of the electrolyte, rather than polarization of the electrodes. A battery of 20 Westinghouse cells has specific power of about 100 W/kg (heat insulation is not taken into account) but the voltage of an individual cell is low.

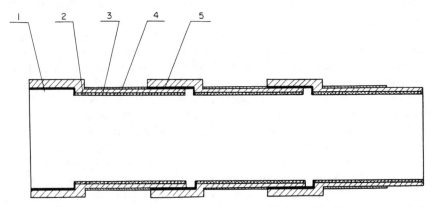

Fig. 103. Schematic of the Westinghouse fuel cell with solid electrolyte: (1) negative terminal; (2) negative electrode; (3) electrolyte; (4) positive electrode; (5) gas-tight interconnection.

The solid electrolyte cells have a number of significant advantages in comparison with the molten electrolyte type, namely, the absence of electrolyte losses and leaks, the absence of corrosive effects of the electrolyte, etc., and are therefore regarded as highly promising. However, extensive development of such cells is still hindered by the rather complicated and expensive technological processes involved in manufacturing thin-wall ceramic electrolytes.

17.11. Prospects for application of fuel cells

The oxygen–hydrogen and hydrazine fuel cells developed by various laboratories have acceptable parameters and reliability. They have been successfully used for various special projects. However, nowhere in the world has mass production of any fuel cell type been started.

Limited use of fuel cells can be explained, partly by the inconvenience of hydrogen and hydrazine for autonomous power units resulting from the difficulty of storing and transporting hydrogen and the toxicity of hydrazine.

It is likely therefore, that even in the future these fuels will be used only in special-application cells. Autonomous power units with hydrogen cells will be extensively used when more effective methods for storing hydrogen in fixed form (in compounds or in the sorbed state) are found. It is often inconvenient to use conversion of liquid hydrocarbons in autonomous power units since this necessitates installing additional devices and auxiliary systems which make the unit more complex and reduce its efficiency and reliability.

Fuel cells are still expensive. They are difficult to manufacture and expensive and scarce materials are needed for them (non-ferrous metals, stainless steels, Teflon, etc.). These difficulties, however, are not insurmountable. The cost of mass-produced fuel cells can be sharply reduced by automation of the technological processes. The expensive materials will be gradually replaced by cheaper ones. In general, the specific metal content of fuel cells is considerably lower than that of other types of chemical power sources.

The most promising type of fuel cell for mass utilization appear to be those with acidic electrolyte working on organic fuel (e.g., ethylene glycol) and air. Development of such cells depends, to a considerable extent, on solving fundamental problems of electrocatalysis associated with finding freely available highly effective catalysts.

If the more complicated problem of development of the low- and medium-temperature fuel cells with direct oxidation of hydrocarbon fuel were solved, this would result in a veritable breakthrough in autonomous power generation, including power units for transport vehicles.

The first serious projects aimed at development of fuel cells for large-scale power generation were started at the beginning of the seventies in the USA. These projects envisage building several 26 MW power plants. The first 4·8 MW pilot plant is soon to be put into operation and used as a part of the New York power system during peak load periods. Heavy oil fractions or natural gas will be used as the starting fuel for this power plant. Conversion of this fuel will produce hydrogen which will be fed into medium-temperature air–hydrogen cells with phosphoric acid electrolyte (Section 17.6). The overall efficiency of the plant is estimated at 38 %. The amount of platinum catalyst in the fuel cells is about 40 g per 1 kW of the rated power. A considerable difficulty encountered in the work on this project is the need for sulphur-free fuel. If such fuel is lacking, the gas fed into the fuel cells will require expensive treatment to remove the sulphur compounds.

The future economic parameters of such power plants with fuel cells are still not clear for two main reasons. Firstly, the need for preliminary conversion of the natural fuel results in considerable energy losses, decreasing the overall efficiency of the power plant. Secondly, even optimistic estimates of the future service life of electrochemical power plants are still far short of the service lives of módern conventional power plants. This leads to higher amortization costs and a corresponding increase in the production costs of electric power.

High-temperature fuel cells will, quite probably, prove to be the most suitable cells for large-scale power generation utilizing natural fuels. The reaction heat in such cells is evolved at fairly high temperatures and can be used to maintain the endothermic conversion reactions. This results in a marked increase of the overall efficiency of the power unit in comparison with the low-temperature cells in which the reaction heat is rejected to the environment. The electrodes in these cells are not poisoned by sulphur-containing fuel gases. Moreover, the products of coal gasification containing large amounts of CO can be used.

Despite their unclear future economic parameters, fuel cells utilizing the natural fuels attract great interest and their development will be continued. It may be expected that at first medium-output (a fraction of a megawatt or several megawatts) power units will be built, probably in remote or sparsely populated areas.

In the coming era of the hydrogen economy large-scale stationary power plants with different types of oxygen–hydrogen cells will be built, making possible the most efficient conversion into electric power of the energy of hydrogen fed through piping systems.

REFERENCES

1. W. Ostwald, Z. *Electrochemie* **1**, 122 (1894).
2. O. K. Davtyan, "Direct Conversion of Chemical Energy of Fuel into Electric Energy", USSR Academy of Sciences, Moscow (1974) (in Russian).
3. R. Wiswall and J. Reilly, Metal hydrides as power sources, *in* Proceedings of 7th Intersociety Energy Conversion Engineering Conference, San Diego, 1972, pp. 1342–1347.

BIBLIOGRAPHY

1. E. W. Justi and A. W. Winsel, "Kalte Verbrennung. Fuel Cells", Steiner, Wiesbaden (1962).
2. Mitchell, W. (ed.), "Fuel Cells". Academic Press, New York and London (1963).
3. M. W. Breiter, "Electrochemical Processes in Fuel Cells". Springer, Berlin (1964).
4. Baker, B. S. (ed.), "Hydrocarbon Fuel Cell Technology", Academic Press, New York and London (1965).

5. W. Vielstich, "Brennstoffelemente". Verlag Chemie, Weinheim (1965).
6. "Advances in Chemistry" series, No. 47. Fuel Cell Systems (1965); No. 90, Fuel Cell Systems II (1969). American Chemical Society, Washington.
7. Williams, K. R. (ed.), "An Introduction to Fuel Cells". Elsevier, Amsterdam (1966).
8. A. B. Hart and G. J. Womack, "Fuel Cells. Theory and Application". Chapman and Hall, London (1967).
9. H. A. Liebhafsky and E. J. Cairns, "Fuel Cells and Fuel Cell Batteries". Wiley, New York (1968).
10. Berger, C. (ed.), "Handbook of Fuel Cell Technology". Prentice-Hall, Englewood Cliffs (1968).
11. J. O'M. Bockris and S. Srinivasan, "Fuel Cells: Their Electrochemistry". McGraw Hill, New York (1969).
12. N. V. Korovin, "Electrochemical Generators". Energiya, Moscow (1974) (in Russian).
13. A. McDougal, "Fuel Cells". Macmillan, London (1976).
14. A. Winsel, Galvanische Elemente und Brennstoffzellen, in "Ullmans Encyclopädie der technischen Chemie, S. 114–135, Bd. 12. Verlag Chemie GmbH, Weinheim (1976).
15. K. V. Kordesch, 25 years of fuel cell development, J. Electrochem. Soc. 125, No. 3, 77C–91C (1978).
16. M. Eisenberg, Design and scale-up considerations for electrochemical fuel cells, in "Advances in Electrochemistry and Electrochemical Engineering" (Tobias, C., ed.), Vol. 2, pp. 235–292. Interscience, New York and London (1962).
17. Sandstede, G. (ed.), "From Electrocatalysis to Fuel Cells". University of Washington Press, Seattle (1972).
18. P. R. Prokopins, M. Warshay, S. N. Symons and R. B. King, Commercial phosphoric acid fuel cell system technology development, in "Proceedings 14th Intersociety Energy Conversion Conference, Boston, 1979", pp. 538–543.

Conclusions

Progress in Chemical Power Sources: Review of Problems

A large number of new types of chemical power sources with better characteristics than those of traditional electrochemical cells have been developed during the last two decades. Many of these cell types are manufactured on an industrial scale and are successfully employed in various devices. Nevertheless, the traditional electrochemical cells, that is the lead storage cells and manganese–zinc cells which have been with us for more than a century, still occupy the leading place in large-scale production.

This impressive "longevity" of traditional cells is explained by the fact that the scale on which a specific cell system is applied depends not only on is specific electrical parameters but also on a considerable number of other factors. Among these are:

convenience of operation and maintenance;
cost of the cell and of the electrical energy it delivers;
availability of raw materials for cell production;
stability of characteristics under the conditions of large-scale manufacturing.

An electrochemical cell must satisfy a considerable number of requirements. Some of these are mutually incompatible because they lead to contradictory approaches to design and technology. For example, the tendency to a maximum increase of admissible delivered power results in the use of thinner electrodes and thinner separating layers; on the other hand, the tendency to increase the cycle life, service life and storageability demands that the electrode and separator thickness be increased. The tendency to a general improvement of cell characteristics and to better reliability necessitates that the manufacturing process be organized more thoroughly, the raw material be better screened, a larger number of

371

monitoring and control operations be organized, and so on, and this tendency leads to measures which considerably raise the cost of the product.

In principle, each version of an electrochemical cell can be optimized with respect to a single selected parameter, that is with respect to specific conditions of operation (loading sequence, temperature, etc.). This optimization, however, means a sacrifice of a very important property of electrochemical cells: their universality. A cell optimized with respect to one characteristic, as a rule, becomes limited in its applicability. From the point of view of practical applications, it is advisable to reduce the number of technological and design models of electrochemical cells to a reasonable minimum, and thus manufacture a certain series of standardized models which can be used in a comparatively wide range of conditions.

Stability of cell characteristics in the conditions of mass production, homogeneity of cell properties, and cell reliability are important criteria for selecting a specific cell model. The quality of the final product often undergoes sharp fluctuations as a result of even insignificant variations in the properties of the initial raw material or in the conditions of the production process. Accumulation of experience is the only way to an understanding of the effects of these factors; consequently, instability of cell characteristics is encountered more often in newer cell models than in traditional cell types.

The problem of raw materials now constitutes one of the basic problems in the production of both traditional and some new variants of electrochemical cells. The more widely used cell designs employ considerable amounts of nonferrous metals: lead, nickel, cadmium, manganese, zinc and some others of which the resources are rather limited. This is an obstacle in the way to expansion of cell production output. Hence, much attention must be paid in the future to the development of those types of electrochemical cells which employ readily available raw materials. The sulphur–sodium storage cells, air–iron cells, as well as various versions of fuel cells are examples of such chemical power sources.

The scientific and technical problems related to further progress in chemical power sources have been discussed in Part Two of the book, in the chapters devoted to specific cell systems. In what follows we list some of the general problems whose solution to a large extent determines further progress in the field.

One of the foremost problems is the improvement of the active mass utilization coefficient, and specifically in the case of heavy discharge currents. This parameter is sufficiently high in some cell types but falls below 50% in others (such as lead batteries).

The primary problem for cells with liquid and gaseous reactants (fuel cells and compound cells) is the development of catalysts maintaining high efficiency of catalysis over long periods of time and requiring no components of limited availability or high cost.

Another important problem is the development of new separator materials for use with alkaline and acid electrolytes. An effective solution to this problem would make possible a considerable increase in the cycle life and in electrical characteristics of storage cells, as well as in the parameters of other types of electrochemical cells.

In the field of alkali storage cells, a stable zinc electrode with negligible displacement of the active mass (shape-changing) in the course of cycling would considerably increase the cell cycle life. Another problem of comparable importance consists in developing an oxygen electrode with high reversibility, to be used in air–metal storage cells.

The problem uppermost in the field of "throw-away" primary cells is the requirement for drastic reduction of the production costs and the use of readily available materials, and the reduction of the cost of energy produced. Magnesium and aluminium, which are cheaper than zinc and theoretically have better performance characteristics, are promising anode materials. Manganese–zinc cells, for instance, could be replaced by easy-to-operate air–metal cells if the problem of a stable, effective, and sufficiently cheap air electrode for weakly alkaline solutions was successfully solved. Production of such cells would pose no supply problems.

Considerable progress has been achieved in recent years in the domain of primary cells with non-aqueous electrolytes. The development of high-capacity storage cells on the basis of these systems would constitute a remarkable success.

Much is expected in the future from solid-electrolyte cells. Once the necessary low-temperature high-conductivity solid electrolytes are developed, thin-film techniques could be applied to produce convenient and highly reliable miniaturized cells.

It can be said in conclusion that the characteristics and the scale of application of various cell types depend on a large number of both technical and economic factors. All these factors have to be taken into account in the research effort aimed at elaboration of the traditional cell types and at the development of new cell types. As one authority in this field has put it, "a well designed battery of good performance is a well balanced compromise."†

† K.-J. Euler, Performance limits of primary and secondary batteries, *J. Power Sources*, **4**, No. 3, p. 216 (1979).

Appendices

Notes to Tables A.1, A.2 and A.3

(1) The electrode and current-producing reactions are given for the discharge process. During charge the reactions proceed in the reverse direction (from right to left).

(2) The column headed "Conditions" in Tables A.1 and A.2 gives the pH values or electrolyte compositions in typical cells with the requisite electrodes (in the charged state).

(3) The parameters $g_Q^{(T)}$ and $g_W^{(T)}$ are the theoretical consumptions of the main reactants per 1 Ah or 1 Wh (without taking into account consumption of other electrolyte components, in particular, water molecules or the ions H^+ or OH^-).

(4) The parameters $E°$ and $\mathscr{E}°$ are the thermodynamic standard electrode potentials and the e.m.f. values (the latter for cells with salt electrolytes, i.e. Leclanché cells and manganese–magnesium cells are given for the working pH values); E, E_r and $\mathscr{E}$ are the actual potentials and o.c.v. values for the electrodes and cells in the charged (but not overcharged) state. The values of $E°$ and E are given with reference to the standard hydrogen electrode potential; E_r is given with reference to the reversible hydrogen electrode in the same solution.

(5) The thermodynamic values of the electrode potentials for manganese dioxide vary in a wide range depending on the modification and state. The figures in brackets are nominal mean values.

(6) The symbol (x) in the column "Equil." (equilibrium) denotes the non-equilibrium values of the potential and o.c.v. (due to complexity of stoichiometry of the electrode reactions, strong inhibition of these reactions, and so on).

 (*continued, p. 380*)

Table A.1. Electrode potentials for cells with aqueous acid and salt solutions (25°C)

Reactant	Electrode reaction	Theoretical values		Specifics	Actual values			Chargeability
		$E°$ (V)	$g_O^{(T)}$ (g/Ah)		Equil.	E (V)	E_r (V)	
Negative electrodes								
Mg	$Mg \rightarrow Mg^{2+} + 2e$	-2.363	0.454	pH = 8	(x)	$-1.2-(-18)$	$-0.7-(-1.3)$	$-$
Al	$Al \rightarrow Al^{3+} + 3e$	-1.662	0.336	pH = 5	(x)	$-0.7-(-1.0)$	$-0.4-(-0.7)$	$-$
Zn	$Zn \rightarrow Zn^{2+} + 2e$	-0.763	1.220	pH = 5		-0.76	-0.46	$++$
Pb	$Pb \rightarrow Pb^{2+} + 2e$	-0.126	3.865	40–50% $HClO_4$		$-0.15-(-0.2)$	$-0.25-(-0.3)$	$++$
	$Pb + HSO_4^- \rightarrow PbSO_4 + H^+ + 2e$	-0.357	3.865	36% H_2SO_4		-0.36	-0.39	$+++$
H_2	$H_2 \rightarrow 2H^+ + 2e$	0.000	0.038	25% H_2SO_4		-0.005	0.00	$++$
CH_3OH	$CH_3OH + H_2O \rightarrow CO_2 + 6H^+ + 6e$	0.017	0.199	25% H_2SO_4	(x)	$0.3-0.5$	$0.3-0.5$	$-$
Positive electrodes								
H_2O	$2H_2O + 2e \rightarrow H_2 + 2OH^-$	-0.828	0.672	pH = 7		-0.41	0.00	$+$
CuCl	$CuCl + e \rightarrow Cu + Cl^-$	0.137	3.694	pH = 8		0.14	0.62	$+$
AgCl	$AgCl + e \rightarrow Ag + Cl^-$	0.222	5.348	pH = 8		0.22	0.70	$++$
$CuSO_4$	$Cu^{2+} + 2e \rightarrow Cu$	0.337	2.978	pH = 5		0.34	0.64	$++$
MnO_2	$MnO_2 + H^+ + e \rightarrow MnOOH$	(1.25)	3.244	pH = 5	(x)	$0.8-1.1$	$1.1-1.4$	$+$
O_2	$O_2 + 4H^+ + 4e \rightarrow 2H_2O$	1.299^1	0.298	25% H_2SO_4	(x)	$0.9-1.0$	$0.9-1.0$	$+$
$K_2Cr_2O_7$	$Cr_2O_7^{2-} + 14H^+ + 6e \rightarrow 2Cr^{3+} + 7H_2O$	1.33	1.830	20–40% H_2SO_4		$1.35-1.45$	$1.3-1.4$	$-$
Cl_2	$Cl_2 + 2e \rightarrow 2Cl^-$	1.360	1.323	pH = 5		1.36	1.66	$+++$
PbO_2	$PbO_2 + 4H^+ + 2e \rightarrow Pb^{2+} + 2H_2O$	1.455	4.462	40–50% $HClO_4$		$1.65-1.75$	$1.55-1.6$	$++$
	$PbO_2 + HSO_4^- + 3H^+ + 2e \rightarrow PbSO_4 + 2H_2O$	1.690	4.462	36% H_2SO_4		1.76	1.73	$+++$

1. With air ($p_{O_2} = 0.2$ atm), $E° = 1.219$ V.

Table A.2. Electrode potentials for cells with aqueous alkaline solutions (25°C)

Reactant	Electrode reaction	Theoretical values		Actual values				Chargeability
		$E°$ (V)	$q_O^{(T)}$ (g/Ah)	Specifics	Equil.	E (V)	E_r (V)	
Negative electrodes								
Li	$Li \rightarrow Li^+ + e$	$-3·045$	$0·259$	3M LiOH		$-3·02$	$-2·18$	—
Zn	$Zn + 2OH^- \rightarrow ZnO + H_2O + 2e$	$-1·254^1$	$1·220$	7M KOH + ZnO_2^{2-} [2]		$-1·30$	$-0·426$	++
Zn	$Zn + 4OH^- \rightarrow ZnO_2^{2-} + 2H_2O + 2e$	$-1·216$	$1·220$	7M KOH + ZnO_2^{2-} [3]		$-1·32$	$-0·44$	+
N_2H_4	$N_2H_4 + 4OH^- \rightarrow N_2 + 4H_2O + 4e$	$-1·160$	$0·299$	7M KOH	(x)	$-0·9$-$(-1·0)$	$-0·1$-0	—
CH_3OH	$CH_3OH + 8OH^- \rightarrow CO_3^{2-} + 6H_2O + 6e$	$-0·91$	$0·199$	7M KOH	(x)	$-0·6$-$(-0·8)$	$0·1$-$0·3$	—
Fe	$Fe + 2OH^- \rightarrow Fe(OH)_2 + 2e$	$-0·877$	$1·042$	7M KOH		$-0·96$	$-0·05$	+++
H_2	$H_2 + 2OH^- \rightarrow 2H_2O + 2e$	$-0·828$	$0·038$	7M KOH		$-0·91$	$0·000$	+++
Cd	$Cd + 2OH^- \rightarrow Cd(OH)_2 + 2e$	$-0·809$	$2·097$	7M KOH		$-0·89$	$0·02$	+++
Positive electrodes								
Cu_2O	$Cu_2O + H_2O + 2e \rightarrow 2Cu + 2OH^-$	$-0·358$	$2·669$	6M NaOH		$-0·41$	$0·47$	+
CuO	$2CuO + H_2O + 2e \rightarrow Cu_2O + 2OH^-$	$-0·158$	$2·968$	6M NaOH $\}$	(x)	$-0·3$-$(-0·4)$	$0·5$-$0·6$	+
	$CuO + H_2O + 2e \rightarrow Cu + 2OH^-$	$-0·258$	$1·484$	6M NaOH				
HgO	$HgO + H_2O + 2e \rightarrow Hg + 2OH^-$	$0·098$	$4·041$	7M KOH		$0·05$	$0·926$	+
MnO_2	$MnO_2 + H_2O + e \rightarrow MnOOH + OH^-$	$(0·35)$	$3·244$	7M KOH	(x)	$0·15$-$0·35$	$1·05$-$1·35$	+
Ag_2O	$Ag_2O + H_2O + 2e \rightarrow 2Ag + 2OH^-$	$0·345$	$4·324$	7M KOH		$0·30$	$1·17$	+++
AgO	$2AgO + H_2O + 2e \rightarrow Ag_2O + 2OH^-$	$0·605$	$4·622$	7M KOH		$0·56$	$1·43\Big\}$	+++
	$AgO + H_2O + 2e \rightarrow Ag + 2OH^-$	$0·475$	$2·311$	7M KOH		—	—	+++
O_2	$O_2 + 2H_2O + 4e \rightarrow 4OH^-$	$0·401^4$	$0·298$	7M KOH	(x)	$0·1$-$0·2$	$1·0$-$1·1$	++
NiOOH	$NiOOH + H_2O + e \rightarrow Ni(OH)_2 + OH^-$	$0·49$	$3·422$	7M KOH		$0·41$-$0·45$	$1·32$-$1·36$	+++
H_2O_2	$HO_2^- + H_2O + 2e \rightarrow 3OH^-$	$0·88$	$0·635$	1M KOH	(x)	$0·2$-$0·5$	$1·0$-$1·3$	—

1. For the most stable modification—ZnO; 2. Solution saturated with potassium zincate; 3. Solution not saturated with potassium zincate; 4. For air ($p_{O_2} = 0·2$ atm) $E° = 0·391$ V.

377

Table A.3. E.m.f. and o.c.v. values in different cell systems (25°C)

Cell type	Electrochemical system	Current-producing reaction	Theoretical values			Actual values		Chargeability
			$\mathscr{E}°$ (V)	$\tilde{\mathscr{E}}$ (V)	$g_W^{(T)}$ (g/Wh)	Equil.	$\mathscr{E}$ (V)	
Early cell types								
Water–zinc (Volta cells)	$Zn\|H_2O\|(Ag)$	$2H_2O + Zn \rightarrow H_2 + Zn^{2+} + 2OH^-$	0·349	0·79	3·50	(x)	0·2–0·4	+
Copper–zinc (Daniell cells)	$Zn\|ZnSO_4::CuSO_4\|Cu$	$Cu^{2+} + Zn \rightarrow Cu + Zn^{2+}$	1·100	1·12	3·82		1·10	+
Nitric acid–zinc (Bunsen cells)	$Zn\|H_2SO_4::HNO_3\|(C)$	$2HNO_3 + 3Zn + 6H^+ \rightarrow 2NO + 3Zn^{2+} + 4H_2O$	1·72	1·85	1·39	(x)	1·9	−
Chromic acid–zinc (Grenet cells)	$Zn\|K_2Cr_2O_7, H_2SO_4\|(C)$	$Cr_2O_7^{2-} + 3Zn + 14H^+ \rightarrow 2Cr^{3+} + 3Zn^{2+} + 7H_2O$	2·09	—	1·46		2·0–2·1	−
Primary cells								
Manganese–zinc (Leclanché cells)	$Zn\|NH_4Cl, ZnCl_2\|MnO_2$	$2MnO_2 + Zn + 2H_2O \rightarrow 2MnOOH + Zn(OH)_2$	(1·7)[1]	(1·4)	(2·6)	(x)	1·55–1·85	+
Manganese–zinc alkaline	$Zn\|KOH\|MnO_2$	$2MnO_2 + Zn + H_2O \rightarrow 2MnOOH + ZnO$	(1·6)	(1·4)	(2·8)		1·5–1·7	++
Mercury–zinc	$Zn\|KOH\|HgO$	$HgO + Zn \rightarrow Hg + ZnO$	1·352	1·33	3·92		1·35	++
Copper–zinc alkaline	$Zn\|NaOH\|CuO$	$CuO + Zn + 2OH^- \rightarrow Cu + ZnO_2^{2-} + H_2O$	1·058	1·00	2·48	(x)	0·9–1·0	+
Air–zinc	$Zn\|KOH\|O_2(C)$	$O_2 + 2Zn \rightarrow 2ZnO$	1·638[2]	1·80	0·92	(x)	1·4–1·5	++
Manganese–magnesium	$Mg\|Mg(ClO_4)_2\|MnO_2$	$2MnO_2 + Mg + 2H_2O \rightarrow 2MnOOH + Mg(OH)_2$	(3·1)[1]	(2·9)	(1·2)	(x)	1·9–2·0	−
Silver chloride–magnesium	$Mg\|MgCl_2\|AgCl$	$2AgCl + Mg \rightarrow 2Ag + MgCl$	2·585	2·81	2·24	(x)	1·6	−
Copper chloride–magnesium	$Mg\|MgCl_2\|CuCl$	$2CuCl + Mg \rightarrow 2Cu + MgCl_2$	2·500	2·73	1·66	(x)	1·5	−
Lead dioxide–lead	$Pb\|HClO_4\|PbO_2$	$PbO_2 + Pb + 4H^+ \rightarrow 2Pb^{2+} + 2H_2O$	1·581	1·51	5·27		1·8–1·9	++
Standard Weston cells	$Cd(Hg)\|CdSO_4::HgSO_4\|(Hg)$	$HgSO_4 + Cd + \frac{8}{3}H_2O \rightarrow Hg + CdSO_4 \cdot \frac{8}{3}H_2O$	1·019	1·03			1·01864	+

Table A.3.—cont.

Cell type	Electrochemical system	Current-producing reaction	Theoretical values			Actual values		
			$\mathscr{E}^\circ$ (V)	$\bar{\mathscr{E}}$ (V)	$g_W^{(T)}$ (g/Wh)	Equil.	$\mathscr{E}$ (V)	Chargeability
Storage cells								
Lead acid	$Pb\,|\,H_2SO_4\,|\,PbO_2$	$PbO_2 + Pb + 2H_2SO_4 \rightarrow 2PbSO_4 + 2H_2O$	2·047	2·04[3]	4·07		2·06–2·15	+ + +
Nickel–cadmium	$Cd\,|\,KOH\,|\,NiOOH$	$2NiOOH + Cd + 2H_2O \rightarrow 2Ni(OH)_2 + Cd(OH)_2$	1·30	1·43	4·25		1·30–1·34	+ + +
Nickel–iron	$Fe\,|\,KOH\,|\,NiOOH$	$2NiOOH + Fe + 2H_2O \rightarrow 2Ni(OH)_2 + Fe(OH)_2$	1·37	1·49	3·36		1·37–1·41	+ + +
Silver–zinc	$Zn\,|\,KOH\,|\,AgO$	$AgO + Zn \rightarrow Ag + ZnO$	1·859	1·74	1·90		1·86	+ +
	$Zn\,|\,KOH\,|\,Ag_2O$	$Ag_2O + Zn \rightarrow 2Ag + ZnO$	1·599	1·65	3·47		1·60	+ +
Nickel–zinc	$Zn\,|\,KOH\,|\,NiOOH$	$2NiOOH + Zn + H_2O \rightarrow 2Ni(OH)_2 + ZnO$	1·74	1·86	2·67		1·74–1·78	+ +
Nickel–hydrogen	$H_2(Ni)\,|\,KOH\,|\,NiOOH$	$NiOOH + \tfrac{1}{2}H_2 \rightarrow Ni(OH)_2$	1·32	1·52	2·62		1·32–1·36	+ + +
Chlorine–zinc	$Zn\,|\,ZnCl_2\,|\,Cl_2(C)$	$Cl_2 + Zn \rightarrow ZnCl_2$	2·123	2·52	1·20		2·10–2·15	+ +
Fuel cells								
Oxygen–hydrogen	$H_2(Ni)\,|\,KOH\,|\,O_2(C, Ag)$	$O_2 + 2H_2 \rightarrow 2H_2O$	1·299	1·48	0·27	(x)	1·0–1·1	+
	$H_2(Pt)\,|\,H_2SO_4\,|\,O_2(Pt)$	$O_2 + 2H_2 \rightarrow 2H_2O$	1·229	1·48	0·27	(x)	0·9–1·0	+
Oxygen–hydrazine	$N_2H_4(Ni)\,|\,KOH\,|\,O_2(C, Ag)$	$O_2 + N_2H_4 \rightarrow 2H_2O + N_2$	1·56	1·56	0·38	(x)	1·0–1·2	–
Oxygen–methanol	$CH_3OH(Pt)\,|\,KOH\,|\,O_2(C, Ag)$	$\tfrac{3}{2}O_2 + CH_3OH + 2OH^- \rightarrow 3H_2O + CO_3^{2-}$	1·31	1·43	0·38	(x)	0·8–0·9	–
	$CH_3OH(Pt)\,|\,H_2SO_4\,|\,O_2(Pt)$	$\tfrac{3}{2}O_2 + CH_3OH \rightarrow 2H_2O + CO_2$	1·21	1·25	0·41	(x)	0·5–0·6	–

1. At pH 7; 2. air ($p_{O_2} = 0\cdot2$ atm); 3. with 36% H_2SO_4; strongly depends on acid concentration due to high heat of dilution.

(*Notes to tables A.1, A.2 and A.3, continued*)

(7) The notation in the column "Chargeability":

(−): total absence of chargeability (the reverse electrochemical reaction does not occur under any conditions);

(+): poor chargeability (the reverse reaction occurs only under certain conditions);

(++): limited chargeability (considerable polarization, morphological changes and so on);

(+++): good chargeability (a storage cell with a long cycle life is feasible).

Table A.4. Standardized sizes of the cylindrical dry cells

Nomenclature			Dimensions	
IEC†	USSR	USA	Diameter (mm)	Height (mm)
R 08	—	O	10	3
R 06	283	—	10	22
R 03	286	AAA	10	44
R 4	314	R	13·5	38
R 6	316	AA	13·5	50
R 8	326	A	16	50
R 10	332	BR	20	37
R 12	336	B	20	59
R 14	343	C	24	49
R 20	373	D	32	61
R 22	374	E	32	75
—	375	—	32	81
R 25	376	F	32	91
R 26	—	G	32	105
R 27	—	J	32	150
—	425	—	39·7	100
—	465	—	50·7	125

† International Electrotechnical Commission.

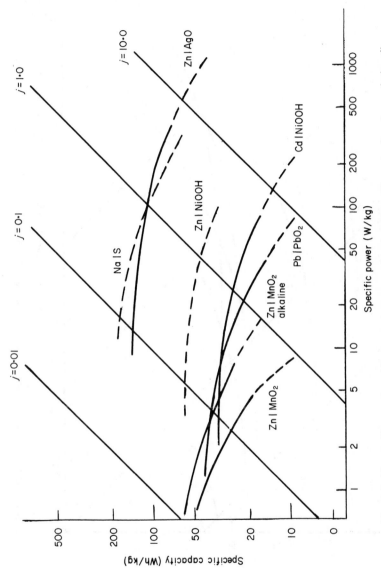

Fig. A.1. The specific capacity with respect to mass as a function of specific power for a heavy drain cell.

381

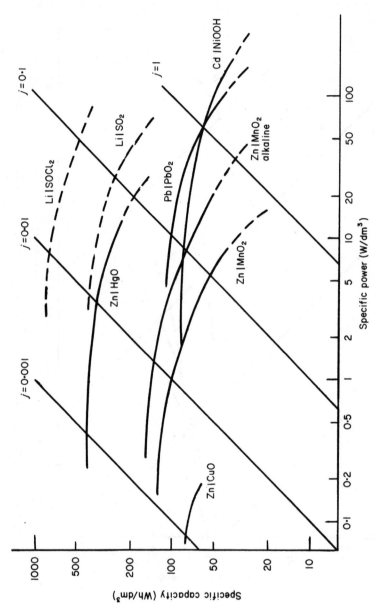

Fig. A.2. The specific capacity with respect to volume as a function of specific power for a light drain cell.

382

Subject Index